Techniques in
Cell Cycle Analysis

Biological Methods

Techniques in Cell Cycle Analysis, edited by **Joe W. Gray** and **Zbigniew Darzynkiewicz,** *1986*

Methods in Molecular Biology, edited by **John M. Walker,** *1984*
   Volume I: *Proteins*
   Volume II: *Nucleic Acids*

Liquid Chromatography in Clinical Analysis, edited by **Pokar M. Kabra** and **Laurence J. Marton,** *1981*

Metal Carcinogenesis Testing: Principles and In Vitro Methods, by **Max Costa,** *1980*

Techniques in
Cell Cycle Analysis

Edited by

Joe W. Gray
and
Zbigniew Darzynkiewicz

Humana Press · Clifton, New Jersey

Library of Congress Cataloging-in-Publication Data

Techniques in Cell Cycle Analysis.

(Biological methods)
Includes index.
1. Flow cytometry. 2. Cell cycle. 3. Cancer cells—Growth.
I. Gray, Joe W. II. Darzynkiewicz, Zbigniew. III. Series.
QH585.5.F56T43   1986   616.99'4071   86-16111
ISBN 0-89603-097-0

© 1987 The Humana Press Inc.
Crescent Manor
PO Box 2148
Clifton, NJ 07015

Printed in the United States of America

Preface

Quantification of the proliferative characteristics of normal and malignant cells has been of interest to oncologists and cancer biologists for almost three decades. This interest stems from (a) the fact that cancer is a disease of uncontrolled proliferation, (b) the finding that many of the commonly used anticancer agents are preferentially toxic to cells that are actively proliferating, and (c) the observation that significant differences in proliferation characteristics exist between normal and malignant cells. Initially, cell cycle analysis was pursued enthusiastically in the hope of generating information useful for the development of rational cancer therapy strategies; for example, by allowing identification of rapidly proliferating tumors against which cell cycle-specific agents could be used with maximum effectiveness and by allowing rational scheduling of cell cycle-specific therapeutic agents to maximize the therapeutic ratio. Unfortunately, several difficulties have prevented realization of the early promise of cell cycle analysis: Proliferative patterns of the normal and malignant tissues have been found to be substantially more complex than originally anticipated, and synchronization of human tumors has proved remarkably difficult. Human tumors of the same type have proved highly variable, and the cytokinetic tools available for cell cycle analysis have been labor intensive, as well as somewhat subjective and in many cases inapplicable to humans. However, the potential for substantially improved cancer therapy remains if more accurate cytokinetic information about human malignancies and normal tissues can be obtained in a timely fashion.

This monograph contains a series of articles describing cytokinetic techniques that have been important to the development of our current cytokinetic data base, as well as those that appear to have substantial promise for future cytokinetic studies in model systems and in the clinic.

Traditional techniques based on the autoradiographic detection of incorporated tritiated thymidine have provided information about the fraction of cells capable of DNA synthesis (a rough indicator in the proliferative activity of the cell population), about the G1-, S-, and G2M-phase durations and dispersions therein, and about the growth fraction (fraction of actively proliferating cells in a population). These techniques, their attributes, and their limitations are described in chapters 1 and 2. Chapter 3 focuses on conventional and developing techniques for estimation of the growth fraction.

Much of our information on the response of tumors to cytotoxic agents has been derived from measurements of the colony-forming ability of cells from model and human tumors grown in vitro. These studies are reviewed in chapter 4.

The majority of this monograph is devoted to flow cytometry and its application to cytokinetics because of the increasing importance of flow cytometry in the field of cytokinetics. Flow cytometry has become the method of choice in many cytokinetic studies within the past decade because of the speed and accuracy with which cellular properties of cytokinetic importance (e.g., DNA content, RNA content, amount of incorporated bromodeoxyuridine, proliferative status, and so on) can be measured. In addition, many cells can be analyzed in each experiment so that rare subpopulations can be studied. Chapter 5 introduces flow cytometry and reviews several common cytokinetic applications, including univariate DNA distribution analysis and bivariate analysis of cellular DNA content and amount of incorporated bromodeoxyuridine. Chapters 6 and 7 discuss the techniques necessary to prepare cells for flow cytometric analysis; chapter 6 reviews techniques for dissociation of solid tissues into suspensions of single cells, and chapter 7 reviews a variety of cell staining techniques that are especially useful for cytokinetic studies. Chapter 8 deals with computer analysis technique for display and cytokinetic analysis of flow cytometric data. Chapter 9 discusses cytochemical techniques to allow flow cytometric discrimination of quiescent and proliferating cells. These techniques complement those described in chapter 3. Chapter 10 deals with the combination of flow cytometry and stathmokinesis for cell cycle analysis.

Most cytokinetically based cancer therapy optimization attempts have presumed the existence of a battery of effective cell cycle-specific agents. Chapter 11 introduces the idea that drug-resistant cells existing prior to therapy and/or developed during the course of therapy must also be considered during the development of cytokinetic therapy strategies. This chapter also suggests several flow cytometric approaches to intracellular drug level quantification.

Much of our information about the cell cycle-specific nature of anticancer agents has come from the application of these agents to synchronized cell populations. Several techniques for synchronization of cells grown in vitro are critically compared in chapter 12.

These 12 chapters encompass many of the techniques that have been especially useful in cell cycle studies or that hold great promise for the future. Emphasis has been placed on the techniques themselves and on critical review of their attributes and limitations. We offer them here in the hope that they will facilitate selection of appropriate techniques for future studies and will stimulate development of new approaches to remove existing limitations so that the true potential of cell cycle analysis in cancer therapy can be realized.

Joe W. Gray
Zbigniew Darzynkiewicz

Contents

CHAPTER 3.   Tumor Growth Fraction Estimation, Perturbation, and Prognostication 47

Paul G. Braunschweiger

CHAPTER 4.   In Vitro Assays for Tumors Grown In Vivo: A Review of Kinetic Techniques 73

Janet S. Rasey

CHAPTER 5.   Flow Cytokinetics *93*
Joe W. Gray, Frank Dolbeare, Maria G. Pallavicini, and Martin Vanderlaan

CHAPTER 6.   Solid Tissue Dispersal for Cytokinetic Analyses *139*
Maria G. Pallavicini

CHAPTER 7.   Multivariate Cell Analysis:
Techniques for Correlated Measurements
of DNA and Other Cellular Constituents *163*

Harry A. Crissman and John A. Steinkamp

CHAPTER 8. Data Analysis in Cell Kinetics Research 207

Phillip N. Dean

CHAPTER 9. Cytochemical Probes of Cycling and Quiescent Cells Applicable to Flow Cytometry 255

Zbigniew Darzynkiewicz

Contributors

PAUL G. BRAUNSCHWEIGER • Department of Experimental Therapeutics, AMC Cancer Research Center and Hospital, Lakewood, Colorado

HARRY A. CRISSMAN • Life Sciences Division, Los Alamos National Laboratory, Los Alamos, New Mexico

ZBIGNIEW DARZYNKIEWICZ • Sloan-Kettering Institute of Cancer Research, Walker Laboratory, Rye, New York

PHILLIP N. DEAN • Biomedical Sciences Division, Lawrence Livermore National Laboratory, University of California, Livermore, California

FRANK DOLBEARE • Biomedical Sciences Division, Lawrence Livermore National Laboratory, University of California, Livermore, California

JOE W. GRAY • Biomedical Sciences Division, Lawrence Livermore National Laboratory, University of California, Livermore, California

DAVID J. GRDINA • Division of Biological and Medical Research, Argonne National Laboratory, Argonne, Illinois

TOD S. JOHNSON • Division of Biological and Medical Research, Argonne National Laboratory, Argonne, Illinois

MAREK KIMMEL • Sloan-Kettering Institute for Cancer Research, Walker Laboratory, Rye, New York

AWTAR KRISHAN • Comprehensive Cancer Center and Department of Oncology, University of Miami Medical School, Miami, Florida

MARVIN L. MEISTRICH • Division of Biological and Medical Research, Argonne National Laboratory, Argonne, Illinois

RAYMOND E. MEYN • Division of Biological and Medical Research, Argonne National Laboratory, Argonne, Illinois

MARIA G. PALLAVICINI • Biomedical Sciences Division, Lawrence Livermore National Laboratory, University of California, Livermore, California

JANET S. RASEY • Department of Radiation Oncology, School of Medicine, University of Washington, Seattle, Washington

PAUL S. RITCH • Department of Medicine, The Medical College of Wisconsin, Milwaukee, Wisconsin

STANLEY E. SHACKNEY • Division of Cancer Treatment, National Cancer Institute, Bethesda, Maryland

LINDA SIMPSON-HERREN • Biochemistry Research Division, Southern Research Institute, Birmingham, Alabama

JOHN A. STEINKAMP • Life Sciences Division, Los Alamos National Laboratory, Los Alamos, California

FRANK TRAGANOS • Sloan-Kettering Institute for Cancer Research, Walker Laboratory, Rye, New York

MARTIN VANDERLAAN • Biomedical Sciences Division, Lawrence Livermore National Laboratory, University of California, Livermore, California

R. ALLEN WHITE • Department of Biomathematics, University of Texas, Houston, Texas

Chapter 1

Autoradiographic Techniques for Measurement of the Labeling Index

Linda Simpson-Herren

1. INTRODUCTION AND HISTORY

Autoradiography is a technique for visualization of radioactive material within an object by registering the charged particles emitted by disintegration of radioactive atoms, and was used by Howard and Pelc to demonstrate the incorporation of ^{32}P into "resting" cells of *Vicia faba* roots (37) and to define the principal phases of the intermitotic period (38). The value of autoradiography as a technique for kinetic studies increased with the synthesis of tritiated thymidine, a specific precursor of DNA, by Taylor et al. (79). High-resolution autoradiography combined with a specific precursor of DNA made feasible identification of cells, initially in S-phase, as they progressed through the cell cycle. With these techniques, Quastler and Sherman (60) performed the first in vivo analysis of cell cycle traverse rates in mouse epithelium.

As interest increased in the cell population kinetics of normal and neoplastic tissues, the techniques for autoradiography, particularly with tritium, developed rapidly. Pilgrim et al. (58) actively developed these techniques to compare the kinetic behavior of several normal tissues. Autoradiography was used in early studies of kinetic changes during carcinogenesis (12,39,45,61) and in experimental tumors (6,10,47–49).

In 1967, Cleaver (17) reviewed the state of the art in *Thymidine Metabolism and Cell Kinetics*. Development of techniques for cloning tumor cells (76) and for cytometry has diversified the study of

1

cell kinetics. Autoradiography continues to be an important tool for studies of individual cells or small, isolated populations, and for studies in which tumor morphology and vascularization are important factors in interpretation of results.

2. PRINCIPLES OF AUTORADIOGRAPHY

Autoradiographs are produced by placing objects or specimens containing the radioactive material in contact with nuclear emulsion, usually in the form of a suspension of crystals of silver bromide in gelatin. Extensive discussions of the principles of autoradiography may be found in books by Baserga and Malamud (7), Gude (32), Rogers (63), Schultze (68), Steel (76), and Evans (26). Briefly, when radiation strikes the emulsion, a change takes place in the silver bromide crystals, forming a latent image. This image is not visible until the emulsion is treated with a developing agent to convert the exposed or changed crystals into grains of metallic silver. The bromide crystals that have not been reduced to metallic silver are dissolved out of the emulsion by the photographic fixative. Thus, the developed image is a pattern of silver grains that indicates the location of radioactive disintegrations within the specimen.

In this chapter emphasis will be placed on techniques for microautoradiography in which the photographic emulsion in the form of a gel or stripping film is placed in direct contact with a slide bearing a sample of cells or a section of tissue containing the radioactive label.

Isotopes of two elements, hydrogen and carbon, are widely used in studies of cells and tissues because of both the ubiquitous presence of these elements in biological materials and the short range of travel of the β-particles emitted. β-Particles emitted from a particular isotope do not all have the same energy, but show a spectrum of energies ranging from a maximum value down to zero. The maximum energy for ^3H is 18.5 keV (kilo electron volts), with a maximum range of about 6 μm in material of unit density. The average range in tissue is approximately 0.5 μm (7). The maximum energy of ^{14}C is 155 keV, with a maximum range of 100 μm in tissue and an average range of 16 μm or less. The range of the particle is inversely related to the density of the material; i.e., the range of particles is longer in air than in water (specific density, 1.0) or tissue (specific density, 1.3) or undeveloped photographic emulsion (specific density, 3.0). The short ranges of the β-particles from ^3H and ^{14}C make it necessary to place the photographic film in direct contact with the sample and make it feasible to determine the source of the disintegration (i.e., specific cell or nucleus).

3. LABELING INDEX

3.1. Definition

Use of tritiated thymidine ($[^3H]$-TdR) and autoradiography is considered the "classic" technique for kinetic studies of DNA-synthesizing cell populations. The validity of measured and calculated kinetic parameters such as the fraction of cells in S-phase and the durations of the G1-, S-, and G2M-phases depends on incorporation of sufficient labeled thymidine by all cells engaged in scheduled DNA synthesis to be detected under the conditions of the experiment and on the specificity of the $[^3H]$-TdR as a precursor of the scheduled synthesis.

A large body of data on both experimental and human normal and abnormal tissues, particularly neoplasms, has been obtained in vivo and in vitro. It is evident after evaluation of the results of many studies that the measured and calculated kinetic parameters do not accurately describe the proliferative kinetics of the tissues under some circumstances. In view of the investment of time required by autoradiography studies, the factors that influence the validity of the observed data should be carefully evaluated for each experimental protocol.

By definition, the labeling index (LI) is a measure of the fraction of a cell population that incorporates a labeled precursor during a pulse exposure, or:

$$LI = \frac{\text{labeled cells}}{\text{labeled cells } + \text{ unlabeled cells}}$$

If the labeled material is a precursor of DNA, then the labeling index is defined as a measure of the fraction of the cell population engaged in DNA synthesis, specifically scheduled synthesis of DNA (TLI is used to designate a labeling index measured after administration of [^{3}H]-TdR). Calculations of other kinetic parameters, such as growth fraction and cell loss factors (76), utilize the observed TLI, and the value of these calculations is limited by the accuracy of the TLI.

When cells are pulse or flash labeled during a specific phase of the cell cycle (such as exposure to [^{3}H]-TdR during S-phase), the radioactivity can serve as a marker to identify these cells as they proceed through mitosis. This "pulse-chase" technique is the basis for the percent labeled mitoses (PLM) method for measuring the phases of the cell cycle. The method is discussed in detail in chapter 2 (71).

Continuous or repeated exposure of cells to [^{3}H]-TdR (or another specific precursor of DNA) will progressively label all categories of actively proliferating cells, including those with limited lifespan, but will fail to label the "resting" or "static" cells that retain capacity to proliferate. The technique has been used as a simple method for estimating growth fraction, but it is subject to errors introduced by the "resting" cells and by the fact that in a growing population the continuous labeling curve will tend to approach 100% as labeled cells divide and produce labeled daughter cells [see ref. (76) for a detailed discussion of the theory of continuous labeling curves].

3.2. Discrete or Continuous Model of Cell Cycle

Interpretation of labeling index data has been subject to considerable uncertainty in view of the differences of two proposed models of the cell cycle and the absence of a definitive basis for a choice between the models in a given system or experiment.

The original model of the cell cycle proposed by Howard and Pelc (38) described a cell cycle consisting of mitosis, a pre- and postmitotic gap, G2 and G1, respectively, separated by a discrete period during which synthesis of genetic or scheduled DNA occurred. According to this model only cells in the S-phase (or entering this phase during exposure to the labeled precursor) would become labeled if [^{3}H]-TdR were available for only a brief period (pulse exposure).

Autoradiography of a pulse-labeled cell population clearly demonstrates a distribution of grains per cell ranging from zero to hundreds, depending on the experimental conditions. The tacit assumption has been made that uptake of [³H]-TdR during the pulse exposure is a measure of the rate of DNA synthesis (*17*); i.e., the autoradiographic grain count produced by cells is proportional to the rate of DNA synthesis within the cell. The validity of this premise was supported by the early studies in mouse L cells in culture by Dendy and Smith (*18*) and by the work of Goldspink and Goldberg (*31*).

The presence of "lightly" labeled cells in the periods of the cell cycle designated as G1 and G2 by the discrete model led to the proposal by Shackney (*69,70*) of a continuous model of the cell cycle. In this model, synthesis of DNA is not confined to a discrete phase, but may occur at some rate throughout the cycle; i.e., at a low rate in G1- and G2-phases. Nicolini (*52*) challenged the validity of the continuous model and suggested that technical problems were responsible for the observations. Maurer (*46*) also questioned the assumption that [³H]-TdR uptake is directly proportional to the rate of DNA synthesis, and recalled earlier observations that fluctuations in the intracellular pools that dilute the exogeneous precursor may increase or decrease incorporation without corresponding changes in the rate of DNA synthesis (*19,74,75*). The validity of the discrete or continuous model of the cell cycle remains a subject of debate, but the practical implications of the lightly labeled cells found in most studies of the labeling index will be discussed in later sections of this chapter.

3.3. Precursors for Measurement of a Labeling Index

The most commonly used labeled precursor for determination of an LI is [³H]-TdR. Thymidine is a nucleoside that consists of a pyrimidine base attached to a deoxyribose moiety (Fig. 1). It is not a normal component of the metabolic pathway producing DNA, but is incorporated into thymidylic acid by phosphorylation with thymidine kinase, an enzyme on the salvage pathway that is rate-limiting for entry of [³H]-TdR into the thymidine monophosphate pool. This endogeneous pool, fed largely by thymidylate synthase reaction on the *de novo* pathway, serves as a diluent for the labeled precursor prior to polymerization into DNA (Fig. 1).

Within 60 min after administration of the precursor in vivo, the radioactivity will have been incorporated into newly formed DNA or into breakdown products. Only a small fraction of the

labeled material injected in vivo is utilized in synthesis of DNA; 0.1–
1% of DNA-thymine synthesized in short-term studies originates

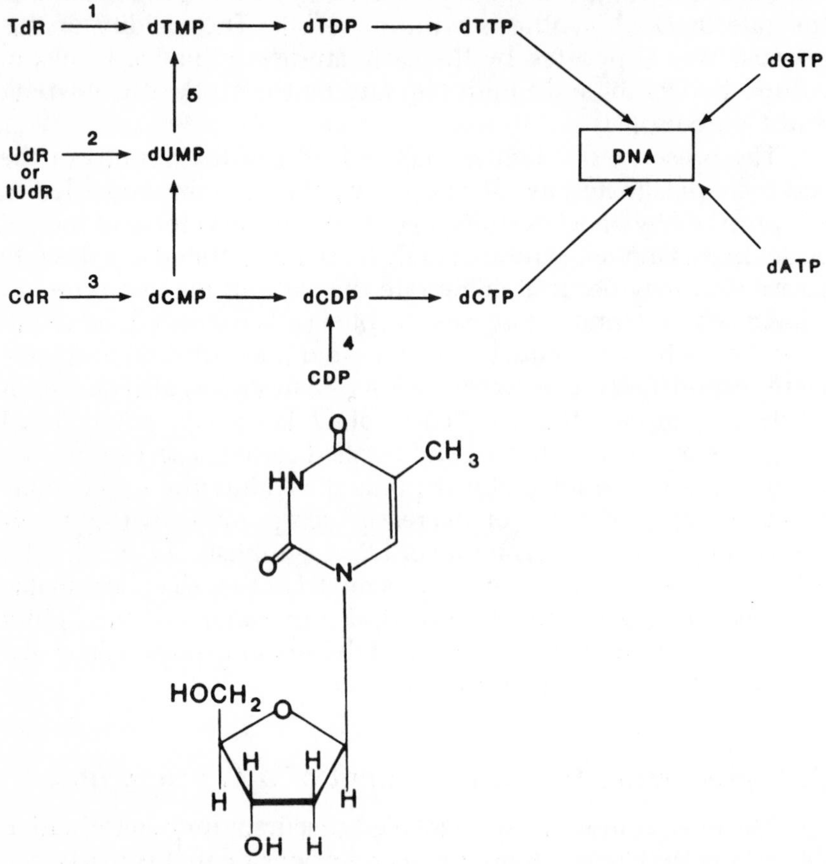

THYMIDINE

Fig. 1. Abbreviations: TdR, deoxythymidine; UdR, deoxyuridine; IUdR, iododeoxyuridine; CdR, deoxycytidine; CDP, cytidine diphosphate; dCMP, deoxycytidine monophosphate; dCDP, deoxycytidine diphosphate; dTMP, thymidine monophosphate; dTDP, thymidine diphosphate; dTTP, thymidine triphosphate; dATP, deoxyadenosine triphosphate; dGTP, deoxyguanosine triphosphate; 1 and 2, thymidine kinase; 3, deoxycytidine kinase; 4, CDP reductase; 5, dTMP synthase.

from exogeneous [³H]-TdR and the upper limit of a tracer dose is about 0.5 μg TdR/g animal weight (42).

Other precursors have also been utilized to measure LIs. [³H]-Deoxycytidine ([³H]-CdR), [¹⁴C]-labeled thymidine ([¹⁴C]-TdR), [¹³¹I]-, [¹²⁵I]-, or [³H]-labeled iododeoxyuridine ([¹²⁵]-IUdR, [¹³¹I]-IUdR, [³H]-IUdR), and [³H]-deoxyuridine ([³H]-UdR) enter the metabolic pathway at different points and have been used to label cells engaged in DNA synthesis (Fig. 1). Differences in utilization of these precursors may result from the rate-limiting enzymes at the point of entry into the pathway or from the dilution that results from the endogeneous nucleotide pools. Based on the data currently available, there is no clear indication that these precursors are superior to [³H]-TdR for measurement of pulse LIs.

3.4. Specificity of Precursors

Numerous investigators (4,25,27,36) have questioned the specificity of DNA precursors for identification cells engaged in DNA synthesis. Pelc (56) reported that nondividing cells of the seminal vesicles, liver, smooth muscle, cardiac muscle, interstitial cells of the testis, and mast cells incorporated low levels of [³H]-TdR (about 2% or less of that incorporated by cells engaged in division). The data from these studies indicated that two distributions of grain counts per cell were present after compensation for background and that no cells with intermediate grain counts were present. Pelc interpreted these data to represent two mechanisms for incorporation of [³H]-TdR, one possibly in nondividing cells. Allison et al. (2) reported the presence of unlabeled cells with S-phase DNA content in [³H]-TdR labeled spheroids of Chinese hamster lung and mouse mammary carcinoma. He suggested that these cells were blocked in S-phase. Allison further suggested that the lightly labeled cells in bone marrow were in G0/G1 or G2 when evaluated by cytometry (1).

Several studies suggest that all cells engaged in DNA synthesis do not utilize various labeled precursors equally. Early studies by Fox and Prusoff (29) suggested that [³H]-TdR was utilized preferentially to [¹²⁵I]-IUdR in synthesis of DNA. In addition, Hamilton and Dobbin (34) reported that [³H]-TdR labels less than half of the DNA-synthesizing cells in the mouse tumor carcinoma NT when compared with [³H]-UdR or [³H]-IUdR. In the Fox and Prusoff studies, the precursors were given on an equimolar basis, but the ratio of specific activity was 4.1 mCi/mmol ([³H]-TdR) and 0.64 mCi/mmol ([¹²⁵I]-IUdR). In the Hamilton and Dobbin study, the precursors were administered as follows:

Precursors	Dose	Specific activity
[³H]-TdR	30 μCi/mouse	5 Ci/mmol
6-[³H]-UdR	50–60 μCi/mouse	15 Ci/mmol
6-[³H]-IUdR	50 μCi/mouse	2.4 Ci/mmol

Labeling cells were evaluated on autoradiographs exposed for 3 wk. The observed differences in LI measured following [³H]-TdR and [³H]-IUdR or [³H]-UdR may be related to the inadequacy of a single period of emulsion exposure (3 wk) to detect the distribution of labeling per cell that occurred following the twofold range of doses, the sixfold range of specific activities, and the variations in uptake of the three precursors.

Results of kinetic studies reported by Dethlefsen (20) suggest that toxic effects were evident when a dose of 0.013 μmol per mouse of IUdR was injected. The toxicity appeared to be primarily a result of chemical effects.

Differential labeling of lymphocytes by [³H]-TdR and [³H]-CdR has been reported (3,33,35). Amano and Everett (3) interpreted this as a difference in ability of short- and long-lived lymphocytes to utilize purine nucleosides for DNA synthesis. Differences in incorporation of [³H]-CdR and [³H]-UdR into cultured lymphocytes were attributed to competition of factors in the culture media (28).

In other studies, the observed LIs as a function of duration of exposure are similar in melanoma (B16) and Ridgway osteogenic sarcoma (ROS) whether the precursor was [³H]-TdR or [³H]-IUdR (Simpson-Herren et al., unpublished data). The LIs of B16 and ROS were measured after doses of 50 μCi/mouse (6.7 Ci /mmol) [³H]-TdR or 50 μCi/mouse (18.0 Ci/mmol) [³H]-IUdR and autoradiographs were exposed for periods ranging from 3 to 96 d (Fig. 2). The differences in LI measured by the two precursors were small and fell within the range of variation between individual tumors.

In a related study, Clausen et al. (16) reported a discrepancy between the observed TLI measured with low doses of [³H]-TdR and the fraction of cells in S-phase determined by flow cytometry in epidermal cells of the mouse. However, higher doses of [³H]-TdR and/or longer emulsion exposures yielded results that confirmed the cytometry data.

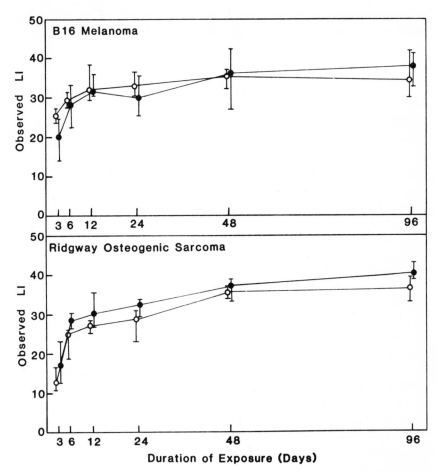

Fig. 2. The observed mean LIs of B16 melanoma and Ridgway osteogenic sarcoma (ROS) following [³H]-TdR (50 μCi/mmol) ○; and [³H]-IUdR (50 μCi/mouse; 18.0 ci/mmol) ●. Bars indicate the range of LIs for individual tumors. Each symbol represents the mean LI of five tumors determined by at least two investigators on duplicate slides at each time point. The slides and autoradiographs were prepared as previously described (70).

The results of studies cited above strongly suggest that the combined factors of specificity of precursors, amount and specific activity of the precursors, effects of drugs (62), and sensitivity of detection (autoradiography, in particular) may lead to confusing or erroneous results in the absence of rigid internal controls. The relationship between observed LI and duration of emulsion exposure will be discussed in detail in section 4 of this chapter.

3.5. Sources of Artifacts

Maurer (46) reviewed the [³H]-TdR techniques to measure cell proliferation and reemphasized the artifacts that may result from careless use of labeled nucleosides for these studies. Many of the problems discussed would be of concern only in specific types of studies, but the following items should be of particular concern in measurement of a labeling index: (1) self-decomposition products of [³H]-TdR stored in aqueous solution that may label cellular fractions other than DNA, (2) drugs that interfere with thymidine transport rather than DNA synthesis (agents capable of interfering with alkylating sulfydryl groups involved in transport reactions), (3) infections from a number of mycoplasma strains (57), which are capable of utilizing and cleaving [³H]-TdR, thus reducing incorporation into cellular DNA, and (4) perturbations of experimental endpoints that result from disintegration of the radioactive atom within the DNA or nucleus of the cell. (This type of artifact is unlikely to occur in studies of a pulse LI or TLI, but may be a serious concern in continuous labeling studies or PLM analysis.)

Specifically labeled nucleic acid precursors are recommended as tracers whenever possible. Tritium atoms are most stable in these compounds if the ³H is in the 2-position of the purine ring or in the 6-position of the pyrimidine ring (27). For further discussion of the importance of purity, stability, and specificity of radioactive precursors, *see* Evans (25,27) and Amano et al. (4).

4. AMBIGUITY OF THE LABELING INDEX

4.1. Experimental Conditions

The value observed for a TLI (or LI) is directly related to the experimental conditions used in the experiment, although historically little attention has been paid to this fact. The duration of exposure required to reach a detectable level of grains per cell is a function of (a) the concentration and specific activity of [³H]-TdR to which the cells are exposed, (b) the duration or efficiency of the autoradiographic process, (c) the labeling characteristics of the cells or tissue under study, and (d) the background threshold chosen to discriminate against those cells with grains originating from sources other than ^{3}H incorporated into DNA (*69,73*).

4.1.1. Dose of Precursor

The dose of [3H]-TdR to which the cells are exposed is usually a compromise between the activity necessary for detection within a reasonable time and the possible adverse effects of excessive levels of radiation. [3H]-TdR may be used at high levels (such as 10 μCi/g body weight) in measurement of pulse TLIs, since radiotoxicity is not a problem in studies of short duration (0.5–1.0 h). If continued normal cell proliferation of the labeled cells is a requirement, as in PLM analysis and in measurement of continuous labeling indices, a compromise must be made between the optimum conditions for the experiment and the possible effects of radiotoxicity (*8,9,14,15,20*).

High-dose levels of labeled precursor increase the possibility of artifacts that result from transfer of the isotope during processing of the samples. Bryant (*13*) reported that during acid hydrolysis with 1.0*N* HCl at 60°C for staining with the Fuelgen technique, a specific reassociation of displaced label could reach significant levels in 6–8 min. Background can also be increased if unincorporated [3H]-TdR or labeled impurities are not removed prior to autoradiography. Treatment of cell smears or sections with weak acid during prestaining or other procedures (*72*) and extraction by solvents during paraffin-embedding of tissues (*21*) effectively remove unincorporated thymidine, radioactive impurities, and degradation products.

4.1.2. Duration of Exposure

The number of autoradiographic grains that results from exposure of a labeled source to photographic emulsion increases in a linear manner with duration of exposure until the image produced in the emulsion begins to fade (*64*). This phenomenon imposes a practical limit on the duration of exposure. The exposure required to reach a given mean grain count in an autoradiograph is inversely related to the level of labeled precursor; i.e., the lower the dose of [3H]-TdR or other precursor, the longer the exposure required to achieve the same observed TLI, provided the amount of available [3H]-TdR is small compared to the size of the nucleotide pools in the cells (Fig. 3).

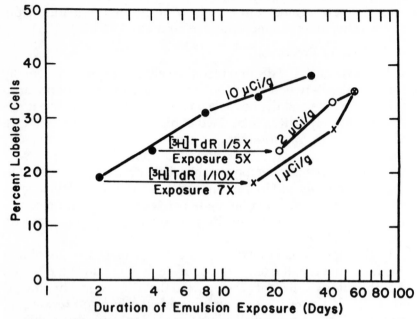

Fig. 3. TI as a function of dose of [³H]-TdR and duration of emulsion exposure in ROS. Arrows connect similar values for observed TIs after different doses of [³H]-TdR and indicate the inverse relationship of duration of emulsion exposure (from ref. 70, with permission of Cancer Research).

The [³H]-TdR incorporated into the DNA of cells is directly related to the specific activity and the dose, but it is also a function of the size of precursor pools that may dilute the label (*see* section 3.2 and Fig. 1 in this chapter). Data obtained under similar experimental conditions, but utilizing different doses of [³H]-TdR or different exposure periods, may be evaluated using an autoradiographic factor (AF); i.e., the product of the dose of [³H]-TdR (μCi/g body weight or μCi/mL) × exposure duration (d) (73). Based on the mathematical model of Lloyd and Simpson-Herren (44), an AF of 100–360 μCi d would be required to detect 95% of the labeled cells for six of seven murine tumors studied. Much longer periods apparently would be needed to detect 95% of the observed labeled cells in Ridgway osteogenic tumors (Table 1) under the experimental conditions used.

TABLE 1
Estimates of the Emulsion Exposure Time Required
to Observe 95% and 99% of the Theoretical Maximum TI
Obtained from the γ-Distributed Uptake Model[a]

Tumor	t_{95}, d	t_{99}, d
C3H	10	28
LL	13	36
AK leukemia-lymphoma	17	33
Plasmacytoma 1	31	74
B16	33	219
Adenocarcinoma 755	36	344
ROS	566	16,307

[a]Reprinted from ref. 42, with permission of Cancer Research.

Allison et al. (2) suggested an autoradiographic exposure unit (AEU) in which one unit corresponds to a 3-h exposure for cytocentrifuge-prepared, single-cell preparations from spheriods exposed to 2 μCi/mL of [^{3}H]-TdR. For the conditions of that study, 8 AEU would be in the middle of the plateau in the plot of observed TLI vs exposure time. (The observed TLIs were similar over a 32-fold range of exposure times.)

4.2. Labeling Characteristics of the Tissue

It should be noted that although the number of grains produced over individual labeled cells increases in a linear manner with time, the increase in fraction of labeled cells detected with increasing exposure is a function of the distribution of ^{3}H (or other label) in the cell population under study (42,70); i.e., the highly labeled cells are detected on short exposure and the lightly labeled cells are detected as labeled only after longer exposure. The changes in observed TLI with time that occurred in seven murine tumor models are shown in Fig. 4 (73), and the times required to detect 95 or 99% of the theoretical maximum TLI, based on a mathematical model (44), are shown in Table 1. This model relates the observed autoradiographic labeling index to the distribution of radioactive atoms among individual cells that originally incorporate labeled material. The model assumes an uptake distribution that fits the observed labeling data and predicts that the labeling index should increase with duration of emulsion exposure. It further indicates

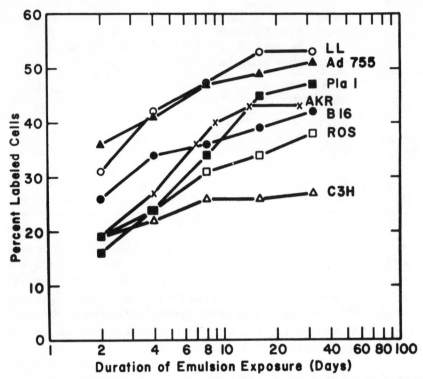

Fig. 4. Effects of duration of emulsion exposure on observed TI of spontaneous and transplanted tumors ([³H]-TdR, 10 μCi/g body weight). Each symbol represents the pooled data from analysis by 2 investigators of duplicate slides from each of five tissues [except four tissues for AK leukemia-lymphoma (AKR) and LL]. The experimental error, estimated as the pooled standard error of the mean for each set of replicates for all exposure periods, was 4.7. Comparison of each pair of TI curves indicated significant differences ($P > 0.05$); either the pairs are not parallel or the TIs are different for the same exposure, or both. Abbreviations: Ad 755, adenocarcinoma 755; Pla 1, plasmacytoma 1; B16, B16 melanoma; C3H, C3H mammary adenocarcinoma (from ref. 70, with permission of Cancer Research).

that the increase is related to the proportion of lightly labeled cells in the population. That the lightly labeled cells detected on long exposure are not an artifact of the autoradioagraphic process is demonstrated by the very low percentages (0–0.033) of labeled mitotic figures found on the same autoradiographs (73). It is apparent from these studies that similar autoradiographic conditions for different tissues may not detect the same fractions of the total labeled cell population.

Low doses of labeled precursor and short exposure periods increase the impact of the distribution of incorporated ^{3}H per cell on the observed TLI. This distribution of ^{3}H per cell results from a variety of factors, including heterogeneity of the cell population; rates of DNA synthesis in individual cells; access of individual cells to the [^{3}H]-TdR (or other precursor), i.e., proximity to the vascular supply in vivo (78); absorption of the β-particle from tritium in the cell or tissue; or sectioning of cell nuclei in preparation of the slide. (For a more complete discussion, *see* refs. cited in ref. 44). Thus, satisfactory levels of [^{3}H]-TdR and exposure durations for a given experiment should be determined from internal controls; i.e., multiple exposures of experimental slides should be used to select conditions to give an observed TLI on the plateau of the TLI vs exposure curve.

Exposure of slides to emulsion for extended periods has two major drawbacks: (1) the background may increase with time, and (2) the cell morphology of the highly labeled cells may be obscured by grains on long exposure. The increase in background is generally not a serious problem in carefully prepared slides on exposures of up to 3 mo. In our experience, high background usually results from bad photographic emulsion, inadequate conditions for protection and storage of slides, and, probably most often, labeled material, either extranuclear or extracellular, that is distributed over the slide prior to application of the emulsion (*see* section 5 for futher details). A more serious problem results from the difficulty in identification of cell morphology of the highly labeled cells. The problem is minimal in a homogeneous cell population or when labeling is primarily confined to the population of interest. An alternate approach to counting long exposures would be the use of a mathematical model (44) to predict the TLI from data obtained on multiple, relatively short exposures where the cell morphology is not obscured by grains.

Efforts have been made to shorten the exposure period of autoradiographs by enhancing or intensifying the process. Panayi and Neill (55) described a technique for immersing slides coated with photographic emulsion into liquid scintillators that developed slides in 24 h. Durie and Salmon (22) modified the technique to use higher specific activity [^{3}H]-TdR, liquid scintillators, and low temperatures for storage. The dried emulsion-covered slides were dipped into a solution of scintillators in dioxane, then dried before storage. Durie and Salmon (22) stated that when the emulsion was impregnated with scintillator, photons were released as the electrons (β-particles) passed through the scintillator, thus activating more silver crystals in the emulsion. A thoughtful review of the efforts to increase autoradiographic efficiency by scintillation techniques suggested that any improvements in efficiency were probably a result of the effective drying of the emulsion by the solvent, thus increasing sensitivity, rather than the production of photons (64). Woodcock et al. (80) reported that the scintillation did not impregnate the emulsion covering of the tissue or cell; thus the scintillator was effectively separated from the β-particles. It is not surprising that use of scintillation enhancement has met with variable success. The techniques described above or modifications thereof have been reported to increase autoradiographic efficiency as much as 30–50 times (43,54,66), whereas others (40,50,51,64,80) reported no increase in efficiency from addition of scintillators, unless added (directly to the tissues) during the histological process.

Addition of gold salts to increase the sensitivity of emulsion, first used in electron microscopy and later applied to autoradiographs for the light microscope, has received more favorable reports (11,41). The technique involves deposit, just prior to development, of a second metal, gold, on minute silver deposits in the emulsion to increase the probability of development during normal development procedures. Additions of the gold increase the size and catalytic potential of each silver deposit without affecting the crystals in the emulsion that have no silver deposits. The treatment transforms sub-images into latent images, but the effect is not simply to achieve equal grain formation in a shorter development time. Even at long development times the total number of grains developed continues to increase. The effect of gold latensification is to increase the probability of development of silver deposits as autoradiographic grains, but may not be directly related to detection of lightly labeled cells after long emulsion exposure, as described in section 4.1.

4.3. Background Threshold

The most popular method of establishing a background threshold, i.e., adoption of a threshold above which cells are scored as labeled, tends to truncate the data by eliminating lightly labeled cells. Quastler (59) described this method as "sweeping questionable cells under the rug." The fraction of cells that may be "swept under the rug" is a function of the distribution of incorporated [^{3}H]-TdR (73).

Alternately, a background value may be determined for each cell from adjacent areas of the autoradiograph not exposed to tissue or cells. If the grain count per cell exceeds the local background by one grain (or more), it may be considered labeled. Thus adjustments are made for variations in the background over the area of the section or cell smear, but the method is subject to error in that the selected local background may not be a valid indicator of background for the particular cell area. This type of error should be random and would not tend to yield data biased toward either low or high values.

In studies of AKR lymphoma (Fig. 5), increased emulsion exposure tended to minimize the difference in observed TLI's obtained with an arbitrary background threshold and a local background threshold. When the grain count/cell for the lightly labeled cells exceeded the arbitrary threshold, this population of cells was no longer discriminated against by the threshold. That the results using the two methods approach the same value on long exposure lends further support to the theory that artificial thresholds and short exposures tend to truncate the data.

Various other methods to classify cells as labeled or unlabeled have been proposed. Stillstrom (77) suggested use of a formula that is based on the number of cells without grains on labeled and unlabeled slides. The assumption is made that the background does not vary from slide to slide. When the background density is high, the number of cells with no grains approaches zero and the method fails. Sawicki et al. (65) proposed a slightly different approach, using the equation described by Stillstrom (77) for evaluation of grain count distributions obtained from labeled and unlabeled slides. Eisen (23) also suggested use of a control slide of unlabeled tissue in estimation of background.

England et al. (24) proposed that the total number of grains on labeled and unlabeled slides be counted, and the total number of cells with no grains on labeled and unlabeled slides be determined. The estimates of background were then based on the assumption that the distribution of grains resulting from background and labeling in cells is Poisson. Application of the method

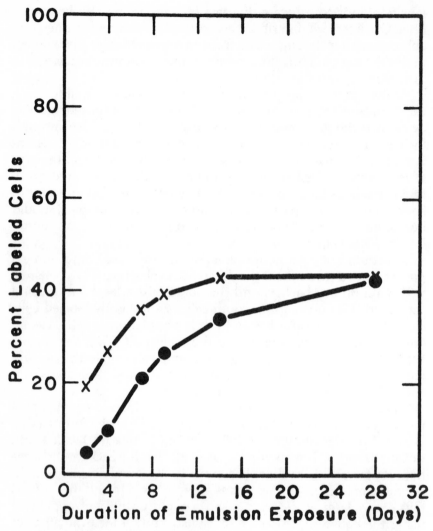

Fig. 5. Effects of duration of emulsion exposure and background threshold on observed TI of AKR spontaneous lymphoma ([³H]-TdR, 10 μCi/g body weight). Each symbol represents the pooled data from the analysis by two investigators of duplicate slides from each of four tissues. x, data obtained using local background, plus 1 grain/cell as threshold of labeled cells; ●, data obtained using an arbitrary five grains/cell as threshold of labeled cells (from ref. 70, with permission of Cancer Research).

would require extensive grain counting when the grain density was high. The assumption of a Poisson distribution for the background grains per cell and a cut-off value used by Barendsen et al. (5), when both the background and labeling density were high, tended to truncate the data in a manner similar to a high arbitrary background threshold.

Schoenfeld and Kallman (67) described another method for estimation of the labeling index based on histograms of grain distributions from slides prepared from cells either labeled after exposure or not labeled to [³H]-TdR. It is based on a model for the frequency distribution of grain counts of those cells with few grains. The labeling index is estimated by a computer program for determining the probability that cells in the range of both labeled and unlabeled cells should be scored as labeled.

This brief review of methods for selection of a background threshold indicates that there is no ideal technique, and that each method represents a compromise between accuracy, labor intensity, and applicability to a given experimental situation. Notably absent are definitive studies that demonstrate that any method of background correction yields comparable results over a wide range of experimental conditions.

5. AUTORADIOGRAPHIC TECHNIQUES

There is extensive literature on autoradiography (7,17,26,30, 32,68,76) with methods applicable to a variety of technical problems. Specific recommendations on many aspects of the technique will not be covered because of the many excellent texts available and because the specific methods chosen should be determined by the requirements of the particular study. Internal controls, designed to establish the validity of the observed labeling data, are a vital part of experimental design and will be emphasized here. The principles are applicable to most studies utilizing autoradiography, regardless of specific method.

5.1. Pretest of Emulsion

Background grains that result from the environment, the experimental materials, and particularly the emulsion should be kept to a minimum to improve identification of labeled cells. These grains cannot be completely eliminated, but can be minimized with careful techniques.

The most frequently used photographic emulsions for measurement of a labeling index are in liquid form and require heating (and often diluting) prior to application. Nuclear research emulsions are commercially available from Eastman Kodak, Rochester, New York and Ilford Ltd., Ilford, Essex, England (available in the USA through Polyscience, Inc., Warrington, PA). Kodak NTB2 is sensitive to low-energy β-particles and Ilford K-2 is especially suitable for detecting α-particles, ^3H, and ^{125}I. Other emulsions produced by these companies have physical properties suitable for other isotopes and experimental conditions.

Prior to use on experimental slides, the background should be determined on each bottle of emulsion. Two or more precleaned slides should be coated with the prepared emulsion, dried, and developed by the standard procedure to evaluate the background microscopically. Although time-consuming, this step will guard against commitment of valuable test slides to inferior or damaged emulsion. (Under optimum conditions, the background should be 1 grain per cell area or less, and if the count exceeds 2 grains per cell area, consideration should be given to discarding the batch of emulsion.) Some investigators recommend that emulsion be heated only once, but we have found that emulsion may be carefully heated several times without measurable deterioration. The background should be rechecked each time.

5.2. Background and Grain Development During Exposure

Even in a carefully executed experiment, occasionally excessive background will develop during exposure, or grains resulting from radioactive disintegrations will not be produced at the expected rate. Multiple factors, ranging from light leaks to improper drying of the emulsion, may introduce these artifacts. These factors should be monitored by two types of controls; (1) precleaned slide(s) (no sample) coated with emulsion to measure background, and (2) slide(s) bearing cells or tissue of known activity, to measure grain production. These control slides should be processed and stored with the experimental slides.

A large number of slides to serve as known positive controls may be prepared from a volume of cultured cells and labeled with the isotope(s) under study. The amount of radioactivity may be adjusted to yield grain counts similar to grain counts in the experimental slides under the chosen experimental conditions. The TLI or LI of the positive control slides can be evaluated over the same range of exposure to be used for the experimental slides. The supply of fixed standard slides should be stored under cool, dry conditions and one or more slides included with each experiment. Although these slides cannot be considered true standards, they do serve as a check on the reproducibility of autoradiographic conditions and are particularly useful when unexpected labeling characteristics are observed on the experimental autoradiographs.

5.3. Negative or Positive Chemography

Negative or positive chemography may result when some factor(s) in the tissue or cells either inhibits activation of silver grains or promotes activation of silver grains in the absence of radiation (64). Chemography is the direct action of reactive groups within the biological material on the photographic emulsion that results in production of a latent image (positive) or, alternatively, may render the emulsion incapable of registering charged particles (negative). Reducing substances in the tissue may produce a latent image or oxidizers may destroy the latent image (53). The rate of development may be increased by alkaline materials or decreased by acid substances.

Negative chemography or inhibition of grains is an insidious problem because it can occur on autoradiographs that appear entirely normal, even ideal, with very low background. The necessity of controls for negative chemography has been demonstrated when sections of tissues, either fixed or frozen, are used, but may be equally necessary in dispersed or single-cell preparations, depending on the procedures employed prior to application of the emulsion.

To detect negative chemography, extra slides are prepared from representative, experimental tissues and are processed with the test slides. Just prior to development, the extra slides are exposed to strong daylight, then processed with the other slides. The exposed slides should be black if the layer of emulsion is uniform and if no chemical reaction occurred between the underlying tissue and the emulsion. In an exposed slide, light areas or streaks from the tissue in the direction that wet slides were hung are indicative of negative chemography. The light areas occur when grain activation is inhibited either in the tissue or section or in the path of material leached from the section. The reduction in grain formation that results is seldom uniform, even over a single slide, and highly erratic data may result.

Identification of negative chemography as a serious problem in autoradiographs of sections of a variety of tissues (fixed in buffered formalin and embedded in paraffin) led to routine treatment in my laboratory of deparaffinized sections with $1N$ HCl at room temperature for 30 min followed by careful washing in water (72) prior to application of the emulsion (Fig. 6). This technique has effectively eliminated the problem. Prestaining of slides, as well as other techniques for preparation of tissues, may accomplish the same purpose, but the absence of negative chemography should be established for each study.

Positive chemography is more readily identifiable because the grains produced are frequently unrelated to individual cells or nuclei and can be monitored by inclusion of a slide or slides with section(s) or smear(s) identical in origin and processing to the experimental slides, but without exposure to radioactive material. This will permit any chemical reaction between the underlying tissue and the emulsion to be monitored.

6. CONCLUSIONS

Classical techniques that utilize a labeled precursor and autoradiography for measurement of kinetic parameters such as labeling indices are highly sensitive and particularly useful in studies of small cell populations or situations in which the location of labeled cells in relation to vasculature, necrosis, or periphery of tissue is necessary for interpretation of data. An important advantage is that cells in solid tissues do not require dispersion into single-cell preparations, with the attendant possibility of loss of fragile cells or specific cell types. The sensitivity of autoradiography permits identification of cells engaged in minimal metabolic activity (such as DNA synthesis) that might be incorrectly classified by other

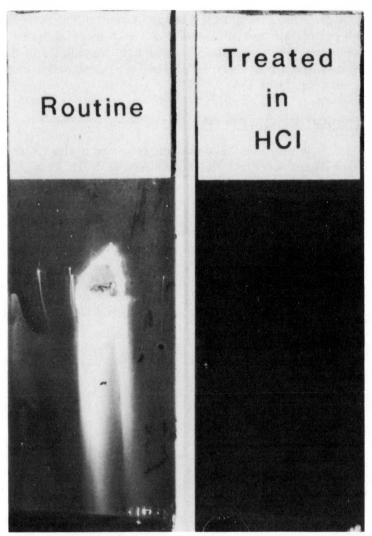

Fig. 6. Autoradiographs that demonstrate the effects of negative chemography after routine histologic processing or no apparent chemography after treatment with 1*N* HCl at room temperature for 30 min. The slides bearing sections of paraffin-embedded murine tumors were processed through xylene, graded alcohols to water (routine) or xylene, graded alcohols to water, followed by HCl treatment and water washes to remove the acid. Slides were held in water during transport to the darkroom for application of the emulsion (*69,70*). After storage, each slide was exposed to light prior to development.

techniques. Information on the range of synthetic activity within the cell population can be obtained. As with most laboratory procedures, controls are required to establish the validity of the observed data and are an essential part of a well-designed experiment to measure an LI or TLI.

ACKNOWLEDGMENTS

Previously unpublished data were obtained under Grant RO1-CA25313-03 and Contract NO1-CM97309, DEB, DCT, NCI, NIH.

REFERENCES

1. Allison, D., Bose, K., Ridolpho, P., and Meyne, J. Lightly [^{3}H]-thymidine-labeled bone marrow cells are in G1-G2. Proc. 8th Ann. Meeting of the Cell Kinetics Society. Cell Tissue Kinet., *17*: 669, 1984.
2. Allison, D. C., Yuhas, J. M., Ridolpho, P. F., Anderson, S. L., and Johnson, T. S. Cytophotometric measurement of the cellular DNA content of [^{3}H] thymidine-labelled spheroids. Cell Tissue Kinet., *16*: 237–246, 1983.
3. Amano, M. and Everett, N. B. Preferential labeling of rat lymphocytes with a rapid rate of turnover by tritiated deoxycytidine. Cell Tissue Kinet., *9*: 167–177, 1976.
4. Amano, M., Messier, B., and Leblond, C. P. Specificity of labelled thymidine as a deoxyribonucleic acid precuror in radioautography. J. Histochem. Cytochem., *7*: 153–155, 1959.
5. Barendsen, G. W., Roelse, H., Hermens, A. F., Madhuizen, H. T., van Peperzeel, H. A., and Rutgers, D. H. Clonogenic capacity of proliferating and non-proliferating cells of a transplantable rat rhabdomyosarcoma in relation to its radiosensitivity. J. Natl. Cancer Inst., *51*: 1521–1526, 1973.
6. Baserga, R. and Kisielaski, W. E. Comparative study of the kinetics of cellular proliferation in normal and tumorous tissues with use of tritiated thymidine. J. Natl. Cancer Inst., *28*: 331–339, 1962.
7. Baserga, R. and Malamud, D. Autoradiography. New York: Harper and Row, 1969.
8. Beck, H.-P. Radiotoxicity of incorporated [^{3}H] thymidine as studied by autoradiography and flow cytometry. Cell Tissue Kinet., *14*: 163–177, 1981.
9. Beck, H.-P. Radiotoxicity of incorporated [^{3}H] thymidine. Cell Tissue Kinet., *15*: 469–472, 1982.

10. Bertalanffy, F. D. and Lau, C. Rates of cell division of transplantable malignant tumors. Cancer Res., *22*: 627–631, 1962.

11. Braunschweiger, P. G., Poulakos, L., and Schiffer, L. M. *In vitro* labeling and gold activation autoradiography for determination of labeling index and DNA synthesis times and solid tumors. Cancer Res., *36*: 1748–1753, 1976.

12. Bresciani, F. Effect of ovarian hormones on duration of DNA synthesis in cells of the C3H mouse mammary gland. Exp. Cell Res., *38*: 13–32, 1965.

13. Bryant, T. R. Hydrolysis in Feulgen autoradiography. I. Loss of label from S-period nuclei and its subsequent association with non-S chromatin. Exp. Cell Res., *56*: 127–133, 1969.

14. Burki, J. H., Bunker, S., Ritter, M., and Cleaver, J. E. DNA damage from incorporated radioisotopes: Influence of the ^{3}H location in the cell. Radiat. Res., *62*: 299–312, 1975.

15. Chan, P. C., Lisco, E., Lisco, H., and Adelstein, S. J. The radiotoxicity of iodine-125 in mammalian cells. II. A comparative study on cell survival and cytogenetic resonses to [125]IUdR, [131]IUdR and [3]HTdR. Radiat. Res., *67*: 332–343, 1976.

16. Clausen, O. P. F., Thorud, E., and Bolund, L. DNA synthesis in mouse epidermis. Virchows Arch. (Cell Pathol.), *34*: 1–11, 1980.

17. Cleaver, J. E. Thymidine Metabolism and Cell Kinetics (Appendix). Amsterdam: North-Holland, 1967.

18. Dendy, P. P. and Smith, C. L. Effects on DNA synthesis of localized irradiation of cells in tissue culture by (i) a U.V. microbeam and (ii) an α-particle microbeam. Proc. Roy. Soc. (B), *160*: 328–344, 1964.

19. Dendy, P. P., Smith, C. L., and Wildy, P. A pool size problem associated with the use of tritiated thymidine. Nature, *194*: 886–887, 1962.

20. Dethlefsen, L. A. ^{3}H-5-Iodo-2′-deoxyuridine toxicity. Cell Tissue Kinet., *7*: 213–222, 1974.

21. Diab, I. M. and Roth, L. J. Autoradiographic differentiation of free bound, pure, and impure thymidine [3]H. Stain Technol., *45*: 285–291, 1970.

22. Durie, B. B. M. and Salmon, W. E. High speed scintillation autoradiography. Science, *190*: 1093–1095, 1975.

23. Eisen, M. A criterion for classifying radioactive cells. Int. J. Appl. Radiat. Isot., *27*: 695–697, 1976.

24. England, J. M., Rogers, A. W., and Miller, R. G. The identification of labelled structures and autoradiographs. Nature, *242*: 612–613, 1973.

25. Evans, E. A. Purity and stability of radiochemical tracers in auto-radiography. J. Microsc., *96*: 165–180, 1971.

26. Evans, E. A. Autoradiography with tritium. *In*: Tritium and Its Compounds. New York: John Wiley, 1974.

27. Evans, E. A., Sheppard, H. C., and Turner, J. C. Validity of tritium tracers. Stability of tritium atoms in purines, pyrimidines, nucleosides and nucleotides. J. Labelled Compounds, *VI*: 76–87, 1970.

28. Forsdyke, D. R. Application of the isotope-dilution principle to the analysis of factors affecting the incorporation of [³H] uridine and [³H] cytidine into cultured lymphocytes. Biochem. J., *125*: 721–732, 1971.

29. Fox, B. W. and Prusoff, W. H. The comparative uptake of I^{125}-labeled 5-iodo-2'-deoxyuridine and thymidine-H^3 into tissues of mice bearing hepatoma-129. Cancer Res., *25*: 234–240, 1965.

30. Gahan, P. B., ed. Autoradiography for Biologists. London: Academic, 1972.

31. Goldspink, D. F. and Goldberg, A. L. Problems in the use of [Me³H] thymidine for the measurement of DNA synthesis. Biochem. Biophys. Acta, *299*: 521–532, 1973.

32. Gude, W. D. Autoradiographic Techniques. Englewood Cliffs: Prentice-Hall, 1968.

33. Hamatani, K. and Amano, M. Different labeling patterns in mouse lymphoid tissues with [³H] deoxycytidine and [³H] thymidine. Cell Tissue Kinet., *13*: 435–443, 1980.

34. Hamilton, E. and Dobbin, J. [³H] Thymidine labels less than half of the DNA synthesizing cells in the mouse tumor, carcinoma NT. Cell Tissue Kinet., *15*: 405–411, 1982.

35. Helpap, B. and Dachselt, U. The pattern of lymphocytes in the thymus and spleen after labeling with ³H-thymidine and ³H-deoxycytidine. Virchows Arch. (Cell Path.), *28*: 287–299, 1978.

36. Hirt, A. and Wagner, H. P. Nuclear corporation of radioactive DNA precursors and progression of cells through S. Cell Tissue Kinet., *8*: 455–466, 1975.

37. Howard, A. and Pelc, S. R. Nuclear incorporation of P^{32} as demonstrated by autoradiographs. Exp. Cell Res., *2*: 178–187, 1951.

38. Howard, A. and Pelc, S. R. Synthesis of deoxyribonucleic acid in normal and irradiated cells and its relationship to chromosome breakage. Heredity (Suppl.), *6*: 261–273, 1953.

39. Iversen, O. H. and Bjerknes, R. Kinetics of epidermal reaction to carcinogens. Oslo: Universitetsforlaget, 1963.

40. Kopriwa, B. M. Quantitative investigation of scintillator intensification for light and electron microscope radioautography. Histochemistry, *68*: 265–279, 1980.

41. Kopriwa, B. M. A comparison of various procedures for fine grain development in electron microscopic radioautography. Histochemistry, 44: 201–224, 1975.

42. Lang, W., Muller, D., and Maurer, W. Prozentuale beteiligung von exogenem thymidin an der synthese von DNA-thymin in geweben der maus und Hela zellen. Exp. Cell Res., 49: 558–571, 1968.

43. Langager, J. M., Howard, G. A., and Baylink, D. J. An improved technique for rapid autoradiography of cells and tissue sections. Histochemistry, 75: 523–531, 1982.

44. Lloyd, H. H. and Simpson-Herren, L. Tumor dependence of observed thymidine index as a function of emulsion exposure. Cancer Res., 43: 1138–1144, 1983.

45. Marques-Pereira, J. P. and Leblond, C. P. Mitoses and differentiation in the stratified epithelium of the rat oesophagus. Am. J. Anat., 117: 73–89, 1965.

46. Maurer, H. R. Potential pitfalls of [^{3}H] thymidine techniques to measure cell proliferation. Cell Tissue Kinet., 14: 111–120, 1981.

47. Mendelsohn, M. L. Autoradiographic analysis of cell proliferation in spontaneous breast cancer of C3H mouse. II. Growth and survival of cells labeled with tritiated thymidine. J. Nat. Cancer Inst., 25: 485–500, 1960.

48. Mendelsohn, M. L. Autoradiographic analysis of cell proliferation in spontaneous breast cancer of C3H mouse. III. The growth fraction. J. Nat. Cancer Inst., 28: 1015–1029, 1962.

49. Mendelsohn, M. L., Dohan, Jr., C. F., and Moorse, Jr., H. A., Autoradiographic analysis of cell proliferation in spontaneous breast cancer of C3H mouse. I. Typical cell cycle and timing of DNA synthesis. J. Nat. Cancer Inst., 25: 477–484, 1960.

50. Meyer, J. S. and Connor, R. E. *In vitro* labeling of solid tissues with tritiated thymidine for autoradiographic detection of S-phase nuclei. Stain Technol., 52: 185–195, 1977.

51. Meyer, J. S. and Facher, R. Thymidine labeling index of human breast carcinoma. Cancer, 39: 2524–2532, 1977.

52. Nicolini, C. The discrete phases of the cell cycle: Autoradiographic, physical and chemical evidences. J. Nat. Cancer Inst., 55(4): 821–826, 1975.

53. Odeblad, E. Artifacts in autoradiography. Acta Radiol., 39: 192–204, 1953.

54. Olszewska, M. J., Bilecka, A., Kuran, H., and Marciniak, K. Application and efficiency of scintillation for autoradiography of plant cells. Microscop. Acta, 85: 133–139, 1981.

55. Panayi, G. S. and Neill, W. A. Scintillation autoradiography—a rapid technique. J. Immunol. Meth., 2: 115–117, 1972.

56. Pelc, S. R. Incorporation of labelled precursors of DNA in non-dividing cells. In: (L. F. Lamerton and R. J. M. Fry, eds.), Cell Proliferation, Philadelphia: F. A. Davis, 1963.

57. Perez, A. G., Kim, J. H., Gelbard, A. S., and Djordjevic, B. Altered incorporation of nucleic acid precursors by mycoplasma-infected mammalian cells in culture. Exp. Cell Res., 70: 301–310, 1972.

58. Pilgrim, C., Erb, W., and Maurer, W. Diurnal fluctuations in the members of DNA synthesizing nuclei in various mouse tissues. Nature, 199: 863–865, 1963.

59. Quastler, H. The analysis of cell population kinetics. In: (L. F. Lamerton and R. J. Fry, eds.), Cell Proliferation, Philadelphia: F. A. Davis, 1963.

60. Quastler, H. and Sherman, G. G. Cell population kinetics in the intestinal epithelium of the mouse. Exp. Cell Res., 17: 420–438, 1959.

61. Reisken, A. B. and Mendelsohn, M. L. A comparison of the cell cycle in induced carcinomas and their normal counterpart. Cancer Res., 24: 1131–1136, 1964.

62. Riccardi, A., Mazzini, G., Montecucco, C., Cresci, R., Traversi, E., Berzuini, C., and Ascari, E. Sequential vincristine, arabinosylcytosine and adriamycin in acute leukemia: Cytologic and cytokinetic studies. Cytometry, 3: 104–109, 1982.

63. Rogers, A. W. Techniques of Autoradiography. Amsterdam: Elsevier/North Holland Biomedical, 1979.

64. Rogers, A. W. Scintillation autoradiography at the light microscope level: A review. Histochem. J., 13: 173–186, 1981.

65. Sawicki, W., Blaton, O., and Rowinski, J. Correction of autoradiographic grain count in respect to precisely calculated background. Histochemie, 26: 67–73, 1971.

66. Sawicki, W. Ostrowski, K., and Platkowska, E. High-speed autoradiography of ³H-thymidine-labelled nuclei. Histochemistry, 52: 341–347, 1977.

67. Schoenfeld, D. and Kallman, R. F. Determining the labeling index in autoradiography. Cell Tissue Kinet., 13: 339–347, 1980.

68. Schultze, B. In: (A. W. Pollister, ed.), Physical Techniques in Biological Research. New York: Academic, 1969.

69. Shackney, S. E. The radiographic transfer function and its implications for radioautographic methodology. J. Nat. Cancer Inst., 55: 811–820, 1975.

70. Shackney, S. E. On the discreteness of the phases of the cell cycle. J. Nat. Cancer Inst., *55*: 826–829, 1975.

71. Shackney, S. E. and Ritch, P. S. Percent labeled mitosis curve analysis *In*: (J. W. Gray and Z. Darzynkiewicz, eds.), Techniques in Cell Cycle Analysis, New Jersey: Humana, 1986.

72. Simpson-Herren, L., Sanford, A. H., and Holmquist, J. P. Cell population kinetics of transplanted and metastatic Lewis lung carcinoma. Cell Tissue Kinet., *7*: 349–361, 1974.

73. Simpson-Herren, L., Sanford, A. H., Holmquist, J. P., Springer, T. A., and Lloyd, H. H. Ambiguity of the thymidine index. Cancer Res., *36*: 4705–4709, 1976.

74. Smets, L. A. Discrepancies between precursor uptake and DNA synthesis in mammalian cells. J. Cell Physiol., *74*: 63–66, 1969.

75. Smets, L. A. and Brohee, H. A hidden pool effect of thymidine incorporation in HeLa Cells. Int. J. Radiat. Biol., *17*: 93–96, 1970.

76. Steel, G. G. Growth Kinetics of Tumours. Oxford: Clarendon Press, 1977.

77. Stillstrom, J. Grain count corrections in autoradiography. Int. J. Appl. Radiat. Isot., *14*: 113–118, 1963.

78. Tannock, I. F. The relation between cell proliferation and the vascular system in a transplanted mouse mammary tumor. Br. J. Cancer, *22*: 258–273, 1968.

79. Taylor, J. A., Woods, P. S., and Hughes, W. L. The organization and duplication of chromosomes as revealed by autoradiographic studies using tritium-labeled thymidine. Proc. Nat. Acad. Sci. USA, *43*: 122–128, 1957.

80. Woodcock, C. L. F., D'Amico-Martel, A., McInnis, C. J., and Annunziato, A. T. How effective is "high-speed" autoradiography? J. Microsc., *117*: 417–423, 1979.

Chapter 2

Percent Labeled Mitosis Curve Analysis

Stanley E. Shackney and Paul S. Ritch

1. EXPERIMENTAL TECHNIQUE

The measurement of percent labeled mitoses (PLM, or alternatively, FLM, for fraction of labeled mitoses) was developed in 1959 by Quastler and Sherman (8) for estimating the durations of the component phases of the cell cycle.

The PLM method involves the use of the technique of autoradiography. In brief, the technique of autoradiography is described as follows: Cells that are actively synthesizing DNA and are exposed briefly to tritiated thymidine ([³H]-TdR) will incorporate this radiotracer exclusively into DNA. Representative samples of cells from the population under study are fixed, placed on glass slides, coated with a sensitive photographic emulsion, and stored in the dark for periods ranging from several days to several months. [³H]-TdR emits weak β particles that travel only 1–2 μm. These β particles activate the emulsion overlying the cell nucleus from which they originated. Thus, when the slide is developed and fixed like a photographic negative, cells in S-phase are identified by the presence of silver grains overlying their nucleii. Tritiated thymidine labeling and the autoradiographic process are discussed in more detail in chapter 1 (12).

For PLM analysis, a population is pulse labeled with [³H]-TdR, and serial samples are obtained at intervals after pulse labeling.

The duration of the phases of the cell cycle and the duration of the cell cycle itself are determined from the changes in the percent labeled mitotic cells over the course of time.

2. PLM ANALYSIS IN KINETICALLY HOMOGENEOUS POPULATIONS

Let us consider a hypothetical cell population in which T_{g1} is 2 h, T_s is 6 h, T_{g2} is 1 h, and T_m (the duration of mitosis) is 1 h, as shown in Fig. 1A$_1$. Their sum, T_c, is 10 h. Let us now consider cell position in the cell cycle on a scale of time required to reach the *next* mitosis instead of time elapsed since the last mitosis. Following a pulse of [^{3}H]-TdR, only cells in S-phase will incorporate labeled tritium, but it will take at least 1 h (the duration of T_{g2}) for any of these labeled cells to move into mitosis. If mitotic cells are examined within 1 h of pulse [^{3}H]-TdR exposure (Fig. 1B$_1$), none will be labeled (Fig. 1B$_2$). From 1 to 7 h after [^{3}H]-TdR exposure, labeled cells will enter mitosis (Fig. 1C$_1$–1G$_1$), and can be detected and quantitated by radioautography (Fig. 1C$_2$–G$_2$). At the time of [^{3}H]-TdR pulse exposure, the G$_1$ cells were not synthesizing DNA; thus, 7–9 h later they will enter the mitotic pool as unlabeled cells (Fig. 1G$_1$–G$_2$). Ten hours after pulse exposure, all the cells will have cycled through mitosis, and the daughter cells will be in the same relative positions that were occupied by their parents at the time of original pulse [^{3}H]-TdR exposure (Fig. 1H$_1$). During the next 10 h after pulse, the PLM curve of the daughter cells should recapitulate that of their parents (Fig. 1I$_1$, I$_2$).

In principle, $T_{g2} + \frac{1}{2}T_m$ can be estimated from the delay in the upstroke of the first wave of labeled mitosis, and T_s from the width of the labeled mitotic wave (Fig. 1I$_2$). Since the interval between successive labeled mitotic waves is that of the cell cycle itself, T_c can also be estimated from the PLM curve. T_{g1} can be calculated from the simple relation that the sum of the durations of the individual phases is equal to the total cell cycle time. That is:

$$T_{g1} + T_s + T_{g2} + T_m = T_c \qquad [1]$$

This analysis assumes that the cell cycle and its phases are of uniform duration. Indeed, early studies in very rapidly proliferating cell populations, such as embryonic mouse tissues (7),

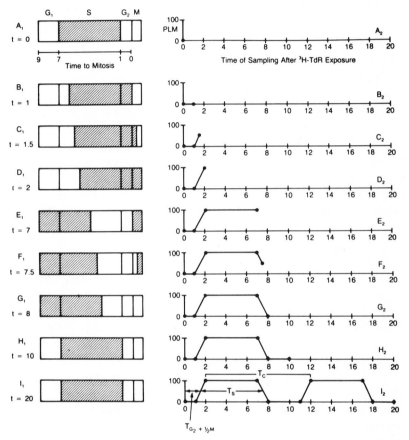

Fig. 1. A schematic representation of the basis for PLM curve analysis.

rat intestinal epithelium (2), and log phase mouse tumor cells in tissue culture (5), produced PLM curves that were similar in many respects to the ideal curve shown in Fig. 1. Thus, early in the 1960s it was generally assumed that T_c and the phase durations were tightly clustered about their respective mean values. The method shown in Fig. 1I_2, or closely related variants (6), was used to analyze a wide variety of PLM curves.

3. PLM ANALYSIS IN KINETICALLY HETEROGENEOUS POPULATIONS

3.1. General Considerations

It soon became apparent that most PLM curves deviate considerably from the ideal, necessitating revisions of the PLM method that would take population kinetic heterogeneity into account.

In reality, it is exceedingly rare for the trough between the first and second wave to fall to zero, and the greater the degree of kinetic heterogeneity, the broader and lower the second wave of the PLM curve. In most human tumors, the second labeled mitotic wave is low, broad, and often difficult to identify at all. In the mid- and late-1960s, computer-based automated curve-fitting methods were developed for PLM curve analysis (14,15). These methods assumed the existence of independent distributions of the cell cycle phase durations. These phase-duration distributions were fitted to the PLM curve data by numerical methods, and the putative distribution of T_c's could then be calculated.

Other methods have also been developed for estimating T_c. We might note, for example, that the [^{3}H]-TdR label acquired during pulse exposure is divided approximately equally between both daughter cells at mitosis. Hence, the mean autoradiographic grain count of labelled cells should fall by half with each successive round of cell division, and T_c can be estimated from the time required for the mean grain count to halve (17).

Each of the various autoradiographic methods used for estimating T_c and the durations of the cycle phases is associated with its own technical pitfalls (12). Overall, the PLM technique provides more information regarding population proliferative characteristics than most other methods, but it alone is insufficient to characterize population proliferative characteristics completely. This can be attributed largely to the difficulties in identifying and characterizing slowly proliferating cells in the presence of rapidly proliferating cells.

3.2. PLM Analysis in Relation to the Growth Fraction and the G0 Cell Pool

It was recognized from the beginning that all cells in a population were not identical with respect to their proliferative behavior.

According to early conceptual formulations, cells were classified as either dividing or nondividing. Dividing cells were treated as having a uniform cell cycle time and uniform cycle phase durations. The nondividing cells were considered to be indistinguishable morphologically from dividing cells, but they did not incorporate [³H]-TdR or contribute to the mitotic cell pool. In the early 1960s, Mendelsohn (4) proposed that the fraction of dividing cells, or the *growth fraction*, could be estimated from the discrepancy between the tritiated thymidine pulse labeling index, or LI, that would be expected for the population if all cells were dividing and the LI that is actually observed, as in the following example. Other aspects of growth fraction estimation are discussed in chapter 3(*1*).

Let us assume that T_c is 10 h and T_s is 6 h. Half of the cells are nonproliferating cells (Fig. $2A_1$). Since the duration of S-phase represents 60% of the total cell cycle time, an LI of 0.6 would be expected if all cells were dividing. Since the nonproliferating cells contribute to the denominator, but not to the numerator of the observed LI, the observed LI is only 0.3. The following equation is easy to verify:

$$\text{growth fraction} = \frac{\text{observed LI}}{\text{expected LI}} \qquad [2]$$

When PLM curve data are avaialble, one can obtain T_s and T_c separately (Fig. $2A_2$), and calculate the LI that would be expected for the dividing cells from the ratio T_s/T_c. Given the observed LI for the entire population, one can then calculate the growth fraction using Eq. [2].

The nondividing cell compartment was thought to contain cells of two basic types: terminally differentiated cells that would never divide again, and nondividing cells that retain the potential for proliferation (7). Quastler suggested that these potentially proliferative cells produced new cells either at a low rate or only upon stimulation. He proposed the term G0 for this potentially reversible nondividing state.

The distinction between a truly nonproliferative G0 state and a very long G1-phase remains a controversial issue to this day. This distinction is difficult to make on conceptual as well as experimental grounds. A cell that "left" the cell cycle from G1, resided in the G0 state, and then "returned" to the cell cycle at a later time would

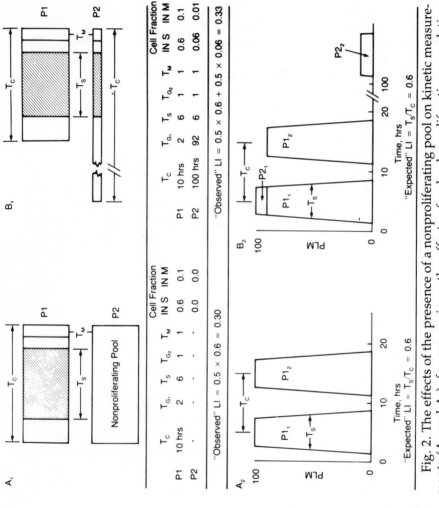

Fig. 2. The effects of the presence of a nonproliferating pool on kinetic measurements (A_1 and A_2); for comparison, the effects of a slowly proliferating population are also shown (B_1 and B_2).

be indistinguishable from a cell that never left the cell cycle at all, but remained in the G1-phase for a long period of time.

Quastler's original definition of G0 was deliberately vague in distinguishing between slowly dividing and truly nondividing cells. His purpose was to distinguish these slowly dividing or potentially proliferative nondividing cells from rapidly proliferating cells on the one hand, and terminally differentiated nondividing cells on the other.

The exclusion of G0 cells from the growth fraction was based on the premise that the cell cycle time of rapidly proliferating cells was uniform, or nearly so (5).

3.3 Problems of Characterizing Broad Cell Cycle Time Distributions by PLM Analysis

In early experimental studies, the average cell cycle times in many mouse tissues were often found to be in the range of 8–12 h; cells with cycle times exceeding 24–36 h were often considered either to be slowly proliferating or in G0 for all practical purposes. Later studies in human tumors indicated mean or median cell cycle times of 2–3 d (13), with considerable variability in cell cycle time, both within and among tumors. More recent computer modeling studies have suggested that the range of cell cycle times is such that one might find some cells dividing every 24 h and others dividing every 40 d in the same tumor cell population (10). This degree of kinetic heterogeneity was not anticipated in early studies, and it should prompt us to reexamine early basic concepts in the light of this new information.

The problem of identifying slowly proliferating cells in the presence of rapidly proliferating cells can be appreciated from the example shown in Figs. 2B$_1$ and 2B$_2$. The population is composed of two subpopulations represented with equal frequency, one with a 10-h cell cycle time and the other with a cycle time of 100 h. Assigned cycle phase durations are shown in the figure. The overall "observed" labeling index for the population is 0.33. In this example, 10% of the rapidly growing cells and 1% of the slowly growing cells are in mitosis at any given time. Thus, although rapidly growing cells and slowly growing cells are equally represented in the *interphase* cell pool, in the *mitotic* cell pool the slowly proliferating cells are outnumbered by 10 to 1. Hence, the second labeled mitotic wave of the PLM curve appears 10 h after the first, and achieves a height exceeding 90% labeled mitoses (Fig. 2B$_2$). Given an S-phase duration of 6 h and a cell cycle time of 10 h (obtained from the PLM curve directly), one can calculate an "expected" labeling index of 0.6. This is almost twice the "observed" value. If half the population were composed of truly nonproliferating cells rather than slowly proliferating cells (Fig. 2A$_1$), there would be a comparable discrepancy between the observed labeling index and that expected from the PLM curve.

This example illustrates two important points. First, the PLM curve is always dominated by the most rapidly proliferating cells in the population; this results in a false impression of cell cycle time uniformity. Second, since only a small fraction of slowly proliferating cells are detected by [^{3}H]-TdR labeling, they contribute little to cell kinetic measurements, and are virtually indistinguishable from truly nonproliferating cells both in theory and in practice.

In the foregoing example, we considered a mixture of two populations with cell cycle times that differed by a factor of 10. In this instance, the distinction between slowly proliferating cells and G0 cells would be of little practical consequence.

However, cells with intermediate cell cycle times are also subject to the biases inherent in the PLM curves method. Such cells are also underrepresented in the PLM curve, but in varying degrees. When large numbers of cells with intermediate cell cycle times are present, the resultant PLM curves have very different properties from those obtained in kinetically homogeneous cell populations, as shown in the following example.

Let us consider a population that contains 10 cell cycle time classes ranging from 10 h to 100 h, as shown in Fig. 3A. In this example, cell cycle time classes are represented equally in interphase, but the most rapidly proliferating cells are better represented in the mitotic cell pool (Fig. 3B). The resultant PLM curve is shown in Fig. 3C, together with the relative contributions of each cycle time class. For simplicity, only the contribution of the 10-h cycle time class is identified explicitly (shaded areas). It is readily apparent that the second mitotic wave is low and broad; it is not sharply peaked, as in the example in Figs. 1 and 2. Although all the cycle time classes contribute to this low, broad second wave, it consists mostly of the repeated divisions of the most rapidly proliferating cells. The *height* of the second wave depends largely on the *relative*

A

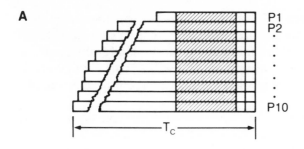

B

	T_C	T_{G_1}	T_S	T_{G_2}	T_M	Interphase Cell Frequency	Mitotic Cell Frequency
P1	10	2	6	1	1	0.1	0.342
P2	20	12	6	1	1	0.1	0.171
P3	30	22	6	1	1	0.1	0.113
P4	40	32	6	1	1	0.1	0.086
P5	50	42	6	1	1	0.1	0.069
⋮	⋮	⋮	⋮	⋮	⋮	⋮	⋮
P10	100	92	6	1	1	0.1	0.034

C

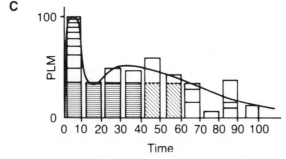

Fig. 3. The effects of cell cycle time heterogeneity on the PLM curve.

abundance of rapidly proliferating cells. Since the amount of label per cell is diluted with successive cell divisions, the *breadth* of the second wave depends largely on the number of divisions that can be detected autoradiographically before cells become too lightly labeled to be detected as such. This, in turn, depends on such factors as initial interphase cell labeling intensity and the grain count detection threshold. For example, if the interphase cell mean grain count of the rapidly proliferating cells is 16 grains/cell initially and the detection threshold is 2 grains/cell, then three successive divisions of labeled rapidly proliferating cells will be detected. If the initial mean grain count of the most rapidly proliferating cells is 64 grains/cell, then the second wave of the PLM curve will include five successive divisions of labeled rapidly proliferating cells (dashed bars, Fig. 3C).

For purposes of exposition, the foregoing example is still highly oversimplified. For example, transitions among cycle time classes during the course of the observation period are not considered; however, more extensive and rigorous computer modeling studies have supported the essential elements of PLM curve analysis described above (10).

The simulated PLM curve in Fig. 3 conforms with observations in kinetically heterogeneous populations. Many experimental tumors and virtually all human tumors lack sharply peaked second waves, yet there is progressive mitotic label dilution that results in light labeling in the middle and late portion of the second PLM wave. This may account for well-recognized discrepancies between human PLM curve data and theoretical PLM curves derived from conventional models (13), since conventional models for PLM curve analysis make no provision for label dilution.

4. COMPREHENSIVE PLM AND GRAIN COUNT HALVING METHODS

If the PLM curve is relatively insensitive to the presence of the more slowly proliferating cells, how *can* the cell cycle time distribution be studied? It is apparent from Fig. 3 that changes in mitotic grain counts over time strongly reflect the kinetic behavior of the most rapidly proliferating cells in the population. On the other hand, since interphase cells are not subject to the biases inherent in mitotic cell studies, changes in *interphase* cell grain counts would reflect the kinetic behavior of the overall cell population. A comparison of differences in the behavior of the mitotic grain count distribution over time and the behavior of the interphase cell grain count distribution over time can be used to reconstruct the overall cell cycle time distribution, using computer simulation techniques. The details of this approach are described elsewhere (10). To date, two human malignancies—adult acute leukemia and human melanoma—have been analyzed extensively using this approach.

The cell cycle time distribution in adult acute leukemia ranges from 16 h to approximately 600 h. In Fig. 4, the cell cycle time

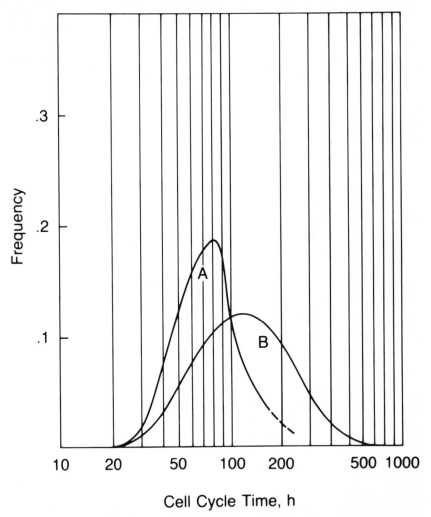

Fig. 4. A comparison of calculated cell cycle time distributions in human malignant melanoma, obtained by two different methods of kinetic analysis.

distribution in human melanoma that was obtained in these studies (curve B) is compared with that obtained by conventional PLM curve analysis (curve A). It should be noted that the abcissa is plotted on a log scale, and that the simulated cell cycle time distribution ranges from 24 h to over 960 h (40 d). Many of the cells with cycle times in the range of 120–960 h would be excluded from the growth fraction by conventional methods of kinetic analysis (*11*).

5. CONCLUSIONS

After all is said and done, why not simply consider these excluded slowly proliferating cells to be G0 cells, as shown in Fig. 5A? Conceptually, it would be simpler to think of growth retardation in terms of a reduction in the growth fraction and entry into the G0 compartment, rather than in terms of a shift in the cell cycle time distribution to longer values (Fig. 5B). By the same token, it would be simpler to think of cell recruitment as a discrete transition from the G0 compartment to the proliferating cell compartment, than as a shift in the cell cycle time distribution from longer to shorter values (Fig. 5C).

Indeed, for most qualitative purposes, the growth fraction-G0, two-compartment model may be quite servicable. However, the two-compartment model is not suited for quantitative applications. In populations that contain large numbers of slowly proliferating cells, when the more slowly proliferating cells are systematically excluded from the cell cycle time distribution, the calculated mean (or median) cell cycle time is spuriously low, and the range of cell cycle times is spuriously narrow (*see* Fig. 4, for example). Published estimates of average cell cycle times in human tumors based on conventional methods of analysis are generally in the range of 2–3 d (*13,16*). The upper limit of the range has been estimated to be about 5–7 d (*16*). In retrospect, the apparent kinetic homogeneity of human tumors may well be a result of conventional methods of PLM curve analysis, which "see" only a narrow region of the cell cycle time distribution containing the rapidly proliferating cells.

This apparent kinetic homogeneity within and among most human tumors has tended to discourage the collection of additional kinetic data in humans. Yet it is among the more slowly proliferating cells that the greatest kinetic differences among tumors are likely to be found. In those human tumors for which sufficient data are available, computer modeling studies that utilize both mitotic and interphase cell data have demonstrated distinctly different patterns

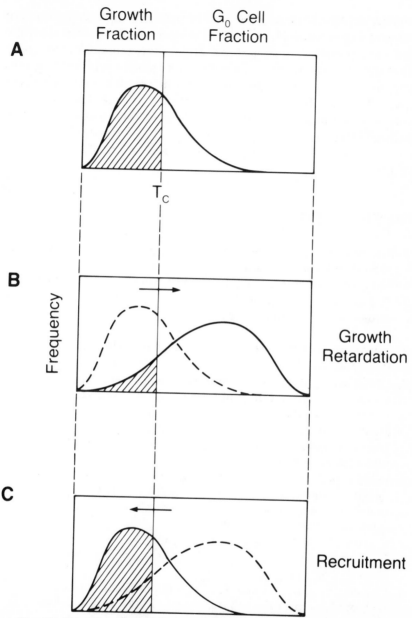

Fig. 5. (A) The relationship among the growth fraction, the non-proliferating pool, and the cell cycle time distribution. (B) The conceptual equivalence of entry into G0 and a shift of the cell cycle time distribution to longer values. (C) The conceptual equivalence of recruitment from G0 into rapid cycle and a shift of the cell cycle time distribution to shorter values.

of DNA synthesis among the slowly proliferating cell fractions of different tumor types (9). It remains to be determined whether these distinctive patterns are subject to therapeutic exploitation. However, the distinctive kinetic features of the more slowly proliferating cells raise interesting possibilities. Cells with intermediate cell cycle times (say, 3–12 d) might be of particular interest as potential targets for treatment optimization with currently available drugs, especially in tumors that are responsive to therapy, but are not often curable (e.g., small cell carcinoma of the lung or small cleaved cell lymphomas).

REFERENCES

1. Braunschweiger, P. F. The tumor growth fraction: Estimation, perturbation, and prognostication? *In*: (J. W. Gray and Z. Darzynkiewicz, eds.), Techniques in Cell Cycle Analysis, Humana, New Jersey, 1986.

2. Cairnie, A. B., Lamerton, L. F., and Steel, G. G. Cell proliferation studies in the intestinal epithelium of the rat. Exp. Cell Res., *39*: 528–538, 1965.

3. Lala, P. K. and Patt, H. M. Cytokinetic analysis of tumor growth. Proc. Nat. Acad. Sci. USA, *56*: 1735–1742, 1966.

4. Mendelsohn, M. L. Autoradiographic analysis of cell proliferation in spontaneous breast cancer of C3H mouse. III. The growth fraction. J. Nat. Cancer Inst., *28*: 1015–1029, 1962.

5. Mendelsohn, M. L. Cell proliferation and tumor growth. *In*: (L. F. Lamerton and R. J. M. Fry, eds.), Cell Proliferation, Blackwell Scientific, Oxford, 1963.

6. Painter, R. B. and Drew, R. M. Studies on deoxyribonucleic acid metabolism in human cancer cell cultures (HeLa). 1. The temporal relationships of deoxyribonucleic acid synthesis to mitosis and turnover time. Lab. Invest., *8*: 278–285, 1959.

7. Quastler, H. The analysis of cell population kinetics. *In*: (L. F. Lamerton and R. J. M. Fry, eds.), Cell Proliferation, Blackwell Scientific, Oxford, 1963.

8. Quastler, H. and Sherman, F. G. Cell population kinetics in the intestinal epithelium of the mouse. Exp. Cell Res., *17*: 420–438, 1959.

9. Shackney, S. E. The role of radioautographic studies in clinical investigative oncology and chemotherapy. Cancer Treat. Rep., *60*: 1873–1886, 1976.

10. Shackney, S. E. A cytokinetic model for heterogeneous mammalian cell populations. III. Tritiated thymidine studies. Correlations among multiple kinetic parameters in human tumors. J. Theoret. Biol., *65*: 421–464, 1977.

11. Shirakawa, S., Luce, J. K., Tannock, I., and Frei III, E. Cell proliferation in human myeloma. J. Clin. Invest., *49*: 1188–1199, 1970.

12. Simpson-Herren, L. Autoradiographic techniques for measurement of the labeling index. *In*: (J. W. Gray and Z. Darzynkiewicz, eds.), Techniques in Cell Cycle Analysis, Humana, New Jersey, 1986.

13. Steel, G. G. Growth Kinetics of Tumours. Clarendon, Oxford, 1977.

14. Steel, G. G. and Hanes, S. The technique of labeled mitoses: Analysis by automatic curve-fitting. Cell Tissue Kinet., *4*: 93–105, 1971.

15. Takahashi, M., Hogg, G. D., and Mendelsohn, M. L. The automatic analysis of FLM curves. Cell Tissue Kinet., *4*: 505–518, 1971.

16. Tannock, I. Cell Kinetics and chemotherapy: A critical review. Cancer Treat. Rep., *62*: 1117–1133, 1978.

17. Wimber, D. E. Methods for studying cell proliferation with emphasis on DNA labels. *In*: (L. F. Lamerton and R. J. M. Fry, eds.), Cell Proliferation, Blackwell Scientific, Oxford, 1963.

Chapter 3

Tumor Growth Fraction Estimation, Perturbation, and Prognostication

Paul G. Braunschweiger

1. INTRODUCTION

The concept that solid tumors can be comprised of both actively cycling and noncycling cell populations was originally proposed by Mendelsohn (32) to account for the observation that after [³H]-TdR (tritiated thymidine) labeling in vivo, the fraction of labeled mitotic cells always exceeded that for interphase cells. Since the mitotic population can be considered a pure cycling cohort, it was concluded that a significant portion of the interphase population must be in a noncycling or quiescent (Q) configuration. The term "growth fraction (GF)" was thus coined to distinguish the cell population, in the tumor, that is actively engaged in cell cycle traverse (P cells) from that that is not (Q cells).

In experimental mammary tumors, autoradiographic studies suggested that cell proliferation is greatest in areas immediately adjacent to the tumor vasculature and that the fraction of [³H]-TdR labeled cells decreases with increasing distance from the capillaries (56). Since cell cycle times for proliferating cells remote from the capillaries were similar to those adjacent to the vasculature, the observed proliferation gradient could be explained on the basis of a similarly decreasing growth fraction gradient. Inasmuch as cell killing by chemotherapeutic and clinically relevant radiotherapy doses is generally proliferation-dependent, the tumor GF is an important clinical consideration; especially if Q cells can reenter the division cycle under appropriate stimuli.

In this paper I will review methods to estimate tumor GF, the significance of therapy-induced changes in the growth fraction for the design of therapeutic strategies, and the possible relationship between tumor GF and chemosensitivity.

2. METHODS FOR ESTIMATION OF TUMOR GROWTH FRACTION

2.1. Pulse-Labeling Methods

The first attempt to measure the fraction of Q cells in solid tumors was reported by Mendelsohn in 1962 (32). He assumed that the total tumor population consisted of P and Q subpopulations and suggested that the GF for the tumor could be calculated as the ratio of the labeling index of the total population (LI_{obs}; the fraction of cells in the total population labeled with a short treatment of [^{3}H]-TdR) to the labeling index of the P population, LI_{th}. For the fraction of proliferating cells in the population:

$$GF = \frac{\text{number of cells in S-phase}}{\text{number of cells in P + Q}} \tag{1}$$

$$\div \frac{\text{number of cells in S-phase}}{\text{number of cells in P}}$$

$$= LI_{obs}/LI_{th} \tag{2}$$

$$= P/(P + Q) \tag{3}$$

Mendelsohn proposed that the fraction of labeled proliferating cells; LI_{th}, be estimated from the Fraction of Labeled Mitotic (FLM) cells measured at several times after pulse labeling. Unfortunately, this is not straightforward. At early intervals after pulse labeling, FLM is not a good estimate of the fraction of labeled P cells because the labeled cells are synchronized. Indeed, immediately after pulse labeling, the FLM is zero, since the labeled cells have not yet reached mitosis. After several generations, the labeled mitotic cells redistribute (desynchronize) to become randomly distributed around the cycle. By this time, however, the transition of labeled proliferating cells from P to Q again renders the FLM a poor estimate of the fraction of labeled P cells. To correct for population expansion and P to Q transition, Mendelsohn suggested that the GF could be estimated using Eq. (1) at several times, t (designated GF_t). For an exponentially growing population, $\ln (1 - GF_t)$ describes a straight line whose intercept at $t = 0$ is an estimate of the GF at the time of pulse labeling.

An estimate for LI_{th} can also be obtained by computer analysis of FLM values measured periodically after pulse labeling (*see* chapter 2 and refs. 2,3,48,57 in this chapter for additional details) to provide accurate estimates of the G1-, S-, and G2-phase durations. In this approach (52), LI_{th} is calculated from T_c, T_s (the S-phase duration), and an assumption about the shape of the age distribution of the proliferating cell population.

Accumulation and accurate analysis of FLM data are essential to analyses of this type. This is not always straightforward, however, especially for measurements of experimental model tumors and for clinical material. For example, FLM analyses usually assume that all cells in DNA synthesis will be exposed to, and incorporate, sufficient [³H]-TdR to be detectable autoradiographically throughout the course of the experiment. Thus, if the emulsion exposure times are too short or if the threshold above which cells are scored as labeled is set too high, the estimate for the GF may be in error (47,48). Estimation of the GF based on analyses of FLM curves is also complicated by the fact that FLM analyses are preferentially sensitive to rapidly cycling cells. Finally, clinical application of these methods is limited by the requirement for administration of [³H]-TdR and the need for serial biopsies. Thus, although these GF analyses have led to some insights as to how therapeutic strategies may be improved, they leave much to be desired in the way of accuracy and ease of use.

2.2. Continuous or Repeated [³H]-TdR Labeling

It is theoretically possible to estimate the relative sizes of the P and Q components of a tumor population by continuously exposing the cell population to [³H]-TdR for a period sufficient to label all P cells as they progress through DNA synthesis (24,52). In most studies, continuous labeling has been approximated by repeated

[³H]-TdR injections at intervals shorter than the minimal DNA synthesis time. Tumors are sampled serially, during the labeling period, and labeling indices determined. If all cells that pass through S-phase become labeled, then only those cells that were not proliferating during the labeling period remain unlabeled. The minimum number of injections should be sufficient to cover at least two median cell cycle times, to minimize the probability that cells with S-phase durations of less than the median S-phase duration will go unlabeled. Unfortunately, the plateau LI level does not provide a reliable GF estimate, since the unlabeled cells will be diluted by subsequent growth in the population and labeled P cells may enter the Q pool. Thus, the continuous labeling curves will always approach 100%. The GF can be estimated by extrapolating the more slowly rising component of the continuous labeling curve back to the end of G2. This technique is usually impractical, however, since large errors in GF estimates may be introduced by imprecise data. Further, since cell loss and variation in cell cycle phase durations influence the shape of the continuous labeling curve, these factors must be modeled in the analysis of continuous labeling curves (52,53).

2.3. Morphological Methods

Studies of peripheral lymphocytes (38) and experimental mammary tumors (37) have indicated that nucleolar morphology may provide some indication of the proliferation status of individual tumor cells. [³H]-TdR labeling studies have indicated that cells with dense nucleoli (DN) constitute a rapidly proliferating cohort, whereas cells with trabeculate nucleoli (TN) or ring-shaped nucleoli (RSN) are components of a slowly proliferating or nonproliferating pool. Microfluorometric analyses of mammary tumor cells with RN and TN have indicated that these cells have G1 DNA content and, further, the incidence of RN and TN cells increases with the degree of tumor differentiation and decreases with increasing tumor growth rate (37,39).

Thus lymphoblasts with dense nucleoli may constitute the proliferative fraction (38) and a close correlation has been observed between the fraction of mammary tumor cells with DN and proliferation rates (37). However studies directly comparing the tumor GF, as measured by more conventional procedures, with the fraction of cells containing DN, have not been performed. It is also uncertain whether this method is applicable to tumor systems other than experimental mammary tumors. Potentially, this method offers the important advantage of rapid, point-in-time GF estimates without [³H]-TdR administration, serial sampling, or autoradiography.

2.4. Flow Cytometry

Flow cytometry is an exciting new tool that may permit detailed analysis of nonproliferating Q cells in a variety of cell systems. Differential acridine orange (AO) staining for DNA and RNA (20), chromatin susceptibility to thermal (18) or acid (20) denaturation, and bromodeoxyuridine (BrdUrd) inhibition of fluorescence from AO bound to DNA (46) have been used to subdivide the cell cycle into 12 functionally distinct compartments (21). Quiescent cells, exemplified by unstimulated peripheral lymphocytes, are characterized by highly condensed chromatin and low RNA content. Although a 24-h incubation with the thymidine analog BrdUrd suppresses AO DNA staining in proliferating cell populations, similar incubations have no effect on AO DNA staining in quiescent lymphocytes (46). Studies in other systems have indicated that Q cells may have G1, S, or G2 DNA content (18), and that following PHA stimulation of peripheral lymphocytes, Q to P transition can be monitored by flow cytometry (22). More details on the use of AO staining for discrimination between P and Q cells are presented in chapter 8 (18).

Although flow analysis has been used to identify Q cells in many tissue culture systems and in human leukemic cell populations (21), the present methodology may not be directly applicable to the study of Q cells in solid tumors until tissue dissociation techniques are developed that can yield representative and reproducible single-cell suspensions. Further, since many human tumors consist of multiple and/or heteroploid cell populations, compartmental overlap must be considered in such analyses.

2.5. PDP Assay

Nelson and Schiffer (35) have described a biochemical approach to GF determination in which P cells are identified as those cells containing DNA polymerase and primer template. Stripped nuclei from these cells incorporate [³H]-thymidine triphosphate ([³H]-TTP) and can be identified autoradiographically. The fraction of cells labeled using this technique is termed the PDP index (primer-available, DNA-dependent DNA polymerase index). The PDP index is not affected by treatment of the nuclei with RNase, but is substantially reduced by treatment with DNase. Labeling is also inhibited by actinomycin D and ethidium bromide. DNA polymerase-α is implicated as the enzyme responsible for [³H]-TTP incorporation since [³H]-TTP incorporation can be inhibited by cytosine arabinoside (ara-C) (*28*) and by hydroxymercuri-benzoate (*8*), both potent and preferential inhibitors of DNA polymerase-α. Substantial data have been accumulated to support the utility of the PDP index as an estimate of the GF. These data are reviewed in the remainder of this section.

Changes in DNA polymerase activity in human peripheral lymphocytes after phytohemagglutinin (PHA) stimulation also have been assessed by adding exogenous primer template (DNase-treated calf thymus DNA) to the PDP reaction mixture (*45*). The fraction of labeled nuclei in this assay (referred to as the DDDP index (DNA-dependent DNA polymerase index), as well as the PDP index, increased 8- to 10-fold within 2 h after PHA stimulation whereas [³H]-TdR labeling indices did not increase significantly until 22–24 h after PHA stimulation. Three days after PHA stimulation, the [³H]-TdR labeling indices were maximal, the PDP index was about 0.50, and the DDDP index was about 0.70. These data showing increased DNA polymerase activity soon after PHA stimulation are qualitatively similar to flow cytometric studies that show significant G0 to G1 transition soon after PHA stimulation and a significant Q population 3 d after stimulation.

Both the PDP index and the [³H]-TdR labeling index decrease with increasing tumor mass in S-180 tumors. However, in this tumor and in others (*8*), the PDP index is always higher than the [³H]-TdR labeling index. Because the PDP index was always shown to exceed the [³H]-TdR LI (*8,35,45*), it was suggested that the PDP assay was detecting more of the P cells than just those in S-phase. Subsequential studies with hydroxyurea-synchronized ascites tumor cells indicated that all S-phase cells and most G1 (possibly excepting early postmitotic G1) cells were PDP-positive (*44*). Although not directly assessed, G2 cells were presumed to be PDP-positive since approximately 50% of mitotic cells were identified as PDP-positive. These results support the suggestion that the PDP index may serve as an estimate of the tumor GF.

Lange-Wantzin (29) reported that the PDP assay method, as reported by Nelson and Schiffer (35), could not be generally applied to leukemic blast populations because the agar stripping step required to expose cell nuclei often resulted in unacceptable cell damage. By substituting an ethanol:acetone fixation step (19) for the agar stripping step, nuclear morphology could be preserved without compromising labeling. Direct comparison of PDP labeling indices using agar stripping or ethanol:acetone fixation showed similar results with both leukemic blasts and mouse ascites tumor cells (30).

Figure 1 shows a recent update of published data (43) comparing the PDP index and tumor GF estimated by FLM analysis in a variety of experimental tumor systems. The analysis now includes the comparison of GF values for Lewis lung tumors (49), B16 melanoma (51), RIF-1 fibrosarcoma (59), C3H/HeJ spontaneous mammary tumors (5), 13762 rat mammary tumors (14), C-6 rat glioma (1), and A67 and S-102F mouse mammary tumor models (33). Linear least-squares regression analysis shows a best fit with a slope of 1.002, a y intercept of 0.025, and a correlation coefficient (r) of 0.978.

Grain count studies of nuclei labeled in the PDP assay show the distribution of grains over labeled nuclei to be bimodal. Figure 2, for example, shows the distribution of grains over T1699 mouse mammary tumor cells. In vitro labeling with [^{3}H]-TdR was used to study the correlation between the distribution of grains in the PDP assay and the cell cycle distribution of the labeled cells. In this approach, the solid tumor cells for PDP assay were grown in vitro for a short period in the presence or absence of [^{3}H]-TdR (2.5 μCi/mL for 60 min), as described previously (4). The cells were then carried through the PDP assay and prepared for autoradiographic analysis. An emulsion exposure time of 7 d was used so that the grains over the [^{3}H]-TdR-labeled cells were heavily labeled (over 90% were too numerous to count; TNTC). Thus, in PDP assays with prior [^{3}H]-TdR labeling, cells actively synthesizing DNA could be visually excluded and the grain count distribution of the remaining cells compared to that obtained without prelabeling. The results suggested that most of the labeled nuclei in the PDP assay (without [^{3}H]-TdR labeling) with eight or more grains/cell were S-phase cells.

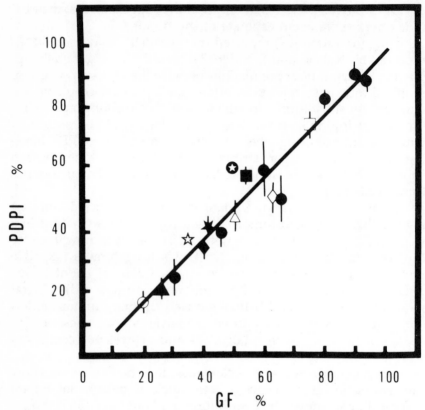

Fig. 1. The correlation between PDP index and tumor GF as measured
by conventional methods for experimental tumor models in mice and rats.
(Solid circle), previously published data (*43*); (hollow star), Lewis lung
(*49*); (hollow diamond), B16 melanoma (*51*); (solid star), RIF-1 sarcoma
(*58*); (hollow circle), C3H spontaneous mammary tumors (*5*); (solid diamond), rat glioma (*1*); (solid triangle), A67 mammary tumor (*33*); (solid
square), S102 mammary tumor (*33*); (hollow square), 13762 rat mammary
tumor (*14*); (hollow star), LoVo colorectal tumor xenograft (*55*). All PDP
assays, except that for LoVo xenografts (*55*), were performed in our
laboratory. References refer to published GF estimates for the respective
tumor lines.

Planometric analysis indicated that the ''S-phase area'' represented
about 45% of the total area in the PDP grain count distribution.
Cells classified as TNTC represented 40% of the total number of
labeled cells in populations subjected to [³H]-TdR labeling and PDP
assay, and the [³H]-TdR LI was 28.4%. Inasmuch as the PDP index in this tumor approximates the tumor growth fraction (*43*), the

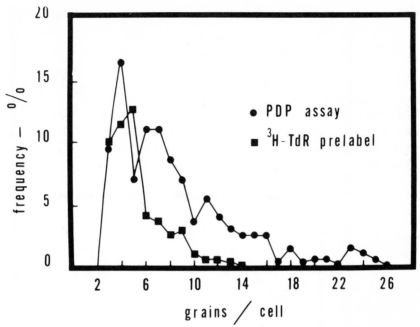

Fig. 2. Autoradiographic grain count distribution for PDP labeled T1699 mammary tumor cells with (solid square) and without (solid circle) prior [³H]-TdR labeling in vitro. Eight hundred (800) labeled cells were evaluated after a 7-d autoradiographic exposure time. Background grain counts were less than 1 grain per equivalent cell area and 3 grains/cell was used as the counting threshold.

labeling index of the P cells, Li_{th}, can be calculated from the PDP grain count distribution as the fraction of cells in the S-phase area.

Although we (5,19,43) and others (55) have provided data to indicate that in unperturbed experimental tumors the PDP index can provide a reasonably good estimate of the GF, there is little evidence, except in xenographs (55), to indicate that it is equally reliable as an estimate of the GF in human tumors. Although it is logistically difficult to directly compare PDP index and the GF as measured by FLM or continuous labeling in clinical material, the PDP index and GF can be compared indirectly through the PDP grain count distribution and [³H]-TdR labeling (9).

In experimental tumor models in which the PDP index was observed to estimate tumor GF, LI_{obs} was determined by in vitro [³H]-TdR labeling and autoradiography (4). LI_{th} was established as T_s/T_c, where T_s was estimated by double labeling in vitro with [³H]-TdR and [¹⁴C]-TdR and double emulsion autoradiography (4,5), and T_c was determined from:

$$T_c = \frac{\ln\,(1\,+\,\text{PDP index})}{\ln 2}\,T_{\text{pot}}$$

The potential doubling time (T_{pot}) was corrected for nonlinear cell age distribution and calculated as described previously (51). If the PDP grain count analysis described above (Fig. 2) reflects the LI_{th}, then LI_{obs}/LI_{th} should be highly correlated with the PDP index if the PDP index is an accurate measure of the GF. The correlation between the PDP index and LI_{obs}/LI_{th} is seen in Fig. 3. Least-

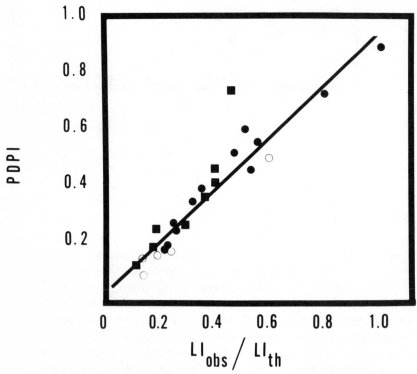

Fig. 3. Correlation between the PDP index and LI_{obs}/LI_{th} for 17 experimental solid tumors where LI_{obs} was determined by in vitro labeling methods (4) and LI_{th} was determined by PDP grain count analysis (hollow circle), or from Ts/Tc (solid circle). Also shown are data derived from the analysis of eight human tumor biopsy samples (solid square), where the LI_{th} was determined from the analysis of PDP grain count distribution.

squares linear regression analysis showed the correlation coefficient to be 0.965 and slope of the regression line to be 0.942. This strong correlation in data from experimental solid tumors and from human tumor biopsy samples suggests that it is reasonable to conclude that, at least in the human tumor samples tested, the PDP index is a reasonable estimate of the GF.

2.6. Estimation of GF in Perturbed Tumors

At the present time estimation of the tumor GF after cytotoxic cytoreduction has been attempted in only a few solid tumor systems, primarily because the analysis of FLM and continuous labeling data is hampered by rapidly changing kinetic parameters. At present, little consensus exists about the recruitment of Q cells following cytoreduction.

Some studies show little recruitment. For example, Hermens and Barendsen (25), using a thymidine-labeling technique, indicated that recruitment of Q cells into proliferation in the Rd-1 rhabdomyosarcoma proceeded very slowly after 1500 rad. Flow cytometric analysis of S-102 mammary tumor cell suspensions and computer simulation indicated little or no Q cell recruitment after 10 mg/kg adriamycin in vivo. However, little or no cell kill was observed either (23).

Other studies support substantial recruitment. Pallavicini et al. (36) used flow cytometry, [³H]-TdR labeling, and computer modeling to quantify the cytokinetic properties in KHT solid tumors treated with 1-β-D-arabinofuranosylcytosine (ara-C). Posttreatment changes in DNA content distribution, [³H]-TdR uptake in surviving cells, and the loss of cells labeled with [³H]-TdR prior to treatment suggested that about 80% of the Q population (initially 57% of the total population) was rapidly recruited into cycle 5–6 h after ara-C treatment. Maximal sensitivity of clonogenic KHT cells to a second ara-C treatment was seen 5–6 h later when labeling studies indicated that most of the newly recruited cells were actively synthesizing DNA.

Serial studies to assess changes in nucleolar morphology in DBAH tumor cell populations were conducted to evaluate P to Q transition after subcurative X-irradiation (40). The results indicated that morphologically defined P and Q populations decreased after irradiation, but that the Q population was more radioresistant. P cell regeneration (i.e., proliferative recovery) was observed to be accompanied by an increased loss of cells from the Q compartment, suggesting that recruitment of surviving Q cells was contributing to repopulation of the P cell pool. The time after irradiation, when increased Q to P transition was maximal, was dose-dependent. The data also indicated that repopulation of the Q compartment was most likely accomplished by P to Q transition.

Changes in the PDP index after therapeutic perturbations were evaluated and compared to that for other kinetic parameters (5,6,10–13,15,42,46) to investigate the utility of the PDP index as an estimate of the GF in perturbed tumors. Figure 4 shows the results from one such study in low growth fraction—C3H/HeJ spontaneous mammary tumors treated with a combination of cyclophosphamide (200 mg/kg, day 0) and adriamycin (2 mg/kg, d 1). These studies showed that the PDP index increased threefold during the first 72 h after treatment and remained elevated for at least another 4 d. Mitotic indices, initially decreased after treatment, were also increased by 72 h. [³H]-TdR LIs fluctuated after treatment with recovery evident by d 4. Although it is not possible to conclude that the absolute values for PDP index between d 3 and 7 represent the tumor growth fraction, the 2.4–4-fold increase seen in the MI and [³H]-TdR LI between d 3 and 7 after treatment support the possibility of an increase in tumor GF during this period.

Figure 5 shows the results from a similar study using the high GF, T1699 mammary tumor model. After cyclophosphamide treatment (100 mg/kg), both the [³H]-TdR LI and PDP index were depressed for 3–4 d. Although the [³H]-TdR LI increased rather rapidly to control values by d 6, PDP indices recovered more slowly. On the surface these results are consistent with a slow Q to P conversion rate. However, if Q cells were recruited early when lethal drug concentrations or unrepaired lesions existed, such Q cells would not contribute to regeneration of the proliferating pool. Alternatively, the P cell pool in the model may have been reestablished by reduced P to Q conversion.

Figure 6 shows the results from a study to assess proliferative changes in residual RIF-1 tumors after partial (¾) surgical resection. Changes in the PDP index are compared to changes in the S-phase clonogenic cell fraction (the fractional reduction in the in vitro plating efficiency following treatment in vivo with a toxic dose

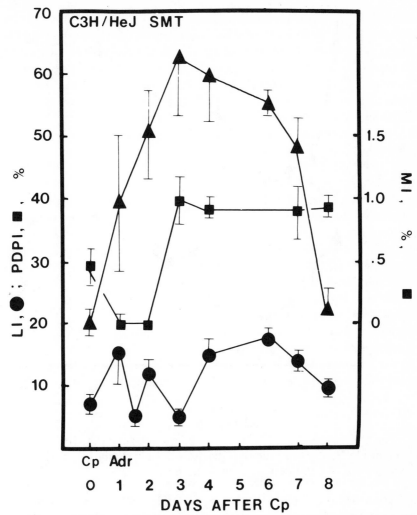

Fig. 4. Changes in the PDP index (solid triangle), [³H]-TdR LI (solid circle), and MI (solid square) in C3H/HeJ spontaneous mammary tumors after combination chemotherapy with 200 mg/kg cyclophosphamide (Cp) and 2 mg/kg adriamycin (Adr). Each symbol is the mean ±SE for four tumors.

of hydroxyurea) (15) and to changes in the [³H]-TdR labeling index as determined by in vitro labeling methods (4). These studies indicate a twofold increase in the PDP index in the residual tumor

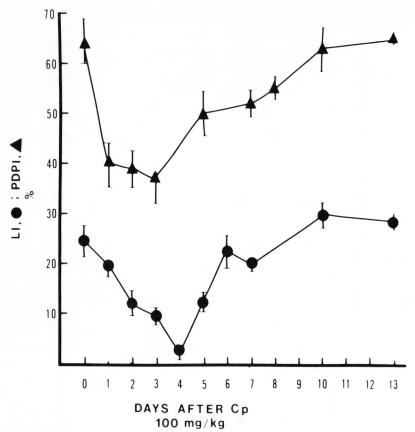

Fig. 5. Changes in the PDP index (solid triangle) and [³H]-TdR LI (solid circle) in T1699 mammary tumors after 100 mg/kg cyclophosphamide (Cp). Each symbol is the mean ±Se for four tumors.

between 48 and 72 h after surgery. [³H]-TdR LIs and S-phase clonogenic cell fractions were also increased about twofold. However, peak values were delayed by about 24 h. Here again, although it is not possible to conclude that the absolute PDP indices between d 3 and 5 represent the tumor GF, the sensitivity of the residual tumor cell population to the proliferation-dependent agents 5-FU (84 mg/kg) and cyclophosphamide (150 mg/kg) was increased 1.5–2-fold, 48 to 72 h after surgery (Fig. 7). Although the changes in drug-induced regrowth delay after surgery cytoreduction could have been caused by changes in drug metabolism and/or drug distribution, it is reasonable to conclude that the increased drug sensitivity is a result of the twofold increase in the size of target population, i.e., the GF.

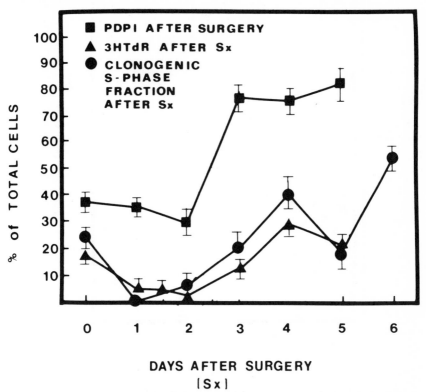

Fig. 6. Changes in the PDP index and S-phase fraction of the clonogenic cell population, assessed by hydroxyurea killing *(15)*, in residual sc RIF-1 tumor tissue after surgical cytoreduction. Each symbol is the mean ±SE for five tumors.

3. GF AS AN INDICATOR OF CHEMOSENSITIVITY

Inasmuch as the killing effects of most chemotherapeutic drugs, including cyclophosphamide, are proliferation-dependent, we speculate that the size of the P cell pool is a contributing factor for the responsiveness of solid tumors to some chemotherapeutic agents. Serial studies after chemotherapeutics *(6,10–15,26,27)* and ionizing radiation *(6,7,26,31)* have indicated that the kinetic

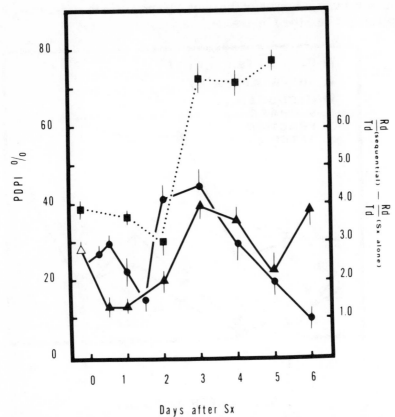

Fig. 7. Changes in the PDP index estimated GF (solid square) and regrowth delays produced in primary, sc RIF-1 tumors by 5-FU (solid triangle) (84 mg/kg) and cyclophosphamide (150 mg/kg) (solid circle) after surgical cytoreduction. Regrowth delays, expressed as multiples of the volumetric doubling, for surgery-only controls were subtracted from those for the surgery-plus-chemotherapy groups to permit evaluation of changing drug sensitivity. Each symbol is the mean ±SE for six to seven tumors.

response to cytotoxic cytoreduction can be characterized by a variable interval of reduced cell proliferation that is modality-, dose-, and tumor-dependent. Unless all clonogenic cells are treated to lethality, repopulation will occur as a result of proliferative recovery. Often this proliferative recovery interval is characterized by a period of hyperproliferation, and numerous studies have indicated that this interval can be an efficacious time to sequence subsequent therapies (6,7,10–15,26,27,31). Inasmuch as the mechanism of cell killing by most alkylating agents, including cyclophosphamide, is proliferation-dependent, we can conclude that unless Q to P transition occurs during intervals of high intracellular drug levels or unrepaired lethal lesions, the GF would represent the potential target population for proliferation-dependent agents. Figure 8 shows the correlation between the GF and the antiproliferative

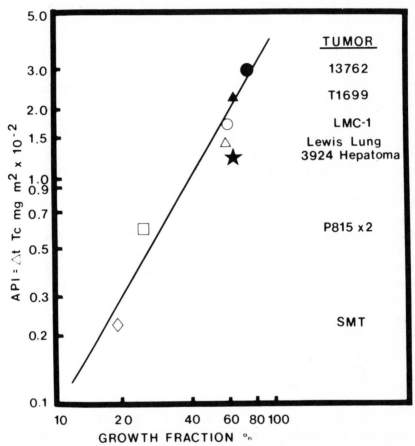

Fig. 8. The correlation between the timing for dose-dependent proliferative recovery after cyclophosphamide in sc experimental solid tumor models of mice and rats and the tumor-specific GF (hollow diamond) C3H/HeJ spontaneous mammary tumor (*10*); (hollow square), P815 × 2 mastocytoma (*41*); (solid star), 3924A rat hepatoma (*31*); (hollow circle), LMC-1 rat mammary tumor (*34*); (solid triangle), T1699 mammary tumor model (*13*); (solid circle), 13762 rat mammary tumor model (*10*); and; (hollow triangle), Lewis lung carcinoma (*26*).

index* (API) for several sc experimental tumor models in mice and rats treated with a single, ip, dose of cyclophosphamide (*10,13,26, 31,34,41*). Since proliferation rates varied in these models, the interval, dT, was expressed as a multiple of the tumor-specific mean cell cycle time and normalized to the cyclophosphamide dose (mg/m^2). As formulated, the API was strongly correlated ($r > 0.96$) with the tumor growth fraction as measured by conventional methods or by PDP assay. These results suggest that at least in these sc experimental tumor models from mice and rats, the cyclophosphamide dose-dependent proliferative recovery is directly related to the size of the actively proliferating component of the cell population and, possibly, the level of cytotoxic cytoreduction in that population. Subsequent studies in other tumor models with cyclophosphamide, adriamycin (*16*), and more recently with *cis*-diamminedichloro platinum II (*17*) have confirmed the relationship between tumor GF and the timing for dose-dependent proliferative recovery. Naturally, correlation does not necessarily prove cause and effect. Phenotypic resistance is an important response limitation. Further, underlying mechanisms controlling the tumor growth fraction, such as tumor vascularity and vascular function, would also be expected to have a major influence on chemoresponsiveness. Studies to further elucidate the biochemical and physiologic mechanisms responsible for or controlling Q to P transition during proliferative recovery could provide insights as to how these phenomenon might be monitored during therapy and how transient intervals of increased chemosensitivity could be exploited for a therapeutic gain.

4. CONCLUSIONS

Although the concept of tumor growth fraction may be a pragmatic way to describe the proliferative behavior of tumor cell populations, it should be recognized that no method has been developed to conclusively demonstrate a distinct boundary between proliferating and nonproliferating compartments in solid tumors.

Using appropriate computer modeling techniques, the growth fraction of unperturbed tumors can be estimated by [^{3}H]-TdR pulse

*APT = $\Delta T/T_c$, where ΔT is the interval between the time of treatment and the time of maximum [^{3}H]-TdR uptake or [^{3}H]-TdR LI.

or continuous (repeated) labeling methodologies. However, uncertainties related to choosing the appropriate analytical models, the accuracy by which FLM data reflect the total proliferating cell population, and the necessity for in vivo [3H]-TdR administration, serial sampling, and autoradiographic analysis render these methods useful only in the well-controlled laboratory setting and generally impractical for clinical application.

Flow cytometric analysis may be the most rapid and the most promising method for identification of Q cells in tissue culture and leukemic cell populations. These studies have allowed for identification and characterization of noncycling populations, based not on their inability to incorporate radiolabeled DNA precursors, but on other criteria of cellular function. The exploitation of this new technology, together with labeling techniques, has led to new information on the nature of noncycling cell populations and their transition to an actively proliferating state. However, problems associated with the preparation of representative single-cell suspensions and the analysis of mixed and/or heteroploid populations must be solved before this technique can be generally applied to the analysis of solid tumors.

Morphological characterization of nucleoli may allow for easy and rapid, point-in-time assessment of the relative size of the proliferating cell pool. Direct comparisons of this method with other methods to estimate GF, however, need to be made.

The PDP index does correlate with the tumor GF as estimated by other methods in a variety of unperturbed animal tumor models and human tumor xenografts. Although this does not confer any more precision or confidence in GF estimation, the PDP assay does offer the advantage of providing point-in-time GF estimates, since neither in vivo [3H]-TdR administration nor serial sampling are necessary. Seven to ten days of autoradiography exposure time may, however, be necessary to ensure adequate labeling for analysis. As with other methods to estimate GF, the PDP index is a function of the total cell population and, thus, may or may not reflect the GF of the clonogenic population, especially if the clonogenic cells are disproportionately distributed in the rapidly cycling cell compartment. Although the PDP assay can provide a reasonably good estimate of GF in experimental tumor models, there is at present only indirect evidence that it can provide equally reasonable estimates of GF in human solid tumors.

Following cytoreduction therapy, pulse and continuous labeling methods are of limited value to serially assess changes in the GF. These methods can be reliably applied only after the cytokinetic perturbation is complete and cells have become randomly distributed through the cell cycle.

Flow cytometric analysis and computer stimulation may be useful to evaluate changes in the Q compartment after cytotoxic treatments in systems that are well-defined and for agents whose mode of action is also well-defined. However, agents that alter the stoichiometry of dye binding and/or its detection in flow systems could lead to erroneous conclusions.

It is uncertain whether the PDP index accurately reflects the tumor GF in perturbed cell populations. PDP indices shortly after drug or radiation treatments may be subject to artifacts related to transient nonlethal modifications of primer template and/or DNA polymerase-α activity. Further, since the PDP index is a function of the total tumor cell population, changes in the PDP index after cytotoxic perturbation may or may not reflect changes in the GF of clonogenic cells. The presence of doomed or dead cells after cytotoxic cytoreduction may also influence PDP index values immediately after cytotoxic treatments. With these limitations it can, however, be stated that the PDP index is a responding cytokinetic parameter after cytotoxic treatment. In some tumor models, the PDP index recovers slowly after cytotoxic cytoreduction, as if repopulation of PDP-positive cells occurs primarily through reduced P to Q transition. In other solid tumor systems, the PDP recovers rapidly prior to the onset of the recovery of the [^{3}H]-TdR LI, suggesting that repopulation could occur through the conversion of Q to P cells (recruitment).

Although it is uncertain whether rapid recruitment of noncycling cells routinely occurs in solid tumors after cytotoxic cytoreduction, the twofold PDP index increase in residual RIF-1 tumors after surgery preceded similar increases in clonogenic cell proliferation, correlated temporally and in magnitude with changes in chemosensitivity, and would be consistent with a relatively rapid Q to P transition in this moderately low GF tumor model. Studies in KHT solid tumors also suggested rapid Q to P transition after ara-C cytoreduction in vivo.

Data derived in a variety of experimental tumor models suggest that pretreatment GF estimates may provide some prognostic information regarding the time course of proliferative recovery after treatment with cyclophosphamide and perhaps other similarly acting agents in vivo.

REFERENCES

1. Barker, M., Hoshino, T., Gurcay, O., Wilson, C. B., Nielson, S. L., Downie, R., and Eliason, J. Development of an animal brain tumor model and its response to therapy with 1, 3-bis (2-chloroethyl 1)-1-nitrosourea. Cancer Res., *33*: 976–986, 1973.
2. Barret, J. C. A mathematical model of the mitotic cycle and its application to the interpretation of percentage labeled mitosis data. J. Natl. Cancer Inst., *37*: 443–950, 1966.
3. Barrett, J. C. Optimized parameters for the mitotic cycle. Cell Tissue Kinet., *3*: 349–353, 1970.
4. Braunschweiger, P. G., Poulakos, L., and Schiffer, L. M. *In vitro* labeling and gold activation autoradiography for determination of labeling index and DNA synthesis time of solid tumors. Cancer Res., *36*: 1748–1753, 1976.
5. Braunschweiger, P. G., Poulakos, L., and Schiffer, L. M. Cell kinetics *in vivo* and *in vitro* for C3H/He spontaneous mammary tumors. J. Natl. Cancer Inst., *59*: 1197–1204, 1977.
6. Braunschweiger, P. G., Schenken, L. L., and Schiffer, L. M. The cytokinetic basis for the design of efficacious radiotheraphy protocols. Int. J. Radiat. Oncol. Biol. Phys., *5*: 37–47, 1979.
7. Braunschweiger, P. G., Schenken, L. L., and Schiffer, L. M. Kinetically directed combination therapy with adriamycin and X-irradiation in a mammary tumor model. Int. J. Radiat. Oncol. Biol. Phys., *7*: 747–753, 1981.
8. Braunschweiger, P. G. and Schiffer, L. M. Nuclear DNA polymerase-α and replicative potential in mammalian cells. Eur. J. Cancer, *13*: 775–779, 1977.
9. Braunschweiger, P. G. and Schiffer, L. M. PDP indices in human tumors: Evidence for proliferative correlations. Proc. Am. Soc. Clin. Oncol., *18*: 276, 1977.
10. Braunschweiger, P. G. and Schiffer, L. M. Therapeutic implications of cell kinetic changes after cyclophosphamide treatment in "spontaneous" and "transplantable" mammary tumors. Cancer Treat. Rep., *62*: 727–736, 1978.

11. Braunschweiger, P. G. and Schiffer, L. M. Cell kinetics after vin-
    cristine treatment of C3H/HeJ spontaneous mammary tumors: Impli-
    cations for therapy. J. Natl. Cancer Inst., *60*: 1043–1048, 1978.

12. Braunschweiger, P. G. and Schiffer, L. M. The effect of methylpred-
    nisilone on the cell kinetic response of C3H/HeJ mammary tumors
    to cyclophosphamide and adriamycin. Cancer Res., *39*: 3812–3815,
    1979.

13. Braunschweiger, P. G. and Schiffer, L. M. Cell kinetic directed se-
    quential chemotherapy with cyclophosphamide and adriamycin in
    T1699 mammary tumors. Cancer Res., *40*: 737–743, 1980.

14. Braunschweiger, P. G. and Schiffer, L. M. Effect of adriamycin on
    the cell kinetics of 13762 rat mammary tumors and implications for
    therapy. Cancer Treatments Rep., *64*: 293–300, 1980.

15. Braunschweiger, P. G., Ting, H. L., and Schiffer, L. M. Receptor-
    dependent antiproliferative effects of corticosteroids in radiation-
    induced fibrosarcomas and implications for sequential therapy.
    Cancer Res., *42*: 1686–1691, 1982.

16. Braunschweiger, P. G. and Schiffer, L. M. Response models for pro-
    liferative recovery in solid tumors after cyclophosphamide and
    adriamycin. Proc. AACR, *24*: 262, 1983.

17. Braunschweiger, P. G. and Schiffer, L. M. The effect of Cisplatinol
    (D-DDp) on cell proliferation in solid tumor models. Cell Tissue
    Kinet., *17*: 672, 1984.

18. Darzynkiewicz, Z. Cytochemical probes of cycling and quiescent cells
    applicable for flow cytometry. *In*: (J. W. Gray and Z. Darzynkiewicz,
    eds.), Techniques in Cell Cycle Analysis. New Jersey: Humana, 1986.

19. Darzynkiewicz, Z. Detection of DNA polymerase activity in fixed cells.
    Exp. Cell Res., *80*: 483–486, 1973.

20. Darzynkiewicz, Z., Traganos, F., Andreeff, M., Sharpless, T., and
    Melamed, M. R. Different sensitivity of chromatin to acid denatura-
    tion in quiescent and cycling cells as revealed by flow cytometry. J.
    Histochem. Cytochem., *27*: 478–485, 1979.

21. Darzynkiewicz, Z., Traganos, F., and Melamed, M. R. New cycle com-
    partments identified by multiparameter flow cytometry. Cytometry,
    1: 98–108, 1980.

22. Darzynkiewicz, Z., Traganos, F., Sharpless, T. R., and Melamed,
    M. R. Cell cycle related changes in nuclear chromatin of stimulated
    lymphocytes as measured by flow cytometry. Cancer Res., *37*: 4635–
    4640, 1977.

23. Dethlefsen, L. A., Riley, R. M., and Roti Roti, J. L. Flow cytometric
    (FCM) analysis of adriamycin-perturbed mouse mammary tumors.
    J. Histochem. Cytochem., *37*: 463–469, 1979.

24. Gerecke, D., Gegsen, A., and Gross, R. Continuous labeling method for autoradiographic analysis of cell cycle parameters in steady-state cell systems. Experientia, 32: 1088–1090, 1976.

25. Hermens, A. F. and Barendsen, G. W. The proliferative status and clonogenic capacity of tumor cells in a transplantable rhabdomyosarcoma of the rat before and after irradiation with 800 rad of X-ray. Cell Tissue Kinet., 11: 83–100, 1978.

26. Houghton, P. J. and Taylor, D. M. Fractional incorporation of ³H-thymidine and DNA specific activity as assays of inhibition of tumor growth. Br. J. Cancer, 35: 68–77, 1977.

27. Kovacs, C. J., Hopkins, H. A., Simon, R. M., and Looney, W. B. Effects of 5-Fluorouracil on the cell kinetic and growth parameters of hepatoma 3924 A. Br. J. Cancer, 32: 42–50, 1975.

28. Lange-Wantzin, G. Effect of cytosine arabinoside on nuclear labeling of leukemic myeloblasts with tritiated thymidine triphosphate. Leukemia Res., 3: 7–13, 1979.

29. Lange-Wantzin, G., Rarle, H., and Killman, S. A. Nuclear DNA polymerase estimation in human leukemic myeloblasts. Br. J. Hematol., 33: 329–334, 1976.

30. Lange-Wantzin, G. Proliferation Characteristics of Human Leukemic Blast Cells *In Vivo* Before and After Cytostatic Drugs, p. 18, Copenhagen, Denmark: Villadsen and Christensen, 1981.

31. Looney, W. B., Hopkins, H. A., Grover, W. H., MacLeod, M. S., Ritenour, E. R., and Hobson, A. S. Solid tumor models for the assessment of different treatment modalities. XIII. Comparison of response and recovery of host and solid tumor to cyclophosphamide and radiation. Cancer, 45: 2793–2798, 1980.

32. Mendelsohn, M. L. Autoradiographic analysis of cell proliferation in spontaneous breast cancer of C3H mouse. III. The growth fraction. J. Natl. Cancer Inst., 28: 1015–1029, 1962.

33. Mendelsohn, M. L. and Dethlefsen, L. A. Cell kinetics of breast cancer: The turnover of non-proliferating cells. Recent Results Cancer Res., 42: 73–86, 1973.

34. Moore, J. V. and Dixon, B. The gross and cellular response of a rat mammary tumor to single doses of cyclophosphamide. Eur. J. Cancer, 14: 91–95, 1978.

35. Nelson, J. S. R. and Schiffer, L. M. Autoradiographic detection of DNA polymerase containing nuclei in sarcoma 180 ascites cells. Cell Tissue Kinet., 6: 45–54, 1973.

36. Pallavicini, M. G., Gray, J. W., and Folstad, L. J. Quantitative analysis of cytokinetic response of KHT tumors *in vivo* to 1-beta-D-arafuranosylcytosine. Cancer Res., 42: 3125–3131, 1982.

37. Potmesil, M. and Goldfeder, A. Identification and kinetics of G1 phase confined cells in experimental mammary carcinomas. Cancer Res., *37*: 857–864, 1977.

38. Potmesil, M. and Goldfeder, A. Nucleolar morphology and cell proliferation kinetics of thymic lymphocytes. Exp. Cell Res., *77*: 31–90, 1973.

39. Potmesil, M. and Goldfeder, A. Nucleolar morphology, nucleic acid synthesis and growth rates of experimental tumors. Cancer Res., *31*: 789–797, 1971.

40. Potmesil, M., Ludwig, D., and Goldfeder, A. Cell kinetics of irradiated experimental tumors: Relationship between the proliferating and the non-proliferating pool. Cell Tissue Kinet., *8*: 369–385, 1975.

41. Schenken, L. L. Proliferative character and growth modes of neoplastic disease as determinants of chemotherapeutic efficacy. Cancer Treat. Rep., *60*: 1761–1776, 1976.

42. Schiffer, L. M., Braunschweiger, P. G. and Stragand, J. J. Tumor cell population kinetics following non-curative treatment. Antibiot. Chemother., *23*: 148–156, 1978.

43. Schiffer, L. M., Markoe, A. M., and Nelson, J. S. R. Estimation of tumor growth fraction in murine tumors by the primer-dependent available DNA-dependent polymerase assay. Cancer Res., *36*: 2415–2418, 1976.

44. Schiffer, L. M., Markoe, A. M., and Nelson, J. S. R. Evaluation of the PDP index as a monitor of growth fraction during tumor therapy. *In*: The Cell Cycle in Malignancy and Immunity, 13th Annual Hanford Biology Symposium, pp. 459–472, US Energy Research and Development Administration, 1975.

45. Schiffer, L. M., Markoe, A. M., Wenkelstein, A., Nelson, J. S. R., and Mikulla, J. M. Cycling characteristics of human lymphocytes in vitro. Blood, *44*: 99–107, 1974.

46. Schuartzendruber, D. E. A bromodeoxyuridine (BUdR)-mithramycin technique for detecting cycling and non-cycling cells by flow cytometry. Exp. Cell. Res., *109*: 439–445, 1977.

47. Shackney, S. A cytokinetic model for heterogeneous mammalian cell populations. II. Tritiated thymidine studies: The percent labeled mitosis (PLM) curve. J. Theor. Biol., *44*: 4990, 1974.

48. Shackney, S. and Ritch, P. Percent labeled mitosis curve analysis. *In*: (J. W. Gray and Z. Darzynkiewicz, eds.), Techniques in Cell Cycle Analysis. New Jersey: Humana, 1986.

49. Simpson-Herren, L., Sanford, A. H., and Holmquist, J. P. Cell population kinetics of transplanted and metastatic Lewis lung carcinoma. Cell Tissue Kinet., *7*: 349–361, 1974.

50. Simpson-Herren, L. L., Sanford, A. H., Holmquist, J. P., Springer, T. A., and Lloyd, H. H. Ambiguity of the thymidine index. Cancer Res., *36*: 4705–4709, 1976.

51. Skipper, H. E. and Schabel, F. M. Quantitative and cytokinetic studies in experimental tumor models. *In*: (J. F. Holland and E. Frei, III, eds.), Cancer Medicine. Philadelphia: Lea and Febiger, 1973.

52. Steel, G. G. Cell loss in experimental tumors. Cell Tissue Kinet., *1*: 193–207, 1968.

53. Steel, G. G. Delayed uptake by tumors of tritium from thymidine. Nature (Lond.), *210*: 806–808, 1966.

54. Steel, G. G. Growth Kinetics of Tumors. Oxford: Oxford University, 1977.

55. Stragand, J. J., Bergerat, J.-P., White, R. A., Hokanson, J., and Drewinko, B. Biological and cell kinetic properties of a human colonic adenocarcinoma (LoVo) grown in athymic mice. Cancer Res., *40*: 2846–2852, 1980.

56. Tannock, I. F. The relationship between cell proliferation and the vascular system in a transplanted mouse mammary tumor. Brit. J. Cancer, *22*: 258, 1968.

57. Takahashi, M., Hogg, G. D., and Mendelsohn, M. L. The automated analysis of FLM curves. Cell Tissue Kinet., *4*: 505–518, 1971.

58. Twentyman, P. R., Brown, J. M., Gray, J. W., Franks, A. J., Scoles, M. A., and Kallman, R. F. A new mouse tumor model system (RIF-1) for comparison of end-point studies. J. Natl. Cancer Inst., *64*: 595–604, 1980.

Chapter 4

In Vitro Assays
for Tumors Grown In Vivo

A Review of Kinetic Techniques

Janet S. Rasey

1. INTRODUCTION

This review of in vitro kinetic techniques for experimental tumors concentrates on methods developed in tumor systems that grow in both the animal and in tissue cultures, and for which cells can be cloned from either growth mode. The ability of cells isolated from these solid tumors to form colonies when plated at low density in tissue culture is widely accepted as a test of their clonogenicity in radiobiology and experimental chemotherapy. It is assumed, but not proven, that a single cell that is capable of proliferating into a large colony in tissue culture would be capable of dividing and repopulating a tumor if left *in situ* in the experimental animal. "Clonogen" is thus defined operationally. This assumption, if valid, allows for the study of the kinetics of the clonogenic tumor cells, rather than the total tumor parenchymal cell population. The clonogenic potential of subpopulations of tumor cells, separated or identified in various ways, and changes in growth kinetics of stem cells during and after a course of antitumor therapy have been major areas of investigation. Techniques have been developed that allow one to infer the cytokinetics of tumor cells *in situ* by directly examining the growth and cloning efficiency of the isolated tumor cells when placed in culture. Several questions may be asked. What is the clonogenicity of S-phase cells, or, alternately, what proportion of the clonogenic cells are in DNA synthesis (S-phase)? What proportion of the stem cells are actively proliferating as defined by the ability of cells to incorporate tritiated thymidine during continuous infusion or multiple injections into the animal? What is the clonogenicity of nonproliferating (Q) vs proliferating (P) cells when both types have a G1 DNA content? What is the clonogenicity of cells in different stages of the cell cycle as determined by DNA content? What is the doubling time of stem cells in the tumor under perturbed vs control growth conditions?

The basic technique of isolating cells from a solid tumor and determining their clonogenicity in tissue culture can be extended to examine one tumor type over time. This allows one to obtain direct information about the doubling time of clonogens or stem cells. Cell sorting and separation techniques based on flow cytometry, elutriation, or density gradient centrifugation allow for more indirect comparisons. Hypoxic cells in tumors may have a different density or size than oxic cells. These differences can form the basis for separation and thus allow the clonogenicity of these two tumor subpopulations to be compared. Thus, one may answer such questions as: Are untreated hypoxic cells as clonogenic as oxic ones? Are hypoxic cells quiescent? Can hypoxic survivors of radiation treatment be recruited into the growth fraction?

Many of the techniques developed have not yet been widely applied. Some of the methods, such as sorting viable cells under aseptic conditions by flow cytometry, are technically difficult and require expensive equipment and sophisticated analysis. Others, such as the hydroxyurea (HU) or high specific activity [^{3}H]-thymidine suicide technique, to selectively kill those stem cells in DNA synthesis, are prone to artifacts that may not be avoidable. The ease of experimental manipulation of the available in vivo–in vitro tumor systems means that techniques used with these systems cannot always be extended to other tumors. In vivo–in vitro tumors also represent a highly selected subset of experimental animal tumors. They have been chosen for reproducible growth and ease of handling in two different growth modes, with the added restriction that individual cells taken from tissue culture, as well as those isolated from the solid tumor, form colonies with a high plating efficiency. The proportion of clonogens in such tumors is almost certainly larger than in spontaneous or other transplantable neoplasms, in which as few as one in 10^4 cells may be a stem cell. Nonetheless, it is possible to obtain useful information about the nature of those cells that can repopulate a tumor and how their kinetics are modified by therapy.

2. REVIEW OF TECHNIQUES

Four techniques have been developed for in vivo–in vitro tumor systems to give essentially direct cytokinetic information about clonogenic tumor cells. These are determination of doubling time of clonogenic tumor cells (2,3,16,18,40); the labeled microcolony method (4,19); the S-phase suicide technique (5,31); and flow cytometry (FCM) sorting and plating of viable cells after vital staining based on DNA content (26). These techniques have one thing in common: The cells forming colonies in the dishes are precisely those for which the cytokinetic information is being obtained. The advantages and disadvantages will be compared. At this point, the advantages are potential, whereas the technical limitations are real and major.

Several indirect methods also will be reviewed briefly. Viable cells from solid tumors may be sorted according to size (21) or density (11) and then plated in cloning tests. These cells may also be analyzed to determine their cell cycle position (actually, DNA content), or residence in the proliferating (P) or nonproliferating (Q) compartment, based on prior labeling with [^{3}H]-thymidine or by use of special staining techniques (6,7,10,42).

2.1. Clonogenic Fraction vs Time

It is possible to measure changes in the number of clonogenic cells per tumor during growth or as a function of time after treatment by simple extension of the basic techniques of in vitro colony-forming assay for cells isolated from tumors growing in vivo.

The objective is to obtain information about the doubling time of tumor stem cells. This may be compared to the doubling time of the total tumor parenchymal cell population. It also is possible to determine if clonogenic cells have different doubling times in small vs large tumors or in tumors regrowing after drug or radiation treatment compared to those not previously treated.

The reason for developing in vivo–in vitro tumor systems was to allow for quantitative study of clonogenic or stem cells in solid tumors. This requires one to accept the operational definition of such a cell, i.e., one that can form a macrocolony in a culture dish when single cell suspensions are plated at low density. For this technique to be meaningful, careful study and optimization of procedures must be done for each tumor type. Parameters to be optimized include developing the best liquid or agar solidified growth medium; use of feeder layers, if necessary; proper choice of enzyme or mechanical disaggregation method to maximize yield of a representative cell population and to minimize cell damage; and a colony growth period sufficiently long to allow for development of all potential clonogens into scorable macrocolonies. Methods for cell dispersal are discussed in more detail in chapter 6 of this volume (24).

To determine the number of clonogens per tumor, the fraction of clonogens determined in plating assays must be multiplied by the number of parenchymal cells per gram of tumor and by the tumor weight, i.e., number of cologens/tumor = (clonogenic fraction)(cells/gram)(grams/tumor).

Parenchymal cell number per unit weight of tumor must be carefully determined by histology to assess the proportion of parenchymal and stromal cells; and/or by measurement of DNA content per cell and DNA content per gram; and/or by careful measurement of cell yield following enzyme disaggregation. Parenchymal-stromal cell balance or average parenchymal cell size may change following cytotoxic treatment (2,3). This must be taken into account to allow for determination of clonogenic doubling time in recurring treated tumors, for comparison to controls.

This approach has several advantages. It is a simple extension of techniques already used with all in vivo–in vitro systems, and the cells cloned are precisely those for which the cytokinetic information is inferred. By examining the number of clonogens over time, the doubling time of clonogens can be compared to the tumor volume doubling time. Much of tumor growth may represent division of cells with limited reproductive potential. The doubling time of stem cells in tumors regrowing during or after treatment can be determined and compared to rates in nontreated controls. Stem cells may have much faster doubling times in a recurrent tumor (2,3,18,39,40).

As a cytokinetic technique, this method has technical and theoretical disadvantages. One cannot determine if a change in clonogen doubling time is caused by an alteration in cell cycle duration, the movement of cells between P and Q compartments, or both. A technical disadvantage is that numerous tumor samples are required over time. Low cell yield and/or selective cell loss can occur if tumors shrink during and after single dose or fractionated cytotoxic therapy, thus limiting the application of this method to tumors regrowing after treatment.

2.2. Labeled Microcolony Technique

The labeled microcolony technique (Fig. 1) was first described by Barendson et al. (4), and has been more extensively developed by Kallman and coworkers (19) and Hermans and Barendsen (16). The objective is to compare the clonogenicity of P vs Q cells by combining the in vitro colony-forming test with autoradiography to detect which proportion of the colony formers had incorporated a radiolabeled DNA precursor. The first step is to label all the P cells in a transplanted tumor by giving multiple injections or continuous infusion of [^{3}H]-thymidine for a time greater than the duration of one cell cycle, just prior to removal of the tumor. Following preparation of a single cell suspension from the labeled tumor, cells are plated into chambers attached to microscope slides and allowed to form microcolonies, usually for 3–5 d. These are then prepared as autoradiographs. A balance must be struck; the cells must be allowed a sufficient number of divisions such that there is a high probability that the microcolony formed would have grown into a macrocolony and thus been scored as a true stem cell. This will allow microcolony plating efficiency to be representative of macrocolony plating efficiency. But the cells must not divide enough times to dilute the [^{3}H]-thymidine label to below autoradiographic detection limits. When applied under carefully controlled conditions, the proportion of labeled microcolonies can be compared to the percent of labeled cells in the tumor cell suspension, and this gives information on the clonogenicity of the P and Q cytokinetic populations. If the percent of labeled microcolonies equals the labeling index in the tumor cell suspension, P and Q cells have the same clonogenicity. It follows from this that one can determine the sensitivity of P vs Q cells to radiation or drugs by treating the tumor

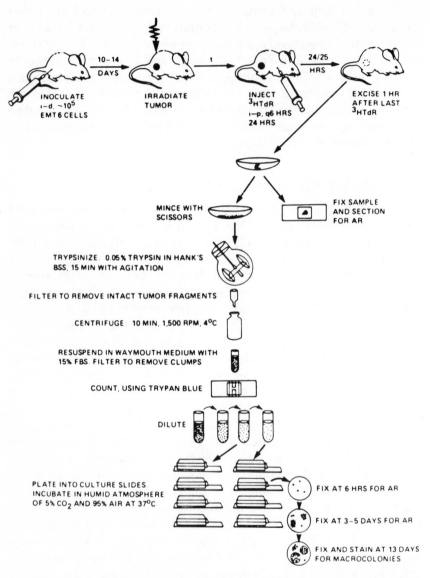

Fig. 1. Schematic representation of the labeled microcolony technique for assaying the clonogenicity of proliferating (P) and nonproliferating (Q) cells in solid tumors. This figure is reproduced from Kallman R. F. and Rockwell S. C., in *Cancer, A Comprehensive Treatise* vol. 6 (F. F. Becker, ed.), 1976, with permission of Plenum Press.

prior to labeling and plating. If P cells are more sensitive, the labeling index in microcolonies will be lower than in nonirradiated controls. The recruitment of cells from Q into P after cell death caused by a cytotoxic treatment can be studied directly by this method and would be seen as an increase in the percent of labeled microcolonies. One must, however, be able to resolve this from initial differences in sensitivity of P vs Q cells and changes in cell cycle duration of proliferating cells that survived treatment.

The problems of applying this technique on a large scale are formidable, and sophisticated equipment is required to analyze carefully the labeling patterns in autoradiographs of large numbers of microcolonies fixed at 3–5 d of age in culture (Fig. 1). Background must be assessed carefully to determine which microcolonies are really labeled. Label is light because cell division dilutes it and the amount of ^3H injected into the animal must be kept small enough to prevent kinetic perturbations from intranuclear radiation. The 3–5-d microcolonies abort with a measurable frequency and some fail to grow to genuine macrocolonies of >50 cells. This frequency can be assessed as a function of cell number/microcolony and thus applied as a correction factor. Overall, this technique is one that gives direct information on clonogenicity of P vs Q cells before, during, or after treatment; the clonogenic and cell kinetic assays are done on the same cell sample at essentially the same time.

2.3. S-Phase Suicide Technique

High specific activity [^3H]-thymidine and HU have been used to preferentially kill those proliferating cells in S-phase and to gain information about cytokinetics of clonogenic cells before and after anticancer therapy (8,44).

The basic approach is to compare the in vitro clonogenicity of two populations of tumor cells; one treated for a limited time with a lethal dose of the S-phase agent immediately prior to plating and the other not so treated. The time allowed for the colonies to develop is the full period normally needed for macrocolony formation with a particular tumor type. The relative plating efficiency for the two groups of cells allows for calculation of the percent killed by the cycle-specific drug. This is assumed to be equal to the percent of clonogenic cells in S-phase (Fig. 2). The S-phase-specific drug may be injected into the tumor-bearing animals shortly prior

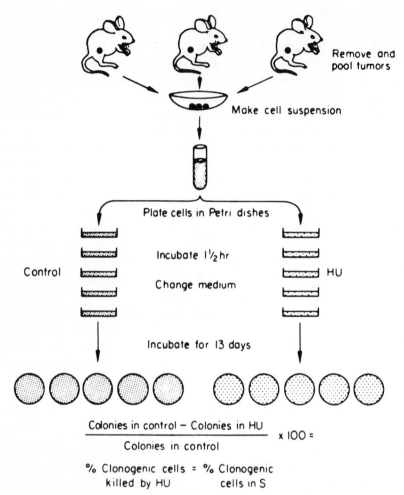

$$\frac{\text{Colonies in control} - \text{Colonies in HU}}{\text{Colonies in control}} \times 100 =$$

% Clonogenic cells   =   % Clonogenic
    killed by HU   cells in S

Fig. 2. Schematic representation of the hydroxyurea suicide technique developed for assaying the proportion of clonigenic tumor cells in S-phase. This figure is reproduced from ref. *31*, with the permission of Blackwell Scientific Publications, Ltd.

to sacrifice, but more commonly the cells are exposed briefly after their isolation from the tumor (*8,31*).

Rockwell et al. (*31*) applied these agents to the in vivo–in vitro EMT-6 tumor to reveal the percent of clonogenic cells in a solid tumor that are in S-phase. Careful preliminary experiments with EMT-6 cells growing in vitro identified incubation times and concentrations of HU or high specific activity [³H]-thymidine that killed a percentage of clonogenic cells equal to the percent in S-phase, as determined by labeling index in autoradiographs prepared after a pulse label with tritiated thymidine. It was determined that HU was far easier to use than [³H]-thymidine; reuse of ³H from the latter produced major radiotoxicity problems and was found not to be applicable to solid tumors.

Figure 2 illustrates one variation of the technique using HU. The tumor cells may be treated with HU after isolation from the tumor by suspending them or plating them briefly in drug-containing medium, as illustrated, or tumor-bearing animals may be injected with HU prior to sacrifice. The optimum concentration of HU in vitro was found to be 3.5 mM, with a 90-min treatment time adequate to kill all S-phase clonogens. Rockwell et al. (31) found that the percent of clonogenic EMT-6 tumor cells killed by in vivo injection of HU prior to removal of the tumor (15 ± 5%) or by plating isolated tumor cells into 3.5 mM HU for 90 min (12 ± 1%) agreed reasonably well with the percent of tumor paren-chymal cells pulse labeled with [³H]-thymidine (17 ± 1%).

This technique has several advantages. The tumor cell population or suspension analyzed for clonogenicity is identical to that analyzed for residence in S-phase, and the tests are done concurrently. Because one counts macrocolonies for treated and untreated cells at the end of a full incubation period (Fig. 2), rather than looking at labeling patterns in microcolonies with a somewhat unknown proliferative future (Fig. 1), this method helps one to avoid a major uncertainty encountered in the labeled microcolony technique. But, like the microcolony technique, this method should allow one to examine recruitment; more clonogens entering the cell cycle should also mean more clonogens in S-phase. The method also allows one to determine if clonogens in S-phase are more or less sensitive to radiation (32) or drugs (5,32), and also to look at reassortment of surviving tumor stem cells into different phases of the cell cycle after cytotoxic treatment (32).

A [³H]-thymidine suicide technique for tumor cell clonogens has several disadvantages. [³H]-Thymidine reutilization and subsequent radiolysis were such major problems that this agent could never be adequately applied to cells isolated from solid tumors, although under very stringent conditions, it could be used to determine the percent of S-phase clonogens for EMT-6 cells growing in vitro (31). The HU suicide technique also has disadvantages. The percent of colony-forming tumor cells killed by plating into HU after preparation of the cell suspension may be slightly or substantially less than the percent that appears to be in S-phase, whether determined by [³H]-thymidine pulse labeling in vivo or FCM analysis of DNA content in isolated cells (31,43). One possible reason for this is that the cell-isolation procedure, which involves enzymatic digestion of the tumor, can temporarily inhibit DNA synthesis (29; Rasey, unpublished). This is a factor when the S-phase-specific agent is applied after cell isolation. In addition, HU may interact with cytotoxic agents or the damage caused by them. Such interaction has been observed with HU plus radiation (37) and HU combined with bleomycin (Rasey, unpublished).

2.4. Viable Cell Sorting Based on DNA Content

The ability to sort viable cells with flow cytometry and the availability of a DNA-specific vital dye Hoechst 33342 (1) opened the way for sorting cells directly on the basis of DNA content and, therefore, position in the cell cycle, followed by plating for determination of clonogenicity of the sorted cell population.

To date, these methods have been applied to only one in vivo–in vitro tumor system—the KHT sarcoma (26). The advantage of this method is that it allows for the direct determination of clonogenicity vs cell cycle position, although it does not allow one to distinguish cycling G1-phase cells from cells with a G1 DNA content that are out of cycle. It also allows for determination of clonogenicity of normal host cells vs tumor parenchymal cells if the latter have a higher DNA content based on increased ploidy. In the KHT tumor, it was used to show that untreated cells with G1, S, or G2 + M DNA contents have equal clonogenicity. Immediately following 1700 rad of X-rays, the cycling position of surviving, presumably hypoxic, cells was determined with this method and the clonogenicity of cells arrested in G2 phase 10 h after this radiation dose also was compared to that for cells in other parts of the cycle (26).

Most disadvantages of this method are technical. FCM sorting is time-consuming and requires expensive, sophisticated equipment. In addition, the vital dye Hoechst 33342 is directly toxic to cells. Pallavicini et al. (26) found that 5 mM was the maximum concentration tolerable for vital staining. Unsorted irradiated KHT cells were more sensitive to the dye than unsorted nonirradiated cells. Sorting plus staining was more toxic than staining alone for nonirradiated cells, but the opposite was true if tumor cells were irradiated *in situ* prior to preparation of a single cell suspension and incubation with the dye. Other limitations of this method are biological rather than technical. Many of the cells in the tumor are not actively proliferating (*38*) and tumor cells may leave the cycle at any stage, not just G1 (*36*). Thus, sorting on the basis of DNA alone cannot distinguish between the P and Q cell compartments. This method cannot distinguish between normal and tumor cells when tumor parenchymal cells have a normal diploid DNA content. However, it may be possible to use FCM to separate malignant from normal cells, using fluorescent antibodies to cell surface markers. The tumor cells could then be further sorted based on DNA content. Viable cell sorting is also discussed in chapter 12 (*13*).

2.5. Purification by Centrifugation

Sorting of viable cells from solid tumors via centrifugal techniques (*11,24*) has been applied to many aspects of tumor biology. Although this is not principally a cytokinetic technique, several investigations with in vivo–in vitro tumors have combined centrifugal separation with [³H]-thymidine labeling or measures of DNA content to derive information about the cell kinetics of various tumor cell subpopulations (*7,9,14,15,20–22,35,36*).

Tumor cell subpopulations have different densities and this forms the basis for separation (*14*). Tumor cells are layered on a sterile density gradient and centrifuged. Viable cells band at different levels in the tube and are removed by inserting a needle through the side of the tube and withdrawing the cells. The cells can then be plated or further characterized in other ways. This technique has principally used density gradients made with the radiographic contrast agent Renografin (*11–15*) and has been applied to the FSa fibrosarcoma of the C3H mouse (*11*). This sarcoma is not an in vivo–in vitro tumor system as usually defined; tumor cell survival for unseparated or separated cell populations is determined by a lung colony rather than a Petri dish assay (*11*). However, the FSa cells will grow in culture (*15*) and the principles developed are the same as with in vivo–in vitro tumors. FSa cells separated on the basis of density have been characterized on the basis of DNA content and, therefore, cell cycle position (*36*). The density fractions have also been studied for DNA synthesizing ability in vivo (*9*) and for the presence of a quiescent nonproliferating population as determined by two-stage acridine orange staining (*9*). For the most part, the different density fractions do not conform to different cell cycle stages (*15*), nor does any one density fraction contain a large pool of quiescent cells (*9*). One might expect that the high-density fraction, which is of lower clonogenicity (*11*), might also contain a high proportion of nonproliferating cells. This fraction, however, does appear to contain most of the chronically hypoxic cells (*12*), and probably includes tumor cells that, *in situ*, are located at distance from a blood vessel (*36*).

Tumor cells also differ in size. Parenchymal cells are often larger than stromal cells or other normal cells such as macrophages that infiltrate a tumor. Cells increase in size as they pass through the cell cycle from the early postmitotic G1 phase to the next mitosis. This forms the basis for cell separation via centrifugal elutriation, which separates cells principally on size, with some influence of cell shape and density as well (*24*). Centrifugal elutriation is based on the principle of velocity sedimentation (*27*), and is a rapid, high-resolution version of this technique that can be applied to large volumes and large total cell numbers (*24*). In this method, a suspension of disaggregated viable cells is loaded into a tapered chamber in a continuous-flow centrifuge rotor while the rotor is spinning. The flow of culture medium through the rotor chamber is counter to the direction of the centrifugal force and only cells with a sedimentation velocity that is less than the minimum flow rate in the chamber will be elutriated into a collection vessel. Increasing the medium flow rate during centrifugation accomplishes the separation (*23*). Centrifugal elutriation is discussed in more detail in chapter 12 of this volume (*13*).

This method of separation of cells from solid tumors has recently been improved and good separation of parenchymal cells from normal cells of low plating efficiency and, more importantly, excellent synchronization of viable parenchymal cells in different phases of the cell cycle were obtained (*21,22,35*). For three widely used in vivo–in vitro tumors, EMT-6/Ro, 9L, and KHT, the quality of synchronization was confirmed by FCM analysis of DNA content. Selected fractions were enriched in specific cell cycle stages, as outlined in Table 1. Although separation by cell cycle stage was much better for the RIF-1 and 9L tumors than for EMT-6/Ro, the potential for use with additional tumor systems seems good (*22*).

The advantages of this technique for studying tumor cell kinetics are considerable. Elutriation is a well-established, nonperturbing method for separation of viable cells. Cell recovery is high,

TABLE 1
Cell Cycle Stages of Elutriation Fractions[a]

	Fractions		
Tumor	I, %	II, %	III, %
9L	93, G1	55, S	78, G2 + M
KHT	94, G1	75, S	73, G2 + M
EMT-6/Rx	91, G1	38, S	40, G2 + M

[a]Taken from ref. 21.

and cell viability is excellent and higher than in unseparated cell populations, which often contain dead cells and debris. The clonogenicity of cells as a function of cell cycle stage can be determined and does not vary for untreated KHT or 9L cells (22). In vivo cell age response to radiation or drugs can be measured by treating tumors prior to cell separation (21). Some of the most exciting developments are those that can couple elutriation to flow cytometry and two-stage acridine orange staining for DNA and RNA. This can distinguish cycling cells with a G1 DNA content from Q cells, the latter usually having a G1 DNA content and a much reduced RNA content (*see* chapter 8 in this volume) (6,10). This method has been successfully used to identify quiescent cells in EMT/Ro spheroids (7), and elutriation can produce a fraction of EMT-6/Ro cells from the solid tumors in which 91% have a G1 DNA content (22). Recently, Wallen et al. (45,46) have successfully identified Q cells for the 66, 67, and 68H mouse mammary tumor cell lines in plateau phase in vitro using the acridine orange staining method. These tumor lines grow in vivo as well as in vitro and can be cloned from both growth modes. Cell size difference as a function of cell cycle position or residence in Q has been shown for cells in vitro. This suggests potential for synchronization of cells from solid tumors by centrifugal elutriation. Because elutriation has worked so well as a method for separating viable cells in good condition, this method, coupled with FCM sorting based on DNA content, and differential staining of Q cells with acridine orange, offers great potential for cytokinetic studies of clonogenic tumor cells.

A major disadvantage of elutriation in these studies is low recovery of cells from some cell cycle stages, i.e., 2% or less of the total cells appear in fractions highly enriched in S or G2/M cells (22). This can become a real problem in studying tumors after cytotoxic treatment, when total cell yield will be reduced.

3. CONCLUSIONS

All of the above methods share the advantages and disadvantages of in vivo–in vitro tumors used in experimental radio- and chemotherapy. The EMT-6, KHT, and RIF-1 mouse sarcomas, the R1 rhabdomyosarcoma of the rat, and the 9L rat brain tumor are the most widely used, and most of these have been studied with some or all of the above techiques. The ability to obtain detailed quantitative information on behavior of tumor stem cells is the real advantage of these tumors. However, there are serious questions as to whether in vitro assays can ever be truly representative of tumor cell kinetics or survival *in situ*, because there is not enough information on tumor microenvironments to duplicate these in vitro (41). The role of the tumor microenvironment in determining cell survival and how heterogeneous these environments are between tumor areas are important considerations. Better techniques for *in situ* measurement of tumor extracellular pH, O_2, glucose levels, and so on, and the development of completely defined tissue culture media to duplicate these environments would make in vivo–in vitro tumor systems a more valuable tool in cell kinetic research.

Obtaining a representative cell sample is a problem for many of these tumor systems (25,29). Although RIF-1 and KHT tumors may yield nearly all of their parenchymal cells in a single cell suspension (30), EMT-6, R-1, and many other tumors produce much smaller yields, typically less than 1.0 to about 20% of the total cells, and these are sometimes not representative samples of the tumor (28,29). Even when isolated cell suspensions have the same [^{3}H]-thymidine labeling index as tumor parenchymal cells in a histological section, this may not be sufficient evidence that this is a random sample, or that the same types of cells are labeled in both preparations.

A problem that is insurmountable in the final analysis is the nature of in vivo–in vitro tumors or most other experimental tumor systems currently used (41). Most are selected for rapid, reproducible growth, and cells from the in vivo–in vitro models are subjected to major selective forces every time they go from one growth mode to another. These tumors have been developed for high plating efficiency, whereas many spontaneous tumors may have only one in 10^4 or 10^5 cells that are clonogenic (34). One might argue for doing these cell kinetic studies with more representative tumor models, such as cell suspensions isolated directly from human tumors.

Human tumor stem cell assays have been extensively developed as a technique for studying those human neoplastic cells that are capable of repopulating the tumor (33). Although the goals of these studies are direct, pragmatic, and laudable, the methods also require much further development before results can be unambiguously interpreted (34). Cell kinetic studies and stem survival assays using in vivo–in vitro tumors can yield useful information if one recognizes the limitations of the models and remembers this caveat: Animal models are to be used to test hypotheses, not to test treatment protocols.

ACKNOWLEDGMENTS

This chapter is a review, but experience with in vivo–in vitro tumor systems provided the background and is based on research sponsored by Grants #CA 12441, CA 19899, CA 27382, and CA 27699, all from NIH, DHHS.

REFERENCES

1. Arndt-Jovin, D. J. and Jovin, T. M. Analysis and sorting of living cells according to deoxyribonucleic acid content. J. Histochem. Cytochem., 25: 585–589, 1977.
2. Barendsen, G. W. and Broerse, J. J. Experimental radiotherapy of a rat rhabdomyosarcoma with 15 MeV neutrons and 300 kV X-rays. I. Effects of single exposures. Eur. J. Cancer, 5: 373–391, 1969.
3. Barendsen, G. W. and Broerse, J. J. Experiemntal radiotherapy of a rat rhabdomyosarcoma with 15 MeV neutrons and 300 kV X-rays. II. Effects of fractionated treatments, applied five times a week for several weeks. Eur. J. Cancer, 6: 89–109, 1969.
4. Barendsen, G. W., Roelse, H., Hermens, A. F., Madhuizen, H. T., Van Peperzeel, H. A., and Rutgers, D. H. Clonogenic capacity of proliferating and nonproliferating cells of a transplantable rat rhabdomyosarcoma in relation to its radiosensitivity. J. Natl. Cancer Instit., 51: 1521–1526, 1974.
5. Bateman, A. E. and Steel, G. G. The proliferative state of clonogenic cells in the Lewis lung tumor after treatment with cytotoxic agents. Cell Tissue Kinet., 11: 445–454, 1978.

6. Bauer, K. D. and Dethlefsen, L. A. Control of cellular proliferation in HeLa-S3 suspension cultures. Characterization of cultures utilizing acridine orange staining procedures. J. Cellular Phys., *108*: 99–112, 1981.

7. Bauer, K. D., Keng, P. C., and Sutherland, R. M. Isolation of quiescent cells from multicellular tumor spheroids using centrifugal elutriation. Cancer Res., *42*: 72–78, 1982.

8. Becker, A. J., McCulloch, E. A., Siminovitch, L., and Till, J. E. The effect of differing demands for blood cell production on DNA synthesis by hemopoietic colony-forming cells of mice. Blood, *26*: 296–308, 1965.

9. Brock, W. A., Swartzendruber, D. E., and Grdina, D. J. Kinetic heterogeneity in density-separated murine fibrosarcoma subpopulations. Cancer Res., *42*: 4999–5003, 1982.

10. Darzynkiewicz, Z. Cytochemical probes of cycling and quiescent cells for flow cytometry. *In*: (J. W. Gray and Z. Darzynkiewicz, eds.), Techniques in Cell Cycle Analysis. New Jersey: Humana, 1986.

11. Grdina, D., Basic, I., Mason, K., and Withers, H. R. Radiation response of clonogenic cell populations separated from a fibrosarcoma. Radiat. Res., *63*: 483–493, 1975.

12. Grdina, D. J., Basic, I., Guzzino, S., and Mason, K. Radiation response of cell populations irradiated *in situ* and separated from a fibrosarcoma. Radiat. Res., *66*: 634–643, 1976.

13. Grdina, D. J., Meistrich, M., Meyn, R., Johnson, T., and White, A. Cell synchrony techniques: A comparison of methods. *In*: (J. W. Gray and Z. Darzynkiewicz, eds.), Techniques in Cell Cycle Analysis. New Jersey: Humana, 1986.

14. Grdina, D. J., Meistrich, M. L., and Withers, H. R. Separation of clonogenic cells from stationary phase cultures by density gradient centrifugation. Exp. Cell Res., *85*: 15–22, 1974.

15. Grdina, D. J., Sigdestad, C. P., and Peters, L. J. Phase-specific cytotoxicity *in vivo* of hydroxyurea on murine fibrosarcoma cells synchronized by centrifugal elutriation. Br. J. Cancer, *39*: 152–158, 1979.

16. Hermens, A. F. and Bardensen, G. W. Effects of ionizing radiation on the growth kinetics of tumor. *In*: 29th Annual Symposium on Fundamental Cancer Research on Growth Kinetics and Biochemical Regulation of Normal and Malignant Cells, Houston, Texas, March 10–12, 1976.

17. Hermens, A. F. and Barendsen, G. W. The proliferative status and clonogenic capacity of tumor cells in a transplantable rhabdomyosarcoma of the rat before and after irradiation with 800 rad of X-rays. Cell Tissue Kinet., *11*: 83–100, 1978.

18. Jung, H., Beck, H. P., Brammer, I., and Zywietz, F. Depopulation and repopulation of the R1H rhabddomyosarcoma of the rat after X-irradiation. Eur. J. Cancer, *17*: 375–386, 1981.

19. Kallman, R. F., Combs, C. A., Franko, A. J., Furlong, B. M., Kelley, S. D., Kemper, H. L., Miller, R. G., Rapacchietta, D., Schoenfeld, D., and Takahashi, M. Evidence for the recruitment of noncycling clonogenic tumor cells. *In*: (R. E. Meyn and H. R. Withers, eds.), Radiation Biology in Cancer Research. New York: Raven, 1980.

20. Keng, P. C., Li, C. K. N., and Wheeler, K. T. Synchronization of 9L rat brain tumor cells by centrifugal elutriation. Cell Biophys., *2*: 191–206, 1980.

21. Keng, P. C., Siemann, D. W., and Wheeler, K. T. Comparison of tumor age response to radiation for cells derived from tissue culture of solid tumors. Br. J. Cancer, *50*: 519–526, 1984.

22. Keng, P. C., Wheeler, K. T., Siemann, D. W., and Lord, E. M. Direct synchronization of cells from solid tumors by centrifugal elutriation. Exp. Cell Res., *134*: 15–22, 1981.

23. Meistrich, M. L., Grdina, D. J., and Meyn, R. E. Application of cell separation methods to the study of cell kinetics and proliferaion. *In*: (B. Drewinko and R. M. Humphrey, eds.), Growth Kinetics and Biochemical Regulation of Normal and Malignant Cells. Baltimore: Williams and Wilkins, 1977.

24. Meistrich, M. L., Grdina, D. J., Meyn, R. E., and Barlogie, B. Separation of cells from mouse solid tumors by centrifugal elutriation. Cancer Res., *37*: 4291–4296, 1977.

25. Pallavicini, M. Solid tissue dispersal for cytokinetic analysis. *In*: (J. W. Gray and Z. Darzynkiewicz, eds.), Techniques in Cell Cycle Analysis. New Jersey: Humana, 1986.

26. Pallavicini, M. G., Lalande, M. E., Miller, R. G., and Hill, R. P. Cell cycle distribution of chronically hypoxic cells and determination of the clonogenic potential of cells accumulated in G2 + M phases after irradiation of a solid tumor *in vivo*. Cancer Res., *39*: 1891–1897, 1979.

27. Pretlow, T. G., Weir, E. E., and Zettergren, J. G. Problems connected with the separation of different kinds of cells. *In*: (G. W. Richter and M. A. Epstein, eds.), International Review of Experimental Pathology. New York: Academic, 1975.

28. Rasey, J. S. and Nelson, N. J. Effect of tumor disaggregation on results of *in vitro* cell survival after *in vivo* treatment of the EMT-6 tumor: Comparison of response to X-rays, cyclophosphamide and bleomycin. In Vitro, *16*: 547–553, 1980.

29. Rasey, J. S. and Nelson, N. J. Response of an *in vivo–in vitro* tumor to X-rays and cytotoxic drugs: Effect of tumor disaggregation method on cell survival. Br. J. Cancer, *41* (suppl. IV): 217–221, 1980.

30. Rasey, J. S. and Nelson, N. J. Discrepancies between patterns of potentially lethal damage repair in the RIF-1 tumor system *in vitro* and *in vivo*. Radiat. Res., *93*: 157–174, 1983.

31. Rockwell, S., Frindel, E., and Tubiana, M. A technique for determining the proportion of the clonogenic cells in S phase in EMT6 cell cultures and tumors. Cell Tissue Kinet., *9*: 313–323, 1976.

32. Rockwell, S., Frindel, E., Valleron, A-J., and Tubiana, M. Cell proliferation in EMT6 tumors treated with single doses of X-rays or hydroxyurea. I. Experimental results. Cell Tissue Kinet., *11*: 279–289, 1978.

33. Salmon, S. E., ed. Cloning of Human Tumor Stem Cells. Progress in Clinical Biological Research. New York: A. R. Liss, 1980.

34. Selby, P., Buick, R. N., and Tannock, I. A critical appraisal of the human tumor stem-cell assay. N. Engl. J. Med., *308*: 129–134, 1983.

35. Siemann, D. W., Lord, E. M., Keng, P. C., and Wheeler, K. T. Cell subpopulations dispersed from solid tumors and separated by centrifugal elutriation. Br. J. Cancer, *44*: 100–108, 1981.

36. Sigdestad, C. P. and Grdina, D. J. Density centrifugation of murine fibrosarcoma cells following *in situ* labeling with tritiated thymidine. Cell Tissue Kinet., *14*: 589–600, 1981.

37. Sinclair, W. The combined effect of hydroxyurea and X-rays on Chinese hamster cells *in vitro*. Cancer Res., *28*: 198–206, 1968.

38. Steel, G. G. *Growth Kinetics of Tumors*. Oxford: Clarendon, 1977.

39. Stephens, T. C., Currie, G. A., and Peacock, J. H. Repopulation of γ-irradiated Lewis lung carcinoma by malignant cells and host macrophage progenitors. Br. J. Cancer, *38*: 573–582, 1978.

40. Stephens, T. C. and Peacock, J. H. Tumor volume response, initial cell kill and cellular repopulation in B16 melanoma treated with cyclophosphamide and 1-(2-chlorethyl)-3-cyclohexyl-1-nitrosurea. Br. J. Cancer, *36*: 313–321, 1977.

41. Sutherland, R. M., Rasey, J. S., and Hill, R. P. Tumor biology. Cancer Treatment Symp., *1*: 49–65, 1984.

42. Traganos, F., Darzynkiewicz, Z., Sharpless, T., and Melamed, M. R. Simultaneous staining of ribonucleic and deoxyribonucleic acids in unfixed cells using acridine orange in a flow cytofluorometric system. J. Histochem. Cytochem., *25*: 46–56, 1977.

43. Twentyman, P. R. The growth of the EMT6 tumor in the lungs of BALB/c mice following intravenous inoculation of tumor cells from culture. Cell Tissue Kinet., *11*: 57–68, 1978.

44. Vassort, F., Winterholer, M., Frindel, E., and Tubiana, M. Kinetic parameters of bone marrow stem cells using *in vivo* suicide by tritiated thymidine or by hydroxyurea. Blood, *41*: 789–796, 1973.

45. Wallen, C. A., Higashikubo, R., and Dethlefsen, L. A. Murine mammary tumor cells *in vitro*: I. The development of a quiescent state. Cell Tissue Kinet., *17*: 65–77, 1984.

46. Wallen, C. A., Higashikubo, R., and Dethlefsen, L. A. Murine mammary tumor cells *in vitro*: II. Recruitment of quiescent cells. Cell Tissue Kinet., *17*: 79–89, 1984.

Chapter 5

Flow Cytokinetics

Joe W. Gray, Frank Dolbeare, Maria G. Pallavicini,
and Martin Vanderlaan

1. INTRODUCTION

Studies of the proliferation characteristics of normal and malignant cells have been important to the biological characterizations of these cell types, to the elucidation of the biochemical mechanisms involved in cell replication, to understanding the mechanisms of action of drugs and radiation, and to the design of more effective cancer therapy strategies. The immediate experimental goal of these studies is usually to determine the rate at which cells change from one proliferative status to another; for example, the rate at which cells move from the G1-phase of the cell cycle to the S-phase, or the rate at which cells are recruited from a noncycling state (e.g., G0) into active proliferation. Alternatively, this information may be presented as the rate at which the cells traverse the various proliferative states or phases. At least four proliferative states have been identified: (1) actively proliferating and capable of extended proliferation (clonogenic), (2) actively proliferating but not clonogenic, (3) temporarily not proliferating (quiescent) but, upon stimulation, capable of proliferation and repopulation, and (4) not proliferating and nonclonogenic (doomed or fully differentiated). In addition, cells in any state can reside in at least four phases; G1, S, G2, and M (12).

During the past 10 years, flow cytometry and supporting cell preparation techniques have developed into powerful tools for cytokinetic studies (23). In this chapter, we review the technology and illustrate its application in cytokinetic analysis of asynchronous and drug treated cells growing in vitro and in vivo. We also illustrate the power of flow cytometry in cytokinetic analyses of subpopulations that can be distinguished during flow cytometry.

2. FLOW CYTOMETRY AND SORTING

The principles of flow cytometry and sorting are illustrated in Fig. 1. In this approach, fluorescently stained cells from the population being analyzed are forced to flow one at a time through the flow cytometer. Thus, the first step in flow cytokinetics is to disperse the cells to form a suspension of single cells. This is straightforward for cells that grow naturally in suspension (e.g., spinner cultures in vitro, peripheral lymphocytes, bone marrow cells, leukemias, and other ascites tumors), but is more complicated for solid normal tissues and tumors. The technical details of cell dispersal for flow cytometry are discussed by Pallavicini (43) in chapter 6. Dispersed cells are stained with one or more fluorescent dyes, one of which is usually DNA-specific; and forced to flow one at a time through a light beam(s), the wavelength of which is adjusted to excite the fluorescent dye(s). The fluorescence emitted as each stained cell crosses the existing light beam is collected by a lens system, projected through a spectral filter(s) passing only the desired wavelengths of light, and onto a photomultiplier. The intensity of the electric signal from the photomultiplier is proportional to the intensity of the emitted fluroescence and is recorded in a pulse height analyzer as a measure of the amount of the biochemical entity in the cell to which the fluorescent dye is bound. For example, the intensity of fluorescence at approximately 520 nm is usually taken to be a measure of the relative DNA content for cells stained with the DNA-specific dye Chromomycin A3 and excited at 458 nm. The insert in Fig. 1 shows a fluorescence distribution for asynchronously growing Chinese hamster ovary cells stained with Chromomycin A3.

The advantages of flow cytometry in cell cycle analysis will be discussed in detail in subsequent sections. However, all derive from the speed, sensitivity, and precision of flow cytometry. Thousands of cells can be processed each second so that in a few minutes a

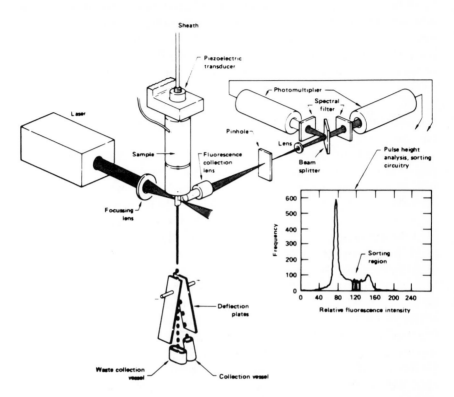

Fig. 1. Schematic diagram illustrating the principles of flow cytometry and sorting. Fluorescently stained cells are introduced into the flow chamber and forced to flow one by one through an intense light beam that excites the fluorescent dye. The resulting fluorescence is collected and recorded as a measure of the amount of bound dye. Since the dye is usually chosen to bind selectively and stoichiometrically to a specific cellular component (DNA in this example), the recorded fluorescence intensity is used as a measure of the amount of the stained cellular component. Cells can be measured at rates of up to 20,000/s with an accuracy approaching 2% (46). After analysis, the cells flow out of a small orifice in a thin jet. The jet is forced to break into uniform droplets by the action of a vibrating piezoelectric transducer mounted on top of the flow chamber. Cells possessing distinctive features (mid-S-phase DNA content in this example) are selected for sorting by applying a charge to the droplets carrying these cells. The charged droplets are separated from the uncharged droplets during passage through an intense electric field and can be collected for further analysis (30–32,39,55).

statistically precise analysis of over 10^5 cells can be recorded in the pulse height analyzer. In addition, rare populations can be detected and analyzed with accuracy. The precision of measurement (i.e., the coefficient of variation of the measured fluorescence intensities of homogeneously stained cells) is high, approaching 1% for highly conserved cellular components such as DNA, so that subtle differences can be detected. Finally, the fluorescence from weakly stained cells (e.g., cells stained with only a few thousand molecules) can be analyzed with accuracy.

Electrostatic sorting (30–32,39,55) allows separation of cells that can be discriminated during flow cytometric analyses. During sorting, the cells are analyzed flow cytometrically, then ejected into air in a thin liquid jet (*see* Fig. 1). The cells flow in the liquid jet in air until they reach the point at which the jet breaks into droplets. Droplet breakup is forced to occur at a fixed time after cell measurement by the vibration of a piezoelectric crystal attached to the flow chamber so that the time between flow cytometric analysis and the arrival of the cell at the droplet breakoff point is constant and can be determined accurately. The instrument is adjusted to induce an electrostatic charge on the jet at the instant any cell to be sorted reaches the end of the jet. Droplets that break off during the charging period are therefore charged as they separate from the jet. Empty droplets and droplets containing unwanted cells are left uncharged. In this way, cells selected according to their cytometric properties (e.g., mid-S phase in Fig. 1) are contained in charged water droplets. All droplets then fall through a high-voltage electric field and the charged droplets containing the cells of interest separate from the uncharged droplets and fall into collection devices. In most sorters, two subpopulations of cells can be sorted simultaneously by charging the droplets containing one subpopulation positively and the droplets containing the other subpopulation negatively.

3. DNA DISTRIBUTION ANALYSIS

One of the most common cytokinetic applications of flow cytometry is the measurement of DNA distributions (24,27,52), the rationale being that such distributions yield information about the cytokinetic status of the cell populations under analysis (i.e., fractions of cells in the G1-, S-, and G2M-phases of the cell cycle). The cytochemical procedures for measuring DNA distributions are straightforward and are described in chapter 7 by Crissman and Steinkamp (10). Briefly, cells to be analyzed are reduced to a monodisperse suspension, fixed, stained with a DNA specific fluorescent dye, and processed through a flow cytometer where the dye is excited. The resulting fluorescence intensities of several hundred thousand cells are recorded and displayed as a DNA distribution. The link between the distribution of cells around the cell cycle and the DNA distribution is illustrated in Fig. 2 and is based on the fact that cells in the G1-, S-, and G2M-phases have distinct DNA contents. Cells in G2M-phase have twice the DNA of cells in G1-phase, and cells in S-phase have intermediate DNA content. The fractions of cells in G1-, S-, and G2M-phase can be estimated from a DNA distribution as described in chapter 8 by Dean (14). For asynchronously growing cell populations with unity growth fraction, the fractions of time spent in the G1-, S-, and G2M-phases are related to the fractions of cells with G1-, S-, and G2M-phase DNA contents, respectively (27). Unfortunately, these relations are complex in situations in which the population is not growing asynchronously or the growth fraction is not unity (23). In addition, it is important to realize that DNA distributions for asynchronously growing cells provide no information about absolute rates of cell cycle traverse. For example, asynchronous populations with G1-, S-, and G2M-phase durations of 5, 5, and 1 h, respectively, will be measured to have exactly the same DNA distributions as populations with G1-, S-, and G2M-phase durations of 50, 50, and 10 h, respectively. Thus, DNA distributions, at best, give only a rough estimate of the cytokinetic status of the population under analysis.

Somewhat more cytokinetic information can be obtained from sequential DNA distributions measured for perturbed or synchronously growing populations (10). This notion is illustrated in Fig. 3, which shows the correlation between the distribution of cells around the cell cycle and the DNA distribution for a population at several times after release from synchrony in G1-phase. It is clear from the sequential DNA distributions that the rate at which the synchronous populations move through the G1-, S-, and G2M-phases can be inferred from the rate of change of the DNA distributions. Several computer analysis programs have been developed to extract G1-, S-, and G2M-phase durations from DNA distribution sequences (14). In analyses of this type, cell cycle traverse is

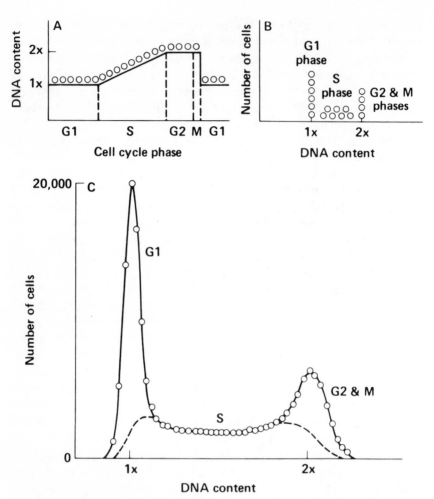

Fig. 2. Relation between the cell cycle and the DNA distribution for an asynchronous population. (A) A hypothetical distribution of cells in the G1-, S-, and G2M-phases of the cell cycle, as well as their DNA contents. (B) Distribution of these same cells as a function of DNA content. Peaks for the G1- and G2M-phase cells are clearly evident. The S-phase cells form a continuum between the G1- and G2M-phase peaks. (C) DNA content distribution actually measured for Chinese hamster ovary cells by flow cytometry. Again, peaks for the G1- and G2M-phase cells are visible as is a continuum formed by the S-phase cells (27).

simulated mathematically assuming an initial distribution for the cell population (in this case in early G1-phase) and the rates of cell cycle traverse and the variability therein. Cell cycle distributions and the corresponding DNA distributions are simulated at several times after the beginning of the analysis corresponding to times at which DNA distributions were measured experimentally. The cell cycle traverse rates in the model are then adjusted until the simulated DNA distributions match the experimental distributions. The G1-, S-, and G2M-phase durations that produce the best match are assumed to be accurate estimates of the actual G1-, S-, and G2M-phase durations and dispersions of the population under analysis. This process is described in more detail in chapter 8 by Dean (*14*). Grdina et al. illustrate the use of perturbed population DNA distribution analysis in chapter 12 on cell synchrony (*29*).

Of course, the analysis of sequential distributions becomes much more difficult for biologically complex populations. Such complexity may occur in tumors during therapy as a result of cell loss, cell cycle traverse inhibition, recruitment of cells into cycle, and/or cell cycle traverse heterogeneity. When these events occur, measurement of a simple sequence of DNA distributions is usually not sufficient for complete analysis, and other supporting data (e.g., rate of DNA synthesis, clonogenicity) must be acquired (*23,44,48*).

A variety of flow cytometric assays has been developed to complement and extend DNA distribution analysis. One approach, described by Darzynkiewicz et al. (*11,12*), allows for discrimination of additional cell cycle states (e.g., quiescent and proliferating cells in the G1-, S-, and G2M-phases). In other approaches, described later in this chapter, other biochemical indicators of cell cycle traverse are measured in addition to DNA content.

4. RADIOACTIVITY PER CELL (RC) ANALYSIS

4.1. Asynchronous Populations

Radioactivity per cell (RC) analysis was developed initially (*25,26*) to allow for accurate estimation of the G1-, S-, and G2M-phase durations and dispersions and the growth fraction for asynchronously growing cell populations. This procedure is analogous in design to the fraction of labeled mitoses procedure described by Shackney and Ritch (*50*) in chapter 2, since the cell cycle phase durations are inferred from measurement of the rate at which a

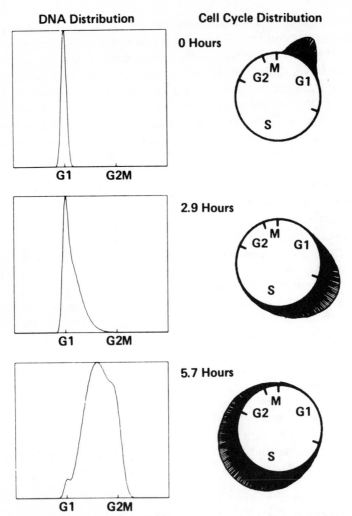

Fig. 3. Relation between the cell cycle distribution and DNA distribution for a synchronous population. DNA distributions and cell cycle distributions are shown for a hypothetical cell population immediately after release from a block in early G1-phase and 2.9 and 5.7 h later. The cohort of cells initially in early G1-phase has moved into early S-phase 2.9 h after release and into late S-phase and G2 phase 5.7 h after release. The rate of movement of the cohort and the loss of synchrony caused by cell-to-cell variability in cell cycle traverse rate can be determined from the DNA distributions.

radioactively labeled cohort of cells traverses the cycle. In the RC method, the S-phase cells are labeled with a radioactive DNA precursor (typically tritiated thymidine; [³H]-TdR) at the beginning of the analysis. The population is then sampled periodically (typically every 2–3 h) for approximately 1.5 times the total cell cycle duration. The samples are dispersed, fixed, stained for DNA content analysis, and processed through a cell sorter. Cells with G1-, mid-S-phase, and/or G2M-phase DNA contents are sorted (typically 10^4 cells for each sample) and the radioactivity per cell (RC) is measured by liquid scintillation counting for each sorted sample (RCG1, RCS_i, and RCG2M, respectively). Figure 4 shows the variation in

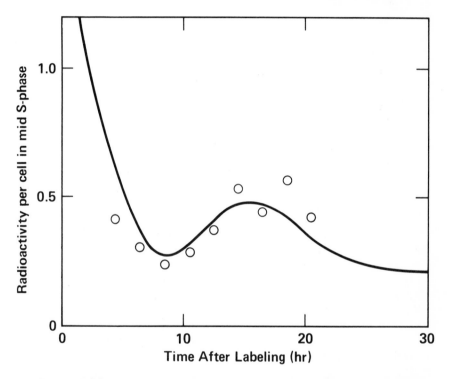

Fig. 4. RCS_i analysis of KHT tumors grown in C3H mice. Replicate mice bearing KHT tumors received injections of [³H]-TdR at time zero. Mice were sacrificed at 2-h intervals thereafter for 20 h. The tumors were removed, dispersed, and stained with the DNA-specific dye Chromomycin A3. These cells were then processed through a cell sorter and cells with mid-S-phase DNA content were sorted into vials. The radioactivity per cell (RCS_i) for each sample was determined by liquid scintillation counting. These data were analyzed by computer to determine the G1, S-, and G2M-phase durations (25), and are shown as open circles. The computer fit to the data is shown as a solid line.

RCS$_i$ values measured for KHT tumors labeled in vivo with [³H]-TdR. Immediately after [³H]-TdR injection, [³H]-TdR-labeled cells are in S-phase, so the RCS$_i$ values are initially high. The RCS$_i$ values fall as the labeled cells move out of S-phase and rise as the labeled cells return to S-phase. Cell cycle phase durations and dispersions can be estimated quantitatively from the variation with time in the RCG1 and RCS$_i$ values (25). Figure 5 shows,

(a)   (b)

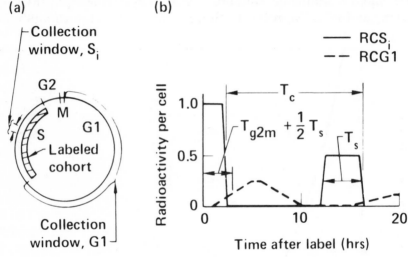

Fig. 5. Schematic diagram showing the relation between RCS$_i$ and RCG1 curves and the G1-, S-, and G2M-phase durations. (A) Cell cycle locations of the mid-S-phase and G1-phase windows from which cells were sorted for RC analysis. (B) Hypothetical RCS$_i$ and RCG1 curves expected for a population in which all cells traverse the G1-, S-, and G2M-phases of the cell cycle in exactly 9, 4, and 1 h, respectively. The relation between the RC curves and the phase durations is illustrated in panel (b) (25). The G1-, S-, and G2M-phase durations are designated T_{g1}, T_s, and T_{g2m}, respectively.

schematically, the connection between the G1-, S-, and G2M-phase durations and the RCS$_i$ and RCG1 curves for a hypothetical population in which all cells traverse the cycle at the same rate. In practice, estimation of phase durations is complicated by cell-to-cell variability in cell cycle traverse rates, and computer procedures must be utilized for proper phase duration analysis (25). Computer analysis of the data in Fig. 5 shows the S-phase and total cell cycle durations to be 12.5 and 16.8 h, respectively.

4.2. Rate of DNA Synthesis

The RC analysis procedure has provided substantial data on the rate of DNA synthesis and its variation across S-phase (8,28). Such data are necessary for accurate analysis of DNA distribution sequences and RC curves. In addition, the rate of DNA synthesis is an important indicator of cell proliferation in drug-treated populations (23,44). To estimate DNA synthesis rates, a radioactive DNA precursor is administered shortly before the cells are harvested for RC analysis. The cells are then prepared for DNA content analysis and processed through a cell sorter in which cells with several S-phase DNA contents are sorted. Figure 6b shows regions of a DNA distribution measured for Lewis lung (LL) tumor cells from which cells were sorted for RC analysis (28). The LL tumor cells were labeled in vitro with [3H]-TdR 30 min prior to harvest. Figure 6a shows the RC values measured for the cells from the various S-phase DNA content regions. Since the [3H]-TdR was administered as a 30-min pulse, each RC value can be considered to be a rough estimate of the rate of DNA synthesis (actually, amount of [3H]-TdR incorporated in 30 min) at the selected DNA content. Of course, the true rate of DNA synthesis can be determined only if [3H]-TdR-uptake rates and endogenous pool sizes are known at each DNA content. This procedure has the advantage that the cell population need not be synchronized in order to estimate the variation across in the rate of DNA synthesis. Thus, it can be applied to drug treated populations and to cells growing in vivo (28).

4.3. Perturbed Populations

The RC analysis technique has also proved useful in the analysis of cell cycle traverse by perturbed populations. Pallavicini et al. (44), for example, used RC analysis to investigate the effect of cytosine arabinoside (ara-C) on the proliferation of KHT tumor cells treated in vivo. In one experiment, the KHT tumors were pulse labeled with [3H]-TdR 30 min before treatment with ara-C so that the cells in S-phase to be killed by the ara-C (an S-phase cytotoxic agent) were radioactively labeled. Tumors were then sampled periodically, dispersed, stained for DNA content, and processed through the flow sorter in which cells from mid-S-phase and G1-phase were sorted into liquid scintillation vials. Figure 7 shows RCS_i and RCG1 values for these cells at several times after ara-C treatment.

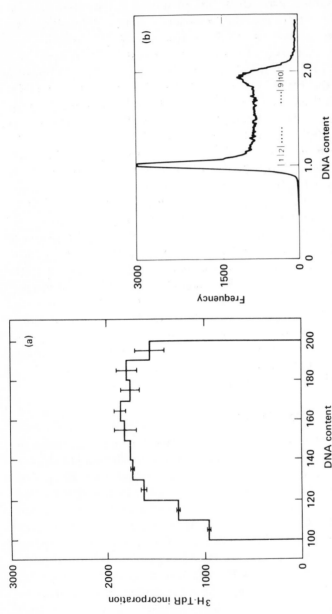

Fig. 6. RC values as a function of DNA content for LL tumor cells isolated from a tumor labeled in vitro with [³H]-TdR 30 min prior to harvest. (a) RC values measured for the cells sorted for the regions indicated in panel (b). In this study, cells were also sorted for autoradiographic determination of the labeling index (a measure of the fraction of G1 and G2M cells contaminating the early and late S-phase sorts, respectively). The RC values shown here were derived by dividing the radioactivity values measured by liquid scintillation counting by the labeling indices measured autoradiographically (28). (b) Regions of the DNA distribution for the LL tumor from which cells were sorted for RC analysis.

The rate of cell cycle traverse and the rate of loss of the radioactively labeled (ara-C-killed) cells was determined by computer analysis of these data. These data show that the S-phase cells progressed slowly or not at all after ara-C treatment and that these ara-C-killed cells were removed from the population at a rate of about 10%/h beginning about 10 h after ara-C treatment. In another experiment, replicate C3H mice bearing KHT tumors were treated in vivo with ara-C. At intervals thereafter, groups of mice were treated with [^{3}H]-TdR for 30 min, sacrificed, and their tumors removed and prepared for RC analysis as described above. The serial samples were processed through a sorter in which the S-phase tumor cells were sorted into liquid scintillation vials. Figure 7b shows the recovery of DNA synthesis in the KHT tumors (i.e., the RCS$_i$ values measured after tritiated thymidine labeling) beginning about 5.5 h after ara-C treatment. Computer analysis of these data also showed that quiescent KHT cells were fully recruited following ara-C treatment, and that the cell cycle traverse rate of the ara-C-treated cells was about 2.5 times the rate prior to treatment (*44*).

5. STATHMOKINETIC ANALYSIS

An absolute estimate of the total time required for cell cycle traverse also can be obtained by measuring the rate at which cells accumulate in the G2M phase of the cell cycle following administration of a stathmokinetic agent (i.e., one that blocks cells in mitosis). In flow-cytometrically-oriented stathmokinetic studies, the rate of accumulation of cells in the G2M phase is determined from sequential DNA distributions measured at several times after administration of the stathmmokinetic agent (e.g., colcemid). Figure 8 shows DNA distributions measured for Chinese hamster ovary (CHO) cells treated in vitro with 0.05 mg/mL colcemid for 0, 2, 4, and 12 h (*18*). By 12 h, essentially all of the cells in the population have been arrested in the G2M phase of the cell cycle. Dosik et al. (*18*) analyzed these data mathematically to estimate the G1-, S-, and G2M-phase durations. They showed good agreement between their estimates and those determined from mathematical analyses of fraction of labeled mitoses measurements. Similar analyses have been performed for peripheral blood lymphocytes grown in vitro (*13*) and for colon and jejunum following exposure to colcemid in vivo (*7*).

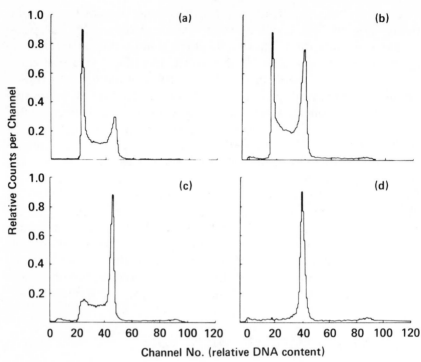

Fig. 7. RCS$_i$ and RCG1 values measured for KHT murine tumors at several times after treatment with ara-C (44). (A) The tumors were labeled in vivo with [^{3}H]-TdR 30 min prior to ara-C treatment. Tumors were harvested periodically thereafter and processed for RCS$_i$ and RCG1 analysis. The closed circles show the RCS$_i$ values and the open triangles show the RCG1 values. The solid line is a computer simulation assuming recovery of cell cycle traverse 5.5 h post-ara-C, a 2.5-fold increase in the cell cycle traverse rate, and full recruitment and loss of cells killed in S-phase beginning 10 h post-ara-C. (B) Tumor-bearing animals were treated at time zero with ara-C. Animals were labeled with [^{3}H]-TdR at several times after the ara-C treatment and sacrificed 30 min later. The [^{3}H]-TdR-labeled tumors were then processed for RCS$_i$ analysis. The solid points show the measured RCS$_i$ values. The solid line is a computer simulation generated using the same assumptions listed for (A).

A multivariate approach to flow stathmokinetic analysis is presented in more detail in chapter 9 by Darzynkiewicz et al. (13).

Stathmokinetic experiments such as these yield, under some circumstances, absolute information about the G1-, S-, and G2M-phase durations and the growth fraction, and can be applied both

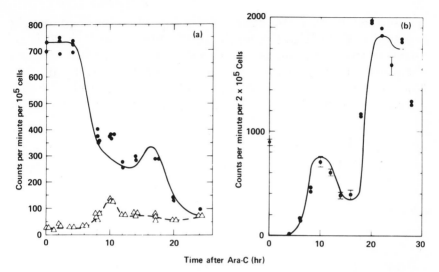

Time after Ara-C (hr)

Fig. 8. DNA distributions measured for CHO cells at several times after administration of colcemid. (A) DNA distribution measured immediately after treatment. (B), (C), and (D) DNA distributions measured after 2, 4, and 12 h, respectively (*18*).

in vitro and to selected systems in vivo. It is important to keep in mind, however, that these estimates are accurate only if (1) the mitotic inhibitor is administered at sufficient strength to block all cells in mitosis, (2) the stathmokinetic agents holds the cells in mitosis throughout the analysis (often difficult in vivo because of colcemid-induced cell death), and (3) the specific loss of mitotic cells is minimal during the dispersal and staining procedures required for flow cytometry. The strengths and limitations of stathmokinetic analyses are reviewed in more detail by Wright and Appleton (*56*).

6. BROMODEOXYURIDINE AS A CYTOKINETIC LABEL

6.1. Bromodeoxyuridine (BrdUrd) Quenching of Hoechst Fluorescence

The DNA-specific dye Hoechst 33258 and its analogs fluoresce intensely when bound to double-stranded DNA (*36,37*). These dyes bind preferentially to adenine-thymine-specific DNA (*37*). However, the fluorescence is quenched considerably when the thymidine is replaced with BrdUrd (35,36). Figure 9 shows that BrdUrd is an analog of thymidine and is incoporated into cellular DNA through the same salvage pathway. The effect of BrdUrd substitution on Hoechst fluorescence can be detected flow cytometrically when a substantial fraction of the thymine in the DNA is replaced by

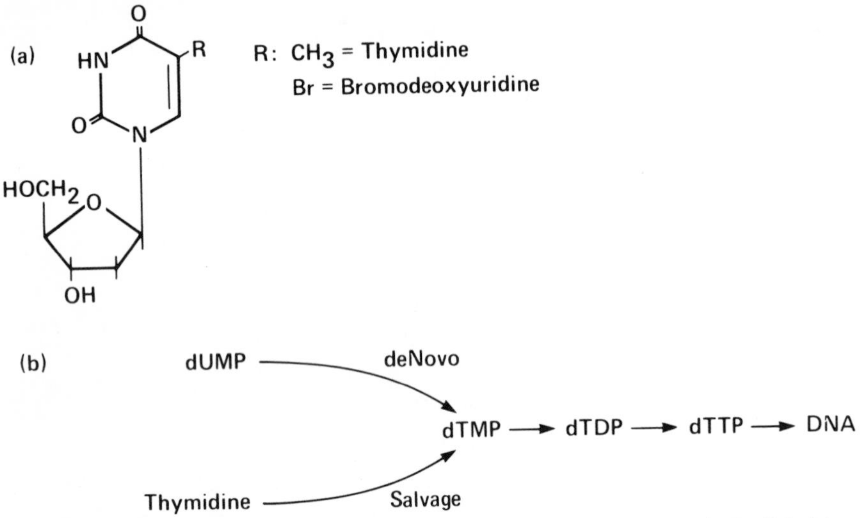

Fig. 9. Molecular structure and metabolic fate of BrdUrd. BrdUrd is an analog of thymidine in which a methyl group is replaced by bromine. BrdUrd is incorporated into DNA through the same salvage pathway as thymidine. Inhibition of the *de novo* pathway may be necessary for maximum BrdUrd incorporation.

BrdUrd (3,4,34,38,47). The extent of the quenching is governed by the extent of BrdUrd incorporation (34). Figure 10 shows Hoechst fluorescence distributions measured for phytohemagglutinin-stimulated human lymphocytes grown for 40–45 h in medium containing BrdUrd at three different concentrations. The quenching induced by the high-concentration BrdUrd ($1.5 \times 10^{-3}M$) is almost twice that induced by the low-concentration BrdUrd ($1.5 \times 10^{-5}M$).

Bohmer and Ellwart (4) have used this technique effectively to estimate G1-, S-, and G2M-phase durations in mouse L cells grown in vitro. The cells of interest were incubated in medium containing 2 mg/mL BrdUrd at the beginning of the analysis. Cell samples were taken periodically after the beginning of the BrdUrd incubation, fixed for subsequent analysis, stained with Hoechst 33258,

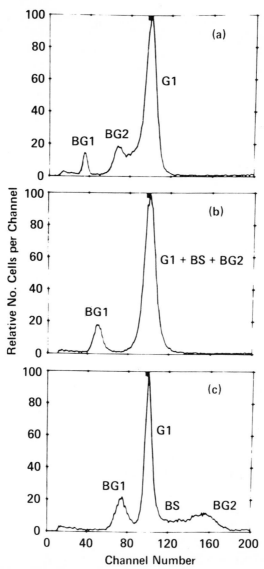

Fig. 10. Hoechst 33258 distributions measured for PHA-stimulated human lymphocytes grown for 40–45 h in medium containing BrdUrd at three different concentrations. (A), (B), and (C) Distributions measured after growth in $1.5 \times 10^{-3}M$, $6.5 \times 10^{-5}M$, and $1.5 \times 10^{-5}M$ BrdUrd, respectively. The prefix B signifies that the cells producing this part of the distribution had replicated in the presence of BrdUrd. For example, BG1 cells indicate G1-phase cells that had completed one round of DNA synthesis in medium containing BrdUrd (34).

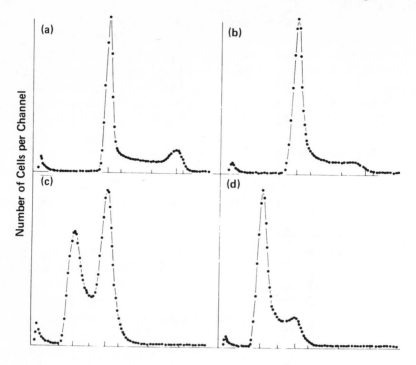

Fluorescence Intensity (channel No.)

Fig. 11. Hoechst 33258 flourescence distributions measured for CHO cells grown for four intervals in medium containing BrdUrd. (A), (B), (C), and (D) Distributions measured immediately and 3, 11, and 19 h after initiation of the BrdUrd labeling (3).

and analyzed flow cytometrically. Immediately after beginning incubation in BrdUrd, the Hoechst fluorescence is proportional to the total cellular DNA content, since little BrdUrd incorporation has occurred. Thus, the fluorescence distribution in Fig. 11a shows the distribution of the cells among the G1-, S-, and G2M-phases. The distribution begins to change in shape, as the BrdUrd labeling induces fluorescence quenching. In this study, the extent of the Hoechst fluorescence quenching was proportional to the BrdUrd labeling duration; that is, cells grown for one generation in BrdUrd were approximately half as fluorescent as those with the same DNA content that had incorporated no BrdUrd. Thus, the fluorescence intensity of cells in G1-, S-, and G2M-phase, when the BrdUrd label was introduced, did not change until the cells divided and halved their fluorescence intensity. Figure 11b shows that the first cells that have divided and halved their flourescence intensity can be seen 3 h after the beginning of the labeling. Figure 11c shows that most of the cells in S and many of the cells in G1 at the time of labeling have divided after 11 h. Figure 11d shows that most of the cells of the population have divided 19 h after label initiation. Bohmer (3) showed that the G1-, S-, and G2M-phase durations can be estimated for kinetically simple cell populations (i.e., asynchronous populations with unity growth fraction, no less loss, and/or recruitment) from the rates at which cells leave the G1-, S-, and G2M-phase regions of the original fluorescence distribution. Rabinovitch (47) has used this technique to study the more complex issue of the cell cycle traverse by human fibroblasts in vitro. He showed in this study that the technique is useful for study of slowly cycling or noncycling cells.

This procedure is especially attractive for analysis of cells in vitro. However, it is essential that sufficient BrdUrd be incorporated so that the Hoechst fluorescence for the newly synthesized DNA be heavily quenched and the extent of BrdUrd incorporation and Hoechst fluorescence quenching be uniform during the course of the analysis. This may be difficult to achieve in some in vivo situations because of heterogeneity in BrdUrd transport (e.g., in solid tumors) and because of the short metabolic half-life of BrdUrd in vivo (33). In addition, BrdUrd-induced toxicity (6,9,19,40), alteration of normal differentiation pathways (1,21), or perturbation of cell cycle traverse may occur after extensive BrdUrd labeling.

6.2. Monoclonal Antibodies Against BrdUrd

Cytokinetic studies are traditionally based on the introduction of a label into a cohort of cells in the population to be studied and on analysis of the rate at which the labeled cohort moves from phase to phase (e.g., S to G2) within the cycle or from state to state (e.g., cycling to noncycling or clonogenic to new clonogenic). The fraction of labeled mitosis method described in chapter 2 (50) is a classic example of this kind of analysis. Such analyses require a nontoxic label to mark the cohort to be followed, a method for detection of the label, and a method to determine the cell cycle location of the labelled cells. Two recent developments now allow all of this to be accomplished quickly and quantitatively using flow cytometry. The first of these developments is a monoclonal antibody against BrdUrd incorporated into cellular DNA (20,22,49,53) and the use of this antibody as a reagent to fluorescently stain cells to reveal small amounts of incorporated BrdUrd (22). The second development is a procedure to simultaneously stain cells for both total DNA content and amount of incorporated BrdUrd (15–17). BrdUrd can be used to label cells in S-phase that are actively synthesizing DNA. BrdUrd has a short metabolic half-life in vivo so only the cells synthesizing DNA immediately after injection incorporate BrdUrd. In vitro, of course, the cells must be washed free of BrdUrd to produce a short labeling period. These labeled cells become a "labelled cohort" whose cell cycle traverse can be analyzed to estimate the cell cycle traverse rates of the labeled population. The cell cycle location of the labeled cells is determined by measuring BrdUrd vs DNA (BrdUrd/DNA) distributions. The information obtained from such distributions is the same as could be obtained by measuring total DNA content and amount of incorporated tritiated thymidine for each cell, and much greater than can be obtained from a [³H]-TdR-labeling index determined autoradiographically (51). As discussed in the following sections, the information on the cytokinetic properties of asynchronous and perturbed populations that can be obtained from such distributions includes estimates for the fractions of cells in the G1-, S-, and G2M-phases, the G1-, S-, and G2M-phase durations, the rate of DNA synthesis, the growth fraction, the cell loss rate, the rate of recruitment of quiescent cells, and the rate of decyling of drug-treated cells.

6.2.1. Antibody Production

Several high-quality monoclonal antibodies against BrdUrd in cellular DNA have now been produced (20,22,49,53) and one, originally developed by Gratzner (22), is available commercially (Becton-Dickinson Monoclonal Antibodies, Sunnyvale, CA). These antibodies recognize BrdUrd in single-stranded DNA (ssDNA). Most of the antibodies against BrdUrd in ssDNA have been produced by immunizing mice with a halogenated pyrimidine–protein complex (e.g., iododeoxyuridine covalently linked to ovalbumin), fusing spleen cells from the immunized mice with myeloma cells, and selecting hybridoma clones that secrete antibodies that selectively bind to cells grown in BrdUrd. The antibodies have been selected by screening for antibodies that bind to BrdUrd-labeled cells whose DNA has been denatured (20,49,53).

6.2.2. BrdUrd/DNA Cytochemistry

Several protocols for simultaneous analysis of cellular DNA content and BrdUrd have been developed in recent years for use with antibodies against BrdUrd in ssDNA (15,17). In one recent protocol (16), BrdUrd labeled cells are fixed in ethanol or another fixative that does not interfere with the binding of common DNA-specific dyes. The fixed cells are treated with RNase to remove dsRNA (the RNase step may be eliminated if the dye used to stain for DNA content does not bind to dsRNA). Proteins are extracted by treatment with 0.1N HCl and the DNA in the cells is then partially denatured (e.g., by heating to 80°C in 50% formamide). These cells are incubated with the monoclonal antibody against BrdUrd in ssDNA. If the anti-BrdUrd antibody is not already fluorescently labeled, the cells are then reacted with fluorescently labeled goat anti-mouse antibody (all monoclonal antibodies against BrdUrd are mouse antibodies). Fluorescein and rhodamine have both been used as fluorescent antibody labels. The cells are then counterstained for DNA. Propidium iodide is the most commonly used dye for this purpose, although Hoechst 33342 has been used as well (45). The double-stained cells are processed through a flow cytometer in which the amounts of each dye are measured for each cell. Cells stained with a fluorescein-labeled anti-BrdUrd antibody and propidium iodide are excited at 488 nm. Green fluorescence is recorded as a measure of the amount of incorporated BrdUrd, and red fluorescence is recorded as a measure of the total cellular DNA content (15–17).

The most troublesome aspect of this protocol is the requirement for partial DNA denaturation required by the antibodies against BrdUrd in ssDNA. Some cell types, notably those in the bone marrow, are fragile and may be lost during the denaturation procedure. If the loss of cells is extensive or selective, the utility of the entire procedure is reduced. The availability of antibodies against BrdUrd in dsDNA will eliminate this problem (20). So far, however, antibodies of this type with proven high specificity and binding affinity are not available.

6.2.3. Pulse and Continuous Labeling Studies

The BrdUrd/DNA analysis procedure has proved applicable both in vitro and in vivo. Figure 12a shows a bivariate BrdUrd/DNA distribution measured for Chinese hamster ovary cells grown for 30 min in 10 mM BrdUrd. The cells with G1- and G2M-phase DNA contents fluoresce weakly at green wavelengths, indicating that they have not incorporated BrdUrd, whereas the cells with S-phase DNA content fluoresce intensely in the green, indicating that they have incorporated substantial amounts of BrdUrd. Some cells with apparent G1- and G2M-phase DNA contents appear to have incorporated intermediate amounts of BrdUrd. These are cells that have moved into early S-phase from G1 or into G2-phase from late-S-phase during the labeling period (17). Figure 12b emphasizes the utility of using the BrdUrd/DNA assay to distinguish between G1-phase and very-early-S-phase cells. This distribution was measured for CHO cells synchronized by elutriation to be in late G1 phase and early-S-phase (15). The S-phase cells are clearly visible at high BrdUrd content; however, they have not synthesized sufficient DNA to appear to have increased DNA content.

In many circumstances, the intensity of the green fluorescence is a quantitative measure of the amount of BrdUrd incorporated (15,16). Figure 13 shows the correlation between BrdUrd incorporation and BrdUrd-linked green fluorescence for CHO cells grown in medium containing tritium labeled BrdUrd for 30 min. The cells were fixed immediately after BrdUrd labeling and processed through a cell sorter. Cells with varying levels of green fluorescence were sorted and collected for liquid scintillation counting to determine the amount of incorporated tritium. Panel (a) shows the regions of the bivariate BrdUrd/DNA from which cells were sorted. Panel (b) shows the correlation between the measured BrdUrd-linked green fluorescence and the measured radioactivity per cell for each sorting window. The two measurements are highly correlated showing the stoichiometry of the BrdUrd staining procedure (17).

One of the more powerful aspects of the BrdUrd staining procedure is its sensitivity. Figure 14 shows the bivariate BrdUrd/DNA distribution measured for CHO cells grown for only 10, 20, and 30 s in medium containing 10 μM BrdUrd. Under these labeling conditions, only about 0.01% of the thymidine molecules in the DNA should be replaced with BrdUrd. This extreme sensitivity is greater than or at least equivalent to that obtained by autoradiographic analysis of tritium labeled cells (51) and illustrates that the technique should be applicable both in vitro and in vivo.

(a)

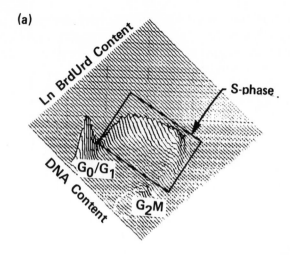

(b)

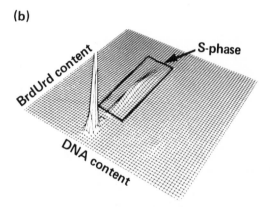

Fig. 12. A bivariate BrdUrd/DNA distribution measured for CHO cells grown for 30 min in medium containing 10 μM BrdUrd. (A) This distribution was measured for asynchronously growing cells. The cells with G1/G0- and G2M-phase DNA content are clearly visible at low BrdUrd values. The BrdUrd-labeled cells with S-phase DNA content are clearly distinct from the G1- and G2M-phase cells. The labeled S-phase cells are enscribed by a rectangular box. (B) This distribution was measured for cells synchronized to be in late-G1-phase and early-S-phase. The S-phase cells have incorporated BrdUrd and are distinctly labeled. The labeled S-phase cells are enscribed by a rectangular box.

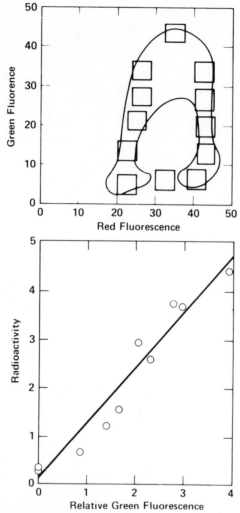

Fig. 13. The correlation between BrdUrd incorporation and BrdUrd-linked green fluorescence. (top panel) Single contour for a bivariate BrdUrd/DNA distribution measured for CHO cells labeled for 30 min with [³H]-BrdUrd. Also shown are square regions from which cells were sorted for RC analysis. (bottom panel) Correlation between the measured RC values and the corresponding BrdUrd-linked green fluorescence (17).

Note that the intensity of BrdUrd-linked green fluorescence increases with labeling duration.

Figure 15 illustrates the in vivo application of the BrdUrd/DNA analysis technique. This figure shows two views of the bivariate

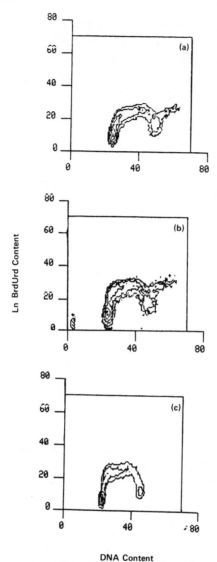

DNA Content

Fig. 14. A bivariate BrdUrd/DNA distribution measured for CHO cells grown in BrdUrd for short periods. (A), (B), and (C) Distributions measured after labeling intervals of 10, 20, and 30 s, respectively. The labeled S-phase cells are clearly visible at elevated BrdUrd levels. These labeling periods should result in incorporation of 10^5–10^6 BrdUrd molecules, assuming complete replacement of all thymine bases with BrdUrd during the labeling period and assuming a 7-h duration for S-phase (Wolfgang Beisker; personal communications).

distribution measured for cells isolated from a near tetraploid murine KHT sarcoma 30 min after ip administration of 100 mg/kg BrdUrd. Two populations are distinguished according to their DNA content; the near tetraploid KHT sarcoma cells and the diploid normal cells located within the tumor mass. The S-phase cells in both the normal and tumor cell populations are distinctly labeled and easily enumerated (14).

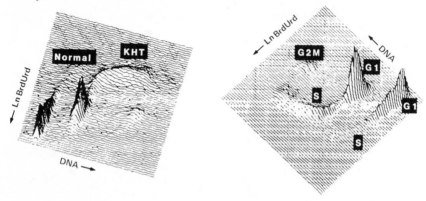

Fig. 15. A BrdUrd/DNA distribution measured for a KHT tumor labeled in vivo for 30 min with 100 mg/kg BrdUrd. (right and left panels) Two views of the same BrdUrd/DNA distribution: the G1-, S-, and G2M-phase regions of the aneuploid KHT tumor cells and of the diploid normal cells are noted in the figure.

The administration of BrdUrd, as a continuous label coupled with BrdUrd/DNA analysis, allows identification of cells that fail to traverse S-phase during the labeling period. Barcellos-Hoff et al. (2) have applied this procedure to the study of noncycling 9L cells grown in vitro as multicellular spheroids. Figure 16 shows the BrdUrd/DNA distribution measured for 9L spheroid cells grown for 24 h in medium containing BrdUrd. A 10% BrdUrd labeling cocktail (3 μM BrdUrd plus 27 μM thymidine) was used in these studies to minimize the possibility of BrdUrd-induced toxicity. Sufficient label (30 μM) was used to ensure labeling throughout the entire 24-h period. The majority of cells in the spheroids have traversed S-phase during the 24-h labeling period, divided, and entered the G1-phase. These G1-phase cells form a peak at high BrdUrd content. Other cells entered the S- and G2M-phases during the labeling period, but have not yet reached G1-phase. These cells fall into the high DNA content/high BrdUrd content part of

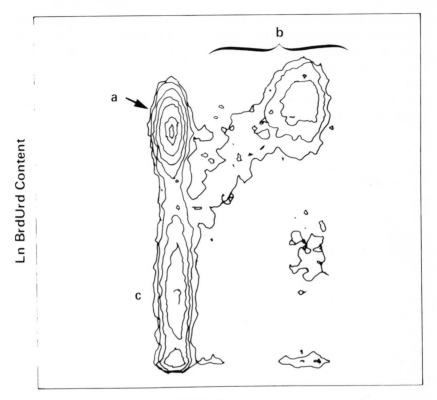

DNA Content

Fig. 16. A BrdUrd/DNA distribution measured for 9L spheroids. The spheroids were grown for 24 h in medium containing BrdUrd plus thymidine. Region (a) contains G1-phase cells that traversed S-phase at least once during the labeling period. Region (b) contains S- and G2M phase cells that initiated DNA synthesis during the labeling period. Region (c) contains G1-phase cells that have not synthesized DNA during the labeling period. These are putative G0 cells (2).

the distribution. Cells that have not entered S-phase during the labeling period are putative G0 cells. They form a peak at low DNA content and low BrdUrd content. Other methods for identification of G0 cells are discussed by Braunschweiger (5) and Darzynkiewicz et al. (12).

6.2.4. Cytokinetic Analysis of Asynchronously Growing Cells

Determination of the G1-, S-, and G2M-phase durations and dispersions (i.e., coefficients of variation of the phase durations) for asynchronously growing cell populations can be accomplished easily using the BrdUrd/DNA methodology. In this assay, the cells to be analyzed are grown for a short period (typically 30 min) in BrdUrd, the BrdUrd is removed (either physically or metabolically), and the cells are then sampled periodically for BrdUrd/DNA distribution analysis as they grow in the absence of BrdUrd. Figure 17 shows a series of BrdUrd/DNA distributions measured for KHT sarcomas in animals labeled in vivo with 100 mg/kg BrdUrd. Pulse labeling in vivo is made simple by the short metabolic half-life of BrdUrd (33). Immediately after the BrdUrd labeling (Fig. 17a), all labeled cells are in S-phase. At subsequent times, the labeled cohort moves through S-phase (panel b), into and through G2M-phase (panel c), into G1-phase (panels d, e, and f), and back into S-phase (panels g and h). At the same time, cells that were in G1-phase during the labeling period move out of G1-phase, through S-, and G2M-phase, and back to G1-phase. Estimates for the G1-, S-, and G2M-phase durations and dispersions can be estimated from the rate at which the labeled and unlabeled cells move through the various phases (17). The cytokinetic information in these sequential BrdUrd/DNA distributions can be reduced to a more manageable form by determining the average green fluorescence (BrdUrd content) per cell in G1 and mid S-phase for each distribution. This is the same information that would be obtained from RC experiments described earlier and can be analyzed using the same computer analysis programs (25).

6.2.5. Cytokinetic Analysis of Perturbed Populations

The BrdUrd/DNA analysis procedure has proved especially useful in the study of perturbed populations. Information typically required for such studies includes the rate of cell cycle traverse of the proliferating cells, rate of DNA synthesis at intervals after perturbation, the fraction of cells in a noncycling state, and so forth. Figures 18 and 19 illustrate the application of the BrdUrd/DNA procedure to the analysis of the affect of ara-C on CHO cells growing in vitro. The BrdUrd labeling was performed in two different ways to maximize the information obtained on the nature of the perturbation caused by the ara-C.

Figure 18 shows BrdUrd/DNA distributions measured for CHO cells that were treated with BrdUrd for 30 min immediately before addition of 100 μM ara-C. The cells in S-phase at the time of treatment appear distinctly labeled, whereas the cells in the G1- and

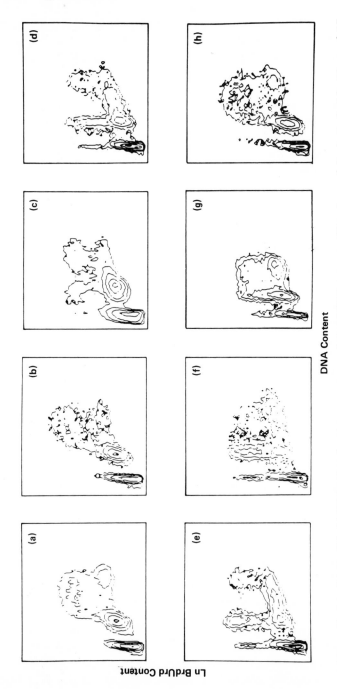

DNA Content

Ln BrdUrd Content

Fig. 17. (a)–(h) BrdUrd/DNA distributions measured for KHT tumors 30 min and 2, 4, 6, 8, 10, 12, and 15 h after ip injection of 100 mg/kg BrdUrd. The cell cycle traverse of the labeled cells (visible at high BrdUrd levels) and the unlabeled cells (visible at low BrdUrd levels) can be followed for both the aneuploid KHT tumor cells and the normal diploid cells.

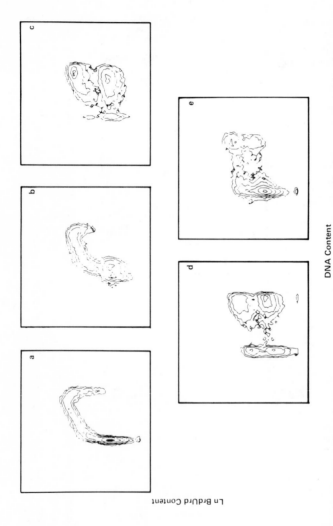

Fig. 18. BrdUrd/DNA distributions measured for CHO cells labeled with BrdUrd 30 min prior to treatment with 100 μM ara-C. (a)–(e) Distributions measured 2, 6, 12, 18, and 24 h, respectively, after treatment with ara-C. The ara-C-killed cells (BrdUrd-labeled) move slowly across and out of S-phase, whereas the cells not in S-phase at the time of ara-C treatment move more rapidly.

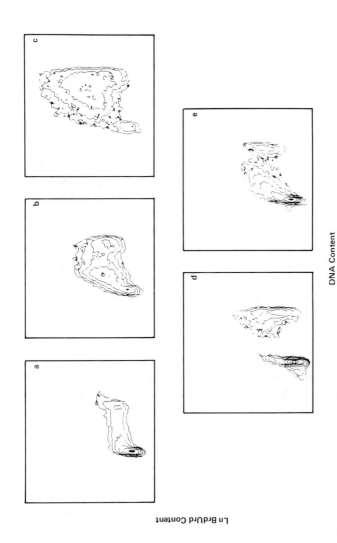

DNA Content

Ln BrdUrd Content

Fig. 19. BrdUrd/DNA distributions measured for CHO cells treated initially with ara-C for 2 h, allowed to grow for times ranging from 2 to 24 h, and then treated with BrdUrd for 30 min. (a)–(e) Distributions, measured after 2, 6, 12, 18, and 24 h, respectively. Recovery of the ability to incorporate BrdUrd begins at 6 h. Rate of incorporation is maximal at 12 h and lower at 24 h.

G2M-phase are unlabeled. The cell cycle traverse characteristics of the labeled S-phase cells are particularly interesting in studies of the effect of ara-C, since ara-C is thought to be maximally toxic to S-phase cells. The cell cycle traverse characteristics of the labeled S-phase cells and the unlabeled G1- and G2M-phase cells can be quantified separately. The distribution in panel (a), measured immediately after treatment, is identical to that measured prior to treatment but after BrdUrd labeling. The distribution in panel (b), measured 6 h after ara-C, shows the relatively slow movement of the labeled cells out of S-phase, as well as the movement by unlabeled cells into S-phase. The distributions in panels (c), (d), and (e) measured 12, 18, and 24 h after ara-C, respectively, show the continued cell cycle traverse of the unlabeled cells and the cell cycle traverse of the labeled cells. Estimates for the posttreatment rates of cell cycle traverse of the cells in S-phase at the beginning of the ara-C treatment (i.e., the BrdUrd-labeled cells) and the cells in G1-plus G2M-phase cells at the beginning of the ara-C treatment (i.e., the BrdUrd-unlabeled cells) can be estimated from the rates of movement of these cells through the G1-, S-, and G2M-phases. The rate of loss of the BrdUrd labeled cells (none in this example) can be quantified as well.

Figure 19 shows BrdUrd/DNA distributions measured for replicate CHO cells populations that were treated with 100 μM ara-C at the beginning of the experiment. Separate populations were labeled with BrdUrd for 30 min at several times after ara-C exposure, fixed, and then prepared for BrdUrd/DNA analysis. These distributions show the capability of the ara-C-treated cells to incorporate BrdUrd (synthesize DNA) at various times after treatment. Panel (a) shows the distribution measured immediately after treatment with ara-C. It is clear that the cells with S-phase DNA content are unlabeled and thus are not synthesizing DNA. Panels (b) through (e) show the recovery of DNA synthesis. Most of the cells with S-phase DNA content at 6 h (panel b) are labeled, and all of the cells with S-phase DNA content are labeled 12 h after treatment (panel c). In addition, these data show that the rate of DNA synthesis (i.e., the extent of BrdUrd incorporation per unit time) varies with time after treatment. BrdUrd incorporation by the cells with S-phase DNA content is highest 12 h after ara-C. These data complement those in Fig. 18, since they show the recovery of DNA synthesis and the rate of DNA synthesis of the ara-C-treated cells.

6.2.6. Critique

The BrdUrd/DNA analysis procedure provides substantial information on the cytokinetic properties of asynchronous and perturbed cell populations. It is sufficiently sensitive that it can be applied in vivo and in vitro. Furthermore, the somewhat subjective discrimination between labeled and unlabeled cells required for analyses based on the use of autoradiography (51) is avoided, so the BrdUrd/DNA analysis procedure is less subjective. The BrdUrd/DNA analysis procedure is fast and quantitative, so that accurate analyses of cell cycle traverse and cell loss can be obtained in a timely fashion. Indeed, a complete analysis of the cell cycle traverse characteristics of a cell population can be completed in a few hours if necessary. Finally, the amount of BrdUrd labeling required for analysis is low, so that the possibility of BrdUrd induced toxicity is minimal. Of course BrdUrd toxicity may still be a problem (1,9,19,40), but it is sufficiently minor that BrdUrd labeling studies have been approved for humans in some circumstances (41,42).

The BrdUrd/DNA analysis procedure suffers from several limitations that should be kept in mind during its application. The principal limitation arises from the fact that the binding of the antibodies used to fluorescently label BrdUrd in DNA [e.g., B-44 (17), IU-1, and IU-2 (54)] seems to depend heavily on the procedures used to denature the DNA and on the BrdUrd-labeling protocol itself. For example, BrdUrd/DNA distributions measured for L1210 cells labeled in vitro with BrdUrd for 30 min show that the distributions measured for the cells denatured by heating in 50% formamide at 80°C for 30 min are different from those measured for cells from the same populations denatured with 2*M* HCl for 30 min (Waldman, personal communication). Such denaturation-dependent staining reduces the confidence that can be placed in estimates of such kinetic parameters as the rate of DNA synthesis. The exact shape of the BrdUrd distribution also depends, to some extent, on the antibody used to label the BrdUrd. BrdUrd/DNA distributions measured for L1210 cells labeled for 30 min with BrdUrd are somewhat different when stained using a commercially available (Becton-Dickinson Monoclonal Antibody Laboratory) fluorescein labeled anti-BrdUrd antibody (B-44) or when using IU-2 and fluorescein labeled goat anti-mouse antibody in a two-step antibody staining procedure.

Another potential limitation of the BrdUrd/DNA analysis procedure arises from the cell loss suffered during the DNA denaturation process. Some denaturation procedures are especially harsh and can be applied only to selected tissues. For example, ethanol-fixed cells from the bone marrow are almost completely destroyed by denaturation in 50% formamide at 80°C. Such tissue-specific cell loss also raises the possibility of selective cell loss during preparation of nonhomogeneous tissues for BrdUrd/DNA analysis. Thus, care should be taken to confirm that BrdUrd distributions are truly representative of the tissue or cell population being studied.

It is also important to be alert for the possibility that BrdUrd may interfere with cell cycle traverse or cause toxicity in some cell systems (21). Halogenated pyrimidines have been shown to interfere with differentiation (1) and fetal development (9,19) in some systems. They have also been shown to be effective radiosensitizing agents (40). In our studies, however, no evidence for such toxicity or cell cycle traverse perturbation has been observed in systems in vitro (CHO, L1210, KHT) or in vivo (murine bone marrow, KHT sarcoma).

7. CYTOKINETIC ANALYSIS OF CYTOMETRICALLY DISTINCT SUBPOPULATIONS

One of the powers of flow cytokinetics is its utility in the analysis of cell subpopulations that can be distinguished flow cytometrically. This capability is illustrated in two studies described in this section. The first study is an RC analysis of the response of normal diploid cells of murine KHT sarcomas to treatment with ara-C. The second study describes separately the responses to ara-C exposure of myeloid- and erythroid-enriched subpopulations in murine bone marrow.

7.1. Diploid Cells in the KHT Sarcoma

Diploid cells in the KHT sarcoma can be distinguished from near tetraploid KHT murine sarcoma cells on the basis of their DNA content, as shown in Fig. 15. Thus, the cytokinetic response of the diploid subpopulation to ara-C can be determined separately from that of the KHT cells, in spite of the fact that these cell types cannot be distinguished readily during microscopic analysis. The cytokinetic response of the diploid cells was accomplished using the RC analysis technique. In this procedure, replicate mice bearing KHT tumors were treated at time zero with 100 mg/kg ara-C. Pairs of mice were selected at 2-h intervals thereafter, injected ip with 40 mCi [³H]-TdR, and sacrificed 30 min later. The cells were dispersed, fixed, stained for DNA content with Chromomycin A3 (*10*), and processed through a cell sorter. Cells with DNA contents midway between the G1 diploid and G1 tetraploid peaks (i.e., diploid cells with S-phase DNA content) were sorted onto filters and the radioactivity per cell in S-phase (RCS_i) was determined by liquid scintillation counting. The first interesting observation is that the diploid cells *do* traverse the cell cycle. Furthermore, they respond dramatically to the ara-C. DNA synthesis is inhibited immediately after ara-C, as indicated by the reduced RCS_i values shown in Fig. 20. DNA synthesis begins to recover 4–5 h after treatment as a synchronous cohort of cells enters S-phase. The RCS_i values remain high as this cohort of cells moves through S-phase. The RCS_i curve is at a maximum at about 10 h post-ara-C, and then falls as the cohort of cells leaves S-phase and enters the G2M- and G1-phases. The synchronous diploid cohort reenters S-phase about 18 h after ara-C. These data suggest a cell cycle duration of about 10 h for the ara-C-treated diploid normal cells. This cell cycle traverse rate is similar to that for the KHT sarcoma cells responding to the same ara-C treatment. It would be difficult to obtain such cytokinetic information on the diploid subpopulation described here by classical cytokinetic techniques. These conclusions have been confirmed using the BrdUrd/DNA analysis procedure.

7.2. BrdUrd/DNA Analysis of Murine Myeloblasts and Erythroblasts

The BrdUrd/DNA analysis procedure is particularly useful in the analysis of subpopulations that can be distinguished flow cytometrically. Pallavicini et al. (*45*), for example, have applied it in the cytokinetic analysis of murine hematopoietic populations distinguished by their affinity for wheat germ agglutinin (WGA), erythroid cells having low WGA affinity and myeloid cells having high WGA affinity. In these studies, WGA was labeled with rhodamine, DNA was stained with Hoechst 33342, and the incorporated BrdUrd was fluorescently labeled using a two-step

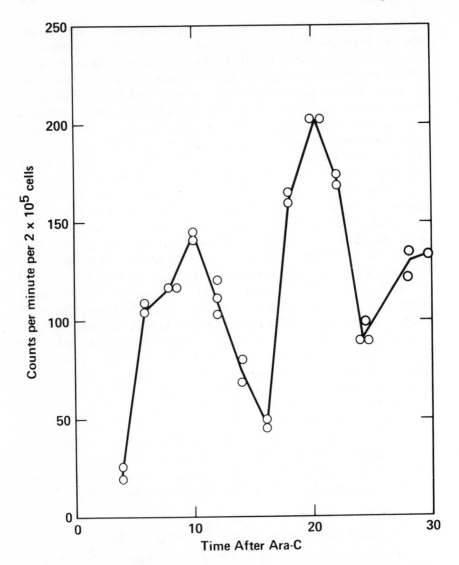

Fig. 20. RC analysis of diploid normal cells in the KHT sarcoma. Animals bearing KHT tumors received 100 mg/kg ara-C at the beginning of the analysis. These animals then received 40 μCi[^{3}H]-TdR at various times thereafter. Thirty minutes after [^{3}H]-TdR labeling, the animals were sacrificed and the tumors were prepared for RCS_i analysis. Only the cells with diploid DNA content were included in this analysis. These data clearly show that the diploid cells do proliferate and that they are perturbed by the ara-C treatment.

monoclonal antibody procedure with the second antibody con-jugated to fluorescein. The cells were then processed through a dual-beam cytometer, where they passed sequentially through laser beams emitting at 488 nm (to excite the rhodamine and fluores-cein) and in the UV (to excite the Hoechst 33342). Data were re-corded in list mode* with red fluorescence recorded as a measure of the amount of bound WGA, green fluorescence recorded as a measure of the amount of incorporated BrdUrd, and blue fluorescence recorded as a measure of cellular DNA content. Figure 21 shows the distribution of the blue vs green fluorescence (DNA vs BrdUrd) for the entire BM sample. Figure 21b shows the distribu-tion of blue vs red fluorescence (DNA vs WGA) for the entire BM. Two populations with distinctly different WGA affinities are visi-ble in Fig. 21b. Polygonal "windows," illustrated in Fig. 21b, were drawn during computer analysis of these data to enscribe the two WGA populations. BrdUrd/DNA distributions were generated separately for the myeloid- (high WGA affinity) and erythroid-enriched (low WGA affinity) populations. These distributions are shown in Figs. 22a and 22b, respectively. Thus, the BrdUrd/DNA analysis procedure has been applied separately to the erythroid and myeloid subpopulations. The cytokinetic properties of the myeloid and erythroid populations appear to be distinctly different in this experiment. Figures 22c and 22d show BrdUrd/DNA distributions for the erythroid and myeloid subpopulations 9 h after treatment with ara-C (the animals were injected ip with BrdUrd 30 min prior to sacrifice so that all cells synthesizing DNA were labeled). It is clear that the erythroid-enriched cells with S-phase DNA content were not synthesizing DNA at this time. On the other hand, a large fraction of the myeloid-enriched cells with S-phase DNA contents were actively synthesizing DNA. A complete analysis of the cytokinetic properties of these subpopulations before and during drug treatment can be obtained from such trivariate data measured periodically for asynchronous populations or after ara-C treatment (45).

*In list mode, the cytometric properties for each cell (e.g., DNA content, BrdUrd content, and amount of bound WGA) are written sequentially into a "list" on magnetic tape. This list is processed one or more times to generate univariate and/or bivariate distributions. During this processing phase of data analysis, distributions of one or two variables (e.g., BrdUrd content and DNA content) may be generated for subpopulations defined by another variable (e.g., WGA binding). More details on list processing can be found in Pallavicini et al. (45).

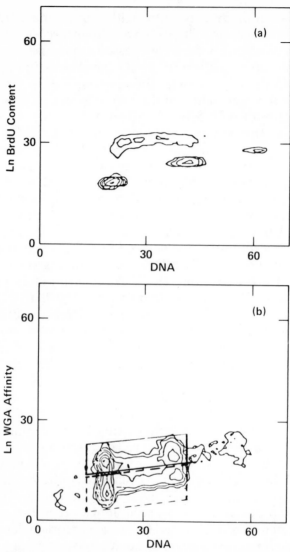

Fig. 21. Trivariate analysis of murine bone marrow cells labeled in vivo for 30 min with 50 mg/kg BrdUrd. The bone marrow cells were stained such that cells binding wheat germ agglutinin were stained with rhodamine, cells containing BrdUrd were stained with fluorescein, and all cells were stained for total DNA content with Hoechst 33258. (A) Bivariate BrdUrd/DNA distribution for the entire bone marrow. (B) Bivariate WGA/DNA distribution for the entire bone marrow; (B) also shows two regions enscribed with rectangular boxes. The upper box (solid lines) enscribes the cells with high WGA affinity and the lower box enscribes the cells with low WGA affinity (45).

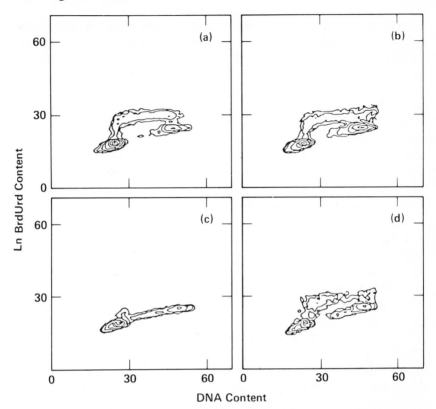

Fig. 22. BrdUrd/DNA distributions of the bone marrow subpopulations defined as indicated in Fig. 21 for control and ara-C-treated mice.
All mice were labeled with BrdUrd 30 min before analysis. (A) and (C)
Distributions measured for cells with low WGA affinity before and 9 h
after treatment with ara-C, respectively. (B) and (D) Distributions
measured for cells with high WGA affinity before and 9 h after treatment
with ara-C. The BrdUrd/DNA distributions for the two WGA populations
appear similar before ara-C treatment. However, (C) and (D) show that
only the cells with low WGA affinity have regained the ability to synthesize DNA (incorporate BrdUrd) 9 h after treatment with ara-C (*45*).

8. CONCLUSIONS

Flow cytometry is becoming increasingly powerful as a tool for
the cytokinetic analysis of those cell populations and subpopulations that can be reduced to representative suspensions of single
cells. Much of the utility of flow cytometry comes from the speed

of analysis and measurement precision. Some systems (46), for example, can process well over a million cells/min with measurement precision of less than 3%. As a result, statistically precise measurements of cell populations can be made in a short time with minimal effort. In the future, flow cytometry should become especially important for analysis of subpopulations occurring at low frequency (e.g., stem cells in the bone marrow or leukemic cells early in relapse) since millions of cells can be processed in a few minutes. In all cases, the results are available immediately to both researchers and clinicians, thereby facilitating their work. In addition, the results may be generated in time to be of use in the design or modification of cancer therapy.

The utility of flow cytometry to cytokinetic studies has been increased significantly in recent years by major cytochemical and procedural developments. Preparation of representative suspensions of single cells is no longer a major problem for most tissues. In addition, new cytochemical advances now allow for identification of numerous proliferative subcompartments (e.g., proliferating and quiescent cells in all phases of the cell cycle), quantification of label incorporation (either [³H]-TdR or BrdUrd) by cells in the G1-, S-, and G2M-phases of the cycle, and discrimination (and hence selective analysis) of subpopulations in complex tissues. These advances are now allowing, for the first time, accurate cytokinetic analyses of therapeutically and biologically relevant populations such as heterogeneous tumors or bone marrow stem cells prior to and in the midst of cancer therapy. These techniques, and those in development, should finally allow for assessment of the true utility of cytokinetics in cancer therapy. In addition, they should greatly facilitate the application of cytokinetics in diverse studies such as the effects of exogenous agents on the DNA synthesis process, the cytokinetic changes occurring during the differentiation, and the response of normal and malignant cells to therapy with cell cycle specific agents.

ACKNOWLEDGMENTS

We deeply appreciate the contributions of unpublished material to this review by Dr. Wolfgang Beisker and Ms. Mary-Helen Barcellos-Hoff. We also appreciate the expert help of Ms. Lil Mitchell and Ms. Angela Riggs in the preparation of this manuscript.

Work has been performed under the auspices of the US Department of Energy by the Lawrence Livermore National Laboratory under contract number W-7405-ENG-48 with support from USPHS grants CA 14533 and CA 28752.

REFERENCES

1. Ashman, C. R., and Davidson, R. L. Induction of Friend erythroleukemic cell differentiation by bromodeoxyuridine: Correlation with amount of bromodeoxyuridine in DNA. J. Cell. Physiol., *102*: 45–50, 1980.
2. Barcellos-Hoff, M., Gray, J. W., Marton, L. J., and Dean, D. F. Cell cycle parameters 9L rat brain tumors in monolayer culture and multicellular spheroids determined with a bromodeoxyuridine labeling technique. Cytometry (submitted.)
3. Bohmer, R.-M. Flow cytometric cell cycle analysis using the quenching of 33258 Hoechst fluorescence by bromodeoxyuridine incorporation. Cell Tissue Kinet., *12*: 101–110, 1979.
4. Bohmer, R.-M., and Ellwart, J. Combination of BUdR-quenched Hoechst fluorescence with DNA-specific ethidium bromide fluorescence for cell cycle analysis with a two-parameter flow cytometer. Cell Tissue Kinet., *4*: 653–658, 1981.
5. Braunschweiger, P. The tumor growth fraction estimation, perturbation and prognostication. *In*: (J. Gray and Z. Darzynkiewicz, eds.), Techniques in Cell Cycle Analysis, New Jersey: Humana, 1986.
6. Burki, J. H., and Aebersold, P. M. Bromodeoxyuridine-induced mutations in synchronous Chinese hamster cells: Temporal induction of 6-thioguanine and oubain resistance during DNA replication. Genetics, *90*: 311–321, 1978.
7. Cheng, H., and Bjerknes, H. Cell production in mouse intestinal epithelium measured by stathmokinetic flow cytometry and Coulter particle counting. Anat. Rec., *207*: 427–434, 1983.
8. Collins, J. Rates of DNA synthesis during the S-phase of HeLa cells. J. Biol. Chem., *253*: 8570–8577, 1978.
9. Craddock, V. M. Shortening of life span caused by administration of 5-bromodeoxyuridine to neonatal rats. Chem. Biol. Intract., *35*: 139–144, 1981.
10. Crissman, H., and Steinkamp, J. Multivariate Cell Analysis: Techniques for Correlated Measurements of DNA and Other Cellular Constituents, *In*: (J. Gray and Z. Darzynkiewicz, eds.), Techniques in Cell Cycle Analysis, New Jersey: Humana, 1986.

11. Darzynkiewicz, Z. Cytochemical probes of cycling and quiescent cells applicable for flow cytometry. *In*: (J. Gray and Z. Darzynkiewicz, eds.), Techniques in Cell Cycle Analysis, New Jersey: Humana, 1986.

12. Darzynkiewicz, Z., Traganos, F., and Melamed, M. New cell cycle compartments identified by multiparameter flow cytometry. Cytometry, *1*: 98–108, 1980.

13. Darzynkiewicz, Z., Traganos, F., and Kimmel, M. Assay of cell cycle kinetics by multivariate flow cytometry using the principle of stathmokinesis. *In*: (J. Gray and Z. Darzynkiewica, eds.), New Jersey: Humana, 1985.

14. Dean, P. N. Data analysis in cell kinetics research. *In*: (J. Gray and Z. Darzynkiewicz, eds.), Techniques in Cell Cycle Analysis, New Jersey: Humana, 1986.

15. Dean, P. N., Dolbeare, F., Gratzner, H., Rice, G., and Gray, J. W. Cell-cycle analysis using a monoclonal antibody to BrdUrd. Cell Tissue Kinet., *17*: 427–436, 1984.

16. Dolbeare, F., Breisker, W., Pallavicini, M. G., Vanderlaan, M., and Gray, J. W. Cytochemistry for BrdUrd/DNA analysis: Stoichiometry and sensitivity. Cytometry, *6*: 521–530, 1985.

17. Dolbeare, F., Gratzner, H., Pallavicini, M., and Gray, J. W. Flow cytometric measurement of total DNA content and incorporated bromodeoxyuridine. Proc. Natl. Acad. Sci. USA, *80*: 5573–5577, 1983.

18. Dosik, G. M., Barlogie, B., White, R. A., Gohde, W., and Drewinko, B. A rapid automated statchmokinetic method for determination of *in vitro* cell cycle transit times. Cell Tissue Kinet., *14*: 121–134, 1981.

19. Franz, J., and Klienebrecht, J. Teratogenic and clastogen effects of BrdUrd in mice. Teratology, *26*: 195–202, 1982.

20. Gonchoroff, N., Greipp, P., Katzman, J., and Kyle, R. A monoclonal antibody reactive with 5-bromo-2-deoxyuridine that does not require DNA denaturation. Cytometry, *6*: 506–512, 1985.

21. Goz, B. The effects of incorporation of 5-halogenated deoxyuridines into the DNA of eukaryotic cells. Pharmacol. Rev., *19*: 249–272, 1978.

22. Gratzner, H. Monoclonal antibody against 5-bromo- and 5-iododeoxyuridine: A new reagent for detection of DNA replication. Science, *218*: 474–475, 1982.

23. Gray, J. W. Quantitative cytokinetics: Cellular response to cell cycle specific agents. Pharmacol. Ther., *22*: 163–197, 1983.

24. Gray, J. W., and Dean, P. Display and analysis of flow cytometric data. Annu. Rev. Biophys. Bioengr., *9*: 509–539, 1980.

25. Gray, J. W., Bogart, E., Gavel, D., George, Y., and Moore II, D. H. Rapid cell cycle analysis: II. Phase durations and dispersions from computer analysis of RC curves. Cell Tissue Kinet., *16*: 457–471, 1983.

26. Gray, J. W., Carver, J. H., George, Y. S., and Mendelsohn, M. L. Rapid cell cycle analysis by measurement of the radioactivity per cell sorted from a narrow window in S-phase (RCSi). Cell Tissue Kinet., 10: 97–109, 1977.

27. Gray, J. W., Dean, P. N., and Mendelsohn, M. L. Quantitative cell cycle analysis. *In*: (M. Melamed, P. Mullaney, and M. Mendelsohn, eds.), Flow Cytometry Sorting, New York: John Wiley, 1979.

28. Gray, J. W., Pallavicini, M. G., George, Y. S., Groppi, V., Look, M., and Dean, P. N. Rates of incorporation of radioactive molecules during the cell cycle. J. Cell. Physiol. 108: 135–144, 1981.

29. Grdina, D. J., Meistrich, M. L., Meyn, R. E., Johnson, T. S., and White, R. A. Cell synchrony: A comparison of methods. *In*: (J. Gray and Z. Darzynkiewicz, eds.), New Jersey: Humana, 1986.

30. Herzenberg, L., Sweet, R., and Herzenberg, L. Fluorescence activated cell sorting. Sci. Am., 234: 108–117, 1976.

31. Horan, P. K. and Wheeles, L. Quantitative single cell analysis and sorting. Science, 198: 149–157, 1977.

32. Horan, P. K., Muirhead, C., and Cram, L. S. (in press) Flow Cytometry: A Primer. Plenum, New York.

33. Kriss, J. P., and Revesz, L. The distribution and fate of bromodeoxyuridine and bromodeoxycytidine in the mouse and rat. Cancer Res., 22: 254–265, 1961.

34. Kubbies, M., and Rabinovitch, P. S. Flow cytometric analysis of factors which influence the BrdUrd-Hoechst quenching effect in cultured human fibroblasts and lymphocytes. Cytometry, 3: 276–281, 1983.

35. Latt, S. A. Microfluorometric detection of DNA replication in human metaphase chromosomes. Proc. Natl. Acad. Sci. USA, 70: 3395–3399, 1973.

36. Latt, S. A., and Stetten, G. Spectral studies on 33258 Hoechst and related bisbenzimidazol dyes useful for fluorescent detection of DNA synthesis. J. Histochem. Cytochem., 24: 24–33, 1976.

37. Latt, S. A., and Wohlleb, J. Optical studies of the interaction of 33258 Hoechst with DNA, chromatin and metaphase chromosomes. Chromosoma, 52: 297–316, 1975.

38. Latt, S. A., George, Y. S., and Gray, J. W. Flow cytometric analysis of BrdU substituted cells stained with 33258 Hoechst. J. Histochem. Cytochem., 25: 927–934, 1977.

39. Melamed, M., Mullaney, P., and Mendelsohn, M. L. (eds.) Flow Cytometry and Sorting, New York: John Wiley, 1979.

40. Morales-Ramires, R., Vallarino-Kelly, T., and Rodriques-Reyes, R. Effects of BrdUrd and low doses of gamma radiation on sister chromatid exchange, chromosome breaks, and mitotic delay in mouse bone marrow cells *in vivo*. Environ. Mutagen., 5: 589–602, 1983.

41. Morstyn, G., Hsu, S.-M., Kinsella, T., Gratzner, H., and Russo, A. Bromodeoxyuridine in tumors and chromosomes detected with a monoclonal antibody. J. Clin. Invest., *71*: 1844–1850, 1983.

42. Morstyn, G., Kinsella, T., Hsu, S.-M., Russo, A., Gratzner, H., and Mitchell, J. Identification of bromodeoxyuridine in malignant and normal cells following therapy: Relationship to complications. Int. J. Radiat. Oncol. Biol. Phys., *10*: 1441–1445, 1984.

43. Pallavicini, M. G. Solid tissue dispersal for cytokinetic analyses. *In*: (J. Gray and Z. Darzynkiewicz, eds.), Techniques in Cell Cycle Analysis, New Jersey: Humana, 1986.

44. Pallavicini, M. G., Gray, J. W., and Folstad, L. Quantitative analysis of the cytokinetic response of KHT tumors *in vivo* to 1-β-D-arabino-furanosylcytosine. Cancer Res., *42*: 3125–3131, 1982.

45. Pallavicini, M. G., Summers, L., Giroud, F., Dean, P. N., and Gray, J. W. Multivariate analysis and list mode processing of murine hematopoietic subpopulations for cytokinetic studies. Cytometry, *6*: 539–549, 1985.

46. Peters, D., Branscomb, E., Dean, P., Merrill, T., Pinkel, D., Van Dilla, M., and Gray, J. W. The LLNL high speed sorter (HISS): Design features, operational characteristics and biological utility. Cytometry, *6*: 290–301, 1985.

47. Rabinovitch, P. S. Regulation of human fibroblast growth rate by both noncycling cell fraction and transition probability as shown by growth in 5-bromodeoxyuridine followed by Hoechst 33258 flow cytometry. Proc. Natl. Acad. Sci. USA, *80*: 2951–2955, 1983.

48. Rasey, J. S. *In vitro* assays for tumors grown *in vivo*: A review of kinetics techniques. *In*: (J. Gray and Z. Darzynkiewicz, eds.), Techniques in Cell Cycle Analysis, New Jersey: Humana, 1986.

49. Raza, A., Preisler, H., Mayers, G., and Bankert, R. Rapid enumeration of S-phase cells by means of monoclonal antibodies. N. Engl. J. Med., *310*:991, 1984.

50. Shackney, S., and Ritch, P. S. Percent labeled mitosis curve analysis. *In*: (J. Gray and Z. Darzynkiewicz, eds.), Techniques in Cell Cycle Analysis, New Jersey: Humana, 1986.

51. Simpson Herren, L. Autoradiographic techniques for measurement of the labeling index. *In*: (J. Gray and Z. Darzynkiewicz, eds.), Techniques in Cell Cycle Analysis, New Jersey: Humana, 1986.

52. Tobey, R., and Crissman, H. Use of flow microfluorometry in detailed analysis of the effects of chemical agents on cell cycle traverse. Cancer Res., *32*: 2726–2732, 1972.

53. Traincard, F., Ternynck, T., Danchin, A., and Aurameas, S. Une technique immunoenzymatique pour la nise en evidence de l'hybridization moleculare entie acides nucleiques. Ann. Immunol. (Inst. Pasture), *134D*: 399–405, 1983.

54. Vanderlaan, M., and Thomas, C. Characterization of monoclonal antibodies to bromodeoxyuridine. Cytometry, *6*: 501–505, 1985.

55. Van Dilla, M. A., Dean, P. N., Melamed, M., and Laerum, O. (eds.) Flow Cytometry: Instrumentation and Data Analysis, New York: Academic, 1985.

56. Wright, N., and Appleton, D. The metaphase arrest technique: A critical review. Cell Tissue Kinet., *13*: 643–664, 1980.

Chapter 6

Solid Tissue Dispersal for Cytokinetic Analyses

Maria G. Pallavicini

1. INTRODUCTION

The dispersal of solid tissues into a single-cell suspension has become an integral component of many current techniques for quantitative cytokinetic analysis of both normal and neoplastic tissues. This is particularly evident in the application of flow cytometric (FCM) procedures for the quantitative analysis of DNA content (see ref. 36), measurements of intracellular components [i.e., bromodeoxyuridine incorporated into DNA (15)], and identification and purification of subpopulations for subsequent cytokinetic studies (42,71). In addition most assay systems for quantitation of the fraction of clonogenic cells, as well as many biochemical/molecular analyses (i.e., incorporation of tritiated thymidine, [3H]-TdR, into DNA), are done on a per cell basis, and thus utilize cells in suspension.

Numerous methods have been described for solid tissue dispersal. These include mechanical disaggregation, enzymatic and/or chemical treatments, and various combinations of these techniques. The choice of a particular method for tissue disaggregation should be dictated by the nature of the intercellular cohesive factors in the solid tissue, and the production of optimal yields of cells representative of those found in the tissue *in situ*. Retention of the structural and functional characteristics of cells comprising the solid tissue is not only highly desirable, but in many cases is a prerequisite for subsequent analyses (i.e., clonogenic assays, cell discrimination on the basis of cell surface markers, and so on). Unfortunately, evaluation of the nature of the cells in suspension is often overlooked in the enthusiasm to obtain data of more interest, and in reality, many investigators use cell suspensions that have not been fully or adequately characterized with regards to the endpoint to be studied. Lack of knowledge of the characteristics of the cells in suspension may have significant impact on data interpretation. For example, failure to appreciate the possibility of subpopulation selection and/or the preferential loss of selected cell types (i.e., S-phase cells) during dispersal of a solid tissue may limit the validity of extrapolation of cytokinetic measurements made on cell suspensions to phenomena in the tissue *in situ*.

Although numerous reviews of cell dispersal procedures (2,51,74) have been published, the objective of this chapter is to emphasize, from a cytokinetic viewpoint, the importance of characterization of the disaggregated cells in suspension. Specifically, I will (1) briefly discuss the biochemical basis of enzymatic and chemical disaggregation methods as the first step in choosing a dispersal technique, (2) suggest a series of guidelines to evaluate dispersal techniques most suitable for preparation of cell suspensions for cytokinetic studies, (3) illustrate application of these guidelines in the dispersal of two model systems (murine sarcoma and small intestine) into suspensions of single cells, and (4) evaluate selected published dispersal protocols with respect to these suggested guidelines.

2. BIOCHEMICAL BASIS OF TISSUE DISAGGREGATION

The dispersal of solid tissue into single cells is dependent upon alteration or removal of the intercellular cohesive materials. The biochemical/molecular characteristics of the cohesive matrix should be one of the primary determinants of a suitable disaggregation technique for an individual tissue system. Although the nature of these cohesive materials is not completely understood, it is known that the composition (connective tissue, proteins, and glycoproteins) varies among tissues (26), and that the effectiveness of selected dispersal techniques is influenced by the nature of these constituents. For example, elastin, a component of fibrous connective tissue, is not digested by trypsin, but can be digested by elastase. Since arteries contain elastic fibers, it is not surprising that elastase treatment is more effective than trypsin for dissociating cardiovascular tissue (30). As our knowledge about the nature of the intercellular cohesive matrix and its local variations (27) within heterogeneous tissues expands, so will our ability to rationally choose the most suitable technique to disaggregate tissues for quantitation of specific cellular endpoints.

It is also known that the method of tissue disruption may transiently or permanently affect the properties of cells in suspension (73). Alteration of cellular properties may be morphologically apparent, such as swollen cells or surface "blebs" (30), or may be more subtle, but equally important, such as alteration of mitochondrial function (20,44), loss of selected surface antigens (37,76), release of sialic acid-containing residues from cell surfaces (11,18,61), protein loss (33), degradation of polysomes (24), and so on. The extent of damage varies between tissues and is influenced by experimental conditions (i.e., temperature, pH, length of exposure, and so on). In some tissues (i.e., small intestine), the techniques developed to obtain single-cell suspensions have not been compatible with retention of cellular function, so efforts to improve tissue dispersal are continually evolving. Tissue dispersal procedures can be broadly categorized into four groups that include enzymatic treatments or exposure to chemical, mechanical, and surface-acting agents. Table 1 provides selected examples of techniques used to disperse both neoplastic and normal tissue systems.

Enzymatic disaggregation procedures are perhaps the most amenable for tissue-dependent selection, and numerous enzymatic methods have been utilized with varying degree of success in different tissue systems (*see* Table 1 for references). These include both proteolytic enzymes and enzymes that degrade selected components of connective tissue. Since proteins and glycoproteins are found in all intercellular materials, dissociation with the aid of proteolytic enzymes is the most common dispersal method. Typical ranges of enzyme concentrations can be found in Bashor (2). Trypsin (which hydrolyzes ester and peptide bonds involving carbonyl groups of arginine and lysine), collagenase (which degrades several molecular species of collagen), lysozymes (which hydrolyze glucosidic linkages of glycoproteins and peptides), papain (a thiol proteinase), pepsin, pronase, and dispase (neutral proteases), will

TABLE 1
Methods Used to Disaggregate Solid Tissue Systems

Tissue systems	Examples of dispersal procedures	References
Tumors		
Animal		
(Mouse/rat)	Trypsin	6,32
	Mechanical/collagenase/elastase	43
	Mechanical/trypsin	50
	Mechanical/pepsin	1,54
	Collagenase/pronase/DNase	7
	Collagenase/trypsin/EDTA	46
	Collagenase/hyaluronidase/DNase	19
	Detergent/trypsin	69,70
	Mechanical	60
	Tetraphenylboron	39
	Neutral protease	40,68
	Mechanical/collagenase	34
Small intestine	Citrate/mechanical/EDTA	29
	Pronase/collagenase	66
	EDTA/dithiothreitol	75
	Mechanical	21
Gastric mucosal cells	Collagenase/pronase/EDTA	31
Lung		
(Type II cells)	DNase/trypsin	25
	Elastase/trypsin	16
	Trypsin/chymotrypsin/elastase	62
Heart	Elastase	30
Muscle	Papain	38
Epidermis	Trypsin/dithiothreitol	3
	Mechanical/trypsin	28
Liver	Collagenase/hyaluronidase	4,57,58
	Lysozymes	23
	Tetraphenylboron	47
Kidney	Mechanical	8
	Trypsin/collagenase/DNase/EDTA	10

release cells in almost all tissues. DNase hydrolyzes DNA complexes with proteins, often found after enzyme digestion, and is used to reduce the cell-trapping gel observed in some cell suspensions.

Elastase digests the fibrous connective tissue glycoprotein elastin, and has been used to dissociate cells from arteries, heart, and liver. Hyaluronidase degrades hyaluronic acid, a constituent of the extracellular matrix, and is often used in combination with other enzymes. Many of the enzymes used for tissue dispersal are crude preparations containing not only a predominant enzyme, but also varying amounts of other contaminants whose activity may contribute to tissue disaggregation and/or adversely affect cellular structure or function.

Chemical methods to disaggregate tissues are based on the role of cations (Ca^{2+} and Mg^{2+}) in maintaining cell surface integrity and the intercellular structural matrix (5). The omission of these compounds from the dissociating medium or the sequestration of these compounds has been used to loosen intercellular bonds. Sequestration is often achieved by exposure to EDTA (ethylene-diaminoacetate) (75) and EGTA [ethylene bis-(β-aminoethyl ether)-N,N'-tetraacetic acid], both of which bind Ca^{2+} and Mg^{2+} to varying degrees. Tetraphenylboron complexes potassium (the latter postulated to be involved in intercellular contacts), and has been used to dissociate liver (47), intestinal crypt cells (21), and solid mammary tumors (39). Sodium citrate is also capable of binding cations and has been used in dissociating solutions (29,75). Hypertonic solutions of disaccharides (sucrose, maltose, lactose, and cellobiose are reported to split gap junctions and zona occludentes (17), whose presence may be responsible for the clusters of cells sometimes remaining after enzymatic tissue digestion.

Exposure of solid tissue fragments to detergents results in lysed cell membranes, thereby producing nuclear suspensions suitable for flow cytometric-cytokinetic analyses. Suspensions of nuclei have been prepared successfully with detergents such as Nonidet P-40 (69,70), and Triton X-100 (65). Aggregation of nuclei, usually a result of incomplete removal of cytoplasmic fragments, and damaged nuclei are two problems commonly associated with preparation of nuclear suspensions. A comparison of nuclei isolation techniques for preparation of samples for FCM analysis (45) suggests that the optimal preparative procedure may vary among tissues.

Mechanical methods of tissue disaggregation utilize shear forces to disrupt cell-to-cell junctions and include, among others, homogenization (45) or passage through nylon or stainless steel mesh (8) or needles of successively smaller gage (35). Often these techniques utilize tissue fragments obtained by mincing with scissors or sharp scalpels (i.e., *45,60*). Although mechanical methods are desirable when it is known that standard enzymatic or chemical exposure will alter the cell constituent endpoint to be measured (e.g., enzyme activity or cell surface components), mechanical disaggregation usually results in low cell yields and damaged cells (*51,73*).

3. SUGGESTED GUIDELINES TO OBTAIN CELL SUSPENSIONS FOR CYTOKINETIC ANALYSES

Characterization of cell suspensions for cytokinetic analyses is best accomplished with a standard set of guidelines to ensure that a cytokinetically representative cell preparation is obtained. A "cytokinetically representative" cell suspension can be defined as one in which (1) the fraction of the cell population present in G1 vs S vs G2M phases is the same as in the tissue *in situ*, and (2) recognizable subpopulations exist at the same frequencies as those found in the tissue *in situ*. The fraction of the cell population residing in S, G2 + M, and the entire proliferating phase of the cell cycle can be compared in the cell suspension and in the solid tissue using autoradiographic techniques after appropriate labeling procedures in vivo. The second criterion of representative subpopulations is more difficult to assess because it requires the availability of subpopulation-specific markers that can be used to compare population frequencies in the solid tissue and in the cell suspension. Depending upon the tissue and subpopulation, the availability of such markers may be limited, although recent advances in molecular and cellular biology and monoclonal antibody technology are making considerable progress in these areas (i.e., *49,56*). However, even if subpopulation-specific markers are not available, preferential loss of high-frequency subpopulations [particularly in tissues with a definite cytokinetic hierarchy (i.e., small intestine)] can be estimated by collective analysis of radioactive labeling patterns of the cells in suspension and the tissue *in situ*. A more detailed discussion of the procedures used to evaluate subpopulation representation in cell suspensions is presented in the following section.

In addition to retaining the cytokinetic characteristics of cells found in the tissue *in situ*, other features are also desirable for definitive cytokinetic studies. Quantitation of DNA content by FCM analysis is often used to monitor drug-induced perturbations; thus high-quality DNA distributions [e.g., displaying minimal cell clumping and low coefficients of variation (CV)] facilitate subsequent phase fraction estimation and mathematical modeling. Maintenance of cell function, as assessed by determinations of cell clonogenicity, intracellular pH, membrane integrity, membrane potential, mitochondrial activity, and so on, is desirable, as techniques to measure these properties become more sophisticated. Since the utilization of structural and/or morphologic characteristics (i.e., lectin affinity, antibody binding, and so on) is becoming more prevalent for subpopulation discrimination (particularly by FCM analyses), retention of these properties will become more important as additional subpopulation-specific markers/probes are developed. Finally, high cell yields are desirable to allow multiple and repeated measurements to be performed on the same cell suspension.

A summary of these guidelines is shown in Table 2. Although an individual dispersal protocol may not completely fulfill all of

TABLE 2
Desirable Features of Cell Suspensions
for Cytokinetic Studies

Cytokinetically representative
Relative phase fractions
Subpopulation frequencies
Cell function maintained
Clonogenicity
Biochemical parameters (i.e., intracellular pH, membrane integrity), membrane potential
High-quality DNA distributions
Retention of cell structure and morphology
Optimal cell yields

these criteria, characterization of the dispersed cells according to these guidelines should provide a firm foundation for more meaningful interpretation of cytokinetic data.

4. APPLICATION OF GUIDELINES FOR TISSUE DISPERSAL

The usefulness of the suggested criteria for characterization of cells disaggregated from solid tissues can be demonstrated by their application to cytokinetic characterization of cell suspensions obtained from two model systems: a murine solid tumor and small intestine. Although the experimental details (such as labeling periods) may vary for individual systems, the methodological approach will be similar for many tissues.

4.1. Preferential Phase-Specific Cell Selection

Labeling in vivo with a radioactive DNA precursor, i.e., [³H]-TdR, and autoradiography of single-cell suspensions and tissue sections can be used to ascertain whether cells are preferentially lost from a particular phase of the cell cycle during the disaggregation procedure. By appropriate experimental conditions, labeled cells residing predominantly in S- or G2 + M-phases or throughout the cell cycle can be generated. By comparing the fraction of labeled cells in the single cell suspension with that in the tissue section after each labeling protocol, phase-specific cell loss can be evaluated.

We have evaluated potential phase-specific cell loss during dispersal of the KHT tumor, a murine sarcoma used extensively as a model system for quantitation of drug/radiation-induced cytokinetic perturbations (22,41,53). Experimental details of these data can be found in Pallavicini et al. (40). The cell cycle phase transit times in the KHT tumor are 3.5, 11.2, and 2.1 h for G1-, S-, and G2 + M-phases, respectively. To produce tumors with labeled cells primarily in S-phase, tumor-bearing mice were injected with [³H]-TdR 30 min prior to sacrifice. A 3-h labeling period was used to produce tumors with labeled cells predominantly in late S-phase and G2M-phases. Labeling of nearly all cycling cells was achieved with two injections of [³H]-TdR—once 13 h before sacrifice and once 6.5 h before sacrifice. At the time of sacrifice the tumor was removed and divided into two aliquots. Half of the tumor was dispersed enzymatically using either neutral protease (40) or a combined mechanical/enzymatic (trypsin/DNase) (63) dispersal technique, and the remainder was fixed in formalin and sectioned. Both the single cells and the tumor sections were placed on glass microscope slides and processed for autoradiography. By comparing the fraction of labeled cells in the disaggregated cell suspension with that of the sectioned tumor, potential phase-specific cell loss was evaluated. These data are shown in Table 3. Since the labeling index (LI) was

TABLE 3
Measurements of Potential Phase-Specific
Cell Loss During Tumor Dispersal[a]

Time interval, h	Labeling index		
	Sections	Single cells	
		Trypsin	Dispase
0.5	27 ± 1 (4)	29 ± 1 (4)·	32 ± 4 (8)
3	31 ± 2 (5)	32 ± 2 (5)	34 ± 2 (6)
13	58 ± 3 (5)	56 ± 1 (5)	55 ± 6 (7)

[a]Tumors were excised at either 0.5, 3, or 13 h after [³H]-TdR injection. Values represent the average ±1 SD of number (n) samples per group [from Pallavicini et al. (40)].

similar for the dispersed cells and the tumor sections with each labeling protocol, we concluded that neither dispersal procedure caused phase-specific cell loss in high-frequency subpopulations. It is unlikely that we would be able to detect phase-specific cell loss in low-frequency subpopulations using such a labeling protocol.

4.2. Preferential Subpopulation-Specific Selection

The small intestine is an example of a tissue with a well-known hierarchy in the production and differentiation of epithelial cells. The most immature cells reside in the bottom of the crypt and migrate up the crypt and onto the villus as differentiation proceeds. Cell division occurs only within defined regions of the crypt and ceases when the cells migrate into the upper crypt and onto the villus. Thus the cytokinetic characteristics of intestinal epithelial cells are dependent upon their spatial location within the epithelium.

Dispersal of the small intestine into a cell suspension has been accomplished by a variety of techniques (*see* Table 1), the majority of which utilize multiple sequential exposures to chemicals or enzymes. The dispersal procedure described by Weiser (75) is commonly used to obtain suspensions of the intestinal epithelial cells. In this technique, the small intestine is excised from the mouse, filled with a buffer solution containing citrate, and subsequently exposed at 37°C to buffer solutions containing EDTA and dithio-threitol (DTH, a mucolytic agent). Cell yields are improved if the EDTA-DTH buffer is periodically replaced with fresh disaggregating solution during the incubation period. We (42) have modified the Weiser procedure to include trypsin digestion following the EDTA-DTH exposures to effect removal of cells at the bottom of the crypt.

We have measured the cytokinetic characteristics of intestinal epithelial cells obtained by the modified Weiser procedure. Collective analysis of FCM bivariate distributions of DNA content and cell length, cell sorting, and autoradiographic analysis of radio-labeled cells was used to determine preferential subpopulation selection occurring during the disaggregation procedure. Intestinal epithelial cells obtained during five consecutive collection periods (0–20, 20–40, 40–60, 60–100, and 100–120 min) were characterized.

Bivariate FCM analysis of the DNA content and cell length of intestinal epithelial cells was used to discriminate mature villus cells, G1 crypt cells, and S-phase crypt cells (*see* Fig. 1). Crypt cells have low cell-length values and are proliferating, whereas villus cells have high cell-length values and a G1 DNA content. We have shown previously that the position of a cell on the cell-length axis reflects its differentiation maturity; cells with high cell-length values are more differentiated than those with low cell-length values (42). Flow cytometric analysis of cells obtained during each of five consecutive collection periods indicated that (1) the relative fraction of crypt and villus subpopulations varied between collection intervals, (2) the relative fraction of the crypt population with an S-phase DNA content differed in each collection period, and (3) the average cell length of the villus population decreased progressively with longer incubation periods, suggesting that the villus subpopulation in the 20–40-min fraction was more immature than the villus subpopulation in the 0–20-min collection period.

Autoradiographic analysis of sorted crypt populations in epithelial cell preparations obtained from mice exposed to [³H]-TdR is shown in Table 4. In these experiments mice were injected with [³H]-TdR 30 min prior to sacrifice to label cells in S-phase. The small intestine was then disaggregated according to the Modified Weiser protocol and cells obtained during each collection period were analyzed with FCM. The crypt subpopulation (G1- and S-phase crypts) was sorted and the LI measured by microscopic scoring of autoradiographs. The LIs for the 0–20-min fraction and 100–120-min

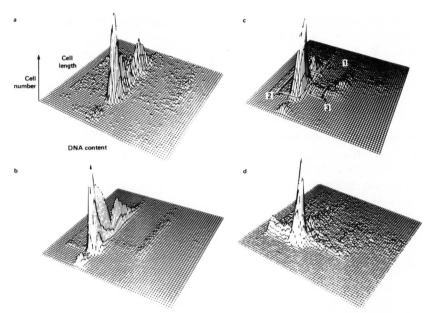

Fig. 1. Bivariate flow cytometric distributions of intestinal epithelial cells collected during the 0–20 min (a), 20–60 min (b), 60–100 min (c), and 100–120 min (d) collection periods. DNA content is represented on the abscissa, cell length on the ordinate, and cell number in the third dimension. Areas 1–3, shown in panel c, are the mature villus, G1 crypt, and S-phase crypt subpopulations, respectively [from Pallavicini et al. (42)].

TABLE 4
Percent Labeling Index of Sorted Cells
From Different Fractions[a]

Fraction, min	Percent labeled cells			
	Villus	All crypt	G1 crypt	S-crypt
0–20	2 (±2)	3 (±1)	3 (±1)	37 (±9)
20–60	2 (±0)	15 (±4)	2 (±2)	71 (±9)
60–100	2 (±1)	13 (±2)	3 (±3)	71 (±12)
100–120	1 (±0)	6 (±2)	2 (±1)	62 (±5)

[a]Values represent the average (±SD) of three or four samples [from Pallavicini et al. (42)].

fraction were less than 6%, whereas LIs of about 15% were obtained for crypt subpopulations harvested in intermediate collection periods. The low LIs of cells in the early and late collection periods were postulated to reflect the age distribution of cells at the top of the crypt (reported to proliferate slowly, if at all) and at the bottom of the crypt (also known to be slowly cycling). Since relatively higher LIs were measured on S-phase crypt cells obtained during intermediate incubation intervals, these cells were postulated to be more rapidly cycling than those with lower LIs. These data, in conjunction with the bivariate FCM distributions, suggest that cytokinetic analysis of dispersed epithelial cells may not yield comparable information to the tissue *in situ* unless cells collected during each incubation interval are combined into one mixed-epithelial cell preparation to minimize apparent subpopulation-specific selection. Similar conclusions were suggested by data of Lawson et al. (*29*).

4.3. Cell Function

Measurements of the fraction of cells surviving exposure to perturbing agents, such as cancer chemotherapeutic drugs, are often used to assess the effectiveness of therapy. The retention of clonogenic capability is a highly desirable feature to consider when selecting a dispersal procedure to provide cells for cytokinetic analyses. For many solid tumor systems (particularly in mice), cell survival can be assessed by colony assays in vitro. In tissues for which such assays are available, plating efficiencies of dispersed cells should be determined for each dispersal protocol. For example, a comparison of the plating efficiencies of KHT solid tumor cells obtained with neutral protease and by the trypsin/DNase disaggregation procedures showed a 1.5-fold higher plating efficiency of cells obtained by the former procedure (*40*). Whether this is a result of less damage induced by neutral protease or a higher yield of the clonogenic cell subpopulation is not known. Regardless, higher plating efficiencies allow for quantitation of greater degrees of cell kill following tissue insult and may imply that a greater fraction of the clonogenic subpopulation is recovered. Data of Twentyman (*67*) and Rasey and Nelson (*48*) indicate that the dispersal procedure used to obtain single-cell suspensions of solid tumors following exposure to cytotoxic agents in vivo can affect the fraction of clonogenic cells assayed by in vitro assays, perhaps suggesting interaction between drug damage and enzyme damage.

4.4. Cell Yields

High yields of dispersed cells allow for multiple and repeat measurements of a particular endpoint and are thus a desirable feature of any dispersal protocol. It is generally estimated that 1 g of tissue contains about 1×10^9 cells, although this number will be affected by the cell size, amount of intercellular matrix, cell packing, and so on, in the tissue. Few dispersal procedures yield more than 1×10^8 cells/g and the majority yield between 1×10^6 and 1×10^7 cells/g. Table 5 compares the cell yields obtained after disaggregating a variety of solid tumors and normal tissues.

TABLE 5
Cell Yield of Disaggregated Solid Tissues

Tissue type	Dispersal method	Approx. cells, $\times 10^{-6}$ g	Reference
Mouse: Tumors			
KHT	Mechanical/DNase/trypsin	10	40
	Neutral protease	50	40
Fibrosarcoma	Mechanical/collagenase/ elastase	90	43
EMT6	Trypsin	5	48
	Collagenase/pronase/DNase	20	7
Small intestine	EDTA	26	42
Rat: Tumors			
Rhadomyo- sarcoma	Mechanical/trypsin	50	50
	Detergent	1	52
Human: Carcinoma, melanoma, sarcoma	Mechanical	27	60
	Mechanical/collagenase/ DNase	48	60
	Mechanical	18	19
	Mechanical/hyaluronidase/ collagenase/DNase	62	19

4.5. DNA Distributions

The extent of cell damage is reflected in the quality of the DNA distributions of the dispersed cells, which in turn can be estimated by quantitating the coefficient of variation (CV) of the G1 peak and the relative fraction of cell clumps. It should be noted, however, that high CV values of the G1 peak may be a result of nonspecific staining of cell constituents with the DNA-fluorochrome, even in the absence of any significant cell damage. CVs of less than 5% have been obtained by numerous investigators (i.e., *1,35,39,40,54, 70*) in a variety of solid tissue systems. Cellular debris is often reflected not only in wide CVs, but also in a fluorescent continuum extending from the origin and underlying the DNA distribution. Although various computer programs are available to account for high levels of debris, cell clumping, and/or poor CVs, cytokinetic analysis is greatly simplified when high-quality DNA distributions are obtained.

4.6. Retention of Cell Morphology

Cell morphology is commonly assessed by light microscopy of either stained or unstained (phase microscopy) cell suspensions. Visible, as well as fluorescent stains specific for nuclear, membrane, and/or cytoplasmic components are utilized. The presence of "blebs" on the outer surface of the cell, nonrefractile cytoplasm, and swollen cells are examples of morphologic characteristics reflecting cellular damage. Observation of cells stained with subpopulation-specific markers is often useful to monitor morphology of selected cell types. The extent of morphology retention required is dependent upon the type of cytokinetic analyses to be performed. For example, in the small intestine, where FCM discrimination of subpopulations is based on cell length differences between the crypt and villus cells, the cell preparation must contain whole cells (villus nuclei would be indistinguishable from G1-cell nuclei of crypts), and the mature epithelial cells are required to retain their oblong "cigar" shape. On the other hand, simple DNA distribution analysis of tumor cell suspensions is not based on morphologic criteria, and thus, retention of cell morphology would not be necessary. In fact nuclear suspensions may provide excellent DNA histograms (*see* Table 1 for references.)

5. EVALUATION OF SELECTED TUMOR DISPERSAL PROCEDURES ACCORDING TO SUGGESTED GUIDELINES

Although it is not feasible to extensively review all commonly used dispersal procedures to obtain cell suspensions for cytokinetic studies, selected examples will serve to illustrate areas in which information is lacking. The majority of disaggregation procedures for solid tumors utilize either enzymes alone or a combined mechanical/enzymatic procedure. Two examples of such procedures in which characterization of the cell suspension was reported are described by Hamburger et al. (*19*) and Slocum et al. (*59,60*). Hamburger et al. (*19*) compared the biologic activities of human tumor cells dispersed mechanically and by exposure to a combination of collagenase, hyaluronidase, and DNase. They found that although the enzymatic method yielded fewer cells than the mechanical procedure, the cells obtained by the enzymatic procedure had higher cloning efficiencies. They also measured tritiated thymidine suicide indices in cells obtained by both procedures to verify that comparable S-phase fractions were obtained. However, the S-phase fraction in the solid tissue could not be quantitated by the thymidine suicide method, so a comparison between S-phase fraction in the dispersed cells and the tissue *in situ* could not be made. Information on the quality of the DNA distributions was not presented.

Slocum et al. (*59,60*) compared mechanical and mechanical plus collagenase and DNase dispersal of human melanoma, sarcoma, and lung tumors. They evaluated cell clonogenicity, nucleotide triphosphate pools, cell yields, and cytologic profile (Papanicolaou and differential staining of monocytes and granulocytes). Although the cytologic profile was similar for cells released by both procedures, enzymatically released cells showed larger median sizes of nucleotide pools. Morphologic analysis of trypan-blue-stained cells indicated that the fraction of dye-excluding cells was greater in enzymatically dispersed cells. However, neither the increase in pool sizes or the higher fraction of trypan-blue-stained cells was reflected in the plating efficiency because similar numbers of colonies were obtained using both dispersal procedures. Higher cell yields were obtained with the enzymatic protocol. Other studies (*72*) by these same authors were directed toward examining subpopulation-specific cell selection. They used karyotypic analysis to show that similar chromosomal aberrations were found in cells released mechanically and enzymatically. In addition they compared four different areas of individual tumor specimens with regard to cell yield, dye exclusion, ATP pool size, and uptake and metabolism of fluoropyrimidines, and showed only small variations between areas. Phase-specific cell selection and FCM analyses of DNA distributions of the cells in suspension were not reported.

McDivitt et al. (34) recently described a method for dissociation of viable human breast cancer cells to produce flow cytometric kinetic information similar to that obtained by thymidine labeling. Using collagenase dissociation they obtained between 1 and 50 million cells/g of tumor tissue. Trypan blue dye exclusion and phase microscopy were used to assess cell viability. Clonogenic cell survival was not reported. By comparing thymidine labeling indices and S-phase fractions, as estimated by FCM, of cells dispersed by a mechanical procedure and by the enzymatic technique, they obtained better correlation of S-phase fractions with labeling indices in the enzymatically dispersed cells. Although this type of analysis does not rule out selection of S-phase cells during enzymatic dispersal of whole tissue, it is an example of evaluation of subpopulation representation in cell suspensions intended for cytokinetic studies.

Russell et al. (55) used differential analysis, zymosan incubation, and direct fluorescence microscopy to characterize the nature of the constituent inflammatory cells in tumor cell suspensions obtained by digestion with trypsin, collagenase, and DNase. By comparing the relative fraction of neutrophils, eosinophils, mast cells (differential analysis), macrophages (zymosan incubation), and B and T lymphocytes (fluorescent microscopy after immunocytochemical labeling) in cell suspensions with similar measurements made in histological sections, they were able to evaluate the most suitable method for their endpoint. They also determined cell yield, DNA recoveries, and fraction of trypan-blue excluding cells in three different murine tumors disaggregated mechanically or by exposure to collagenase, chymotrypsin, papain, pronase, or trypsin. By performing these types of measurements the investigators were able to conclude that for their endpoints (1) complete disaggregation was not required to obtain representative inflammatory cell subpopulations, (2) surface immunoglobulin was stripped from the cell surface most rapidly by pronase and chymotrypsin, slowly by trypsin and papain, and not at all by collagenase, and (3) the combination of enzymes with the least effect on inflammatory cells was trypsin, collagenase, and DNase. Although DNA distributions, cell clonogenicity, and phase-specific cell loss were not evaluated, this study is an excellent example of characterization of cell suspensions for subpopulation representation.

Tissue dispersal procedures that provide suspensions of nuclei are commonly utilized for FCM analysis of DNA content. Although the quality of the DNA distributions is often quite good, nuclear suspensions are nonclonogenic and do not retain cellular morphologic features; thus they cannot be utilized for subpopulation discrimination on the basis of cell size/shape or surface markers. Cytokinetic characterization of nuclear suspensions is generally lacking. For example, Thornwaite et al (64) recently described a technique to obtain nuclear suspensions by exposure of minced tumor fragments to trypsin and Nonidet P-40. These investigators sampled multiple areas of the tumor and obtained similar DNA distributions of nuclei from all areas sampled and concluded that they were dealing with a representative tumor cell suspension. It is unlikely that preferential loss of low frequency subpopulations could be detected without the availability of specific markers. For this particular dispersal technique, as well as numerous others (1,9,12,45,69,70), information about nuclei yields and phase-specific cell/nuclei loss are not available. Retention of mitotic cells in disaggregation procedures yielding whole cells is marginal at best, and the complete absence of such cells in nuclear suspensions is to be expected. However, Darzynkiewicz et al. (13) observed that Triton X-100 treatment in the presence of serum proteins (which most likely stabilize the membrane components) and low pH makes cells permeable, but does not lyse them. Since the fraction of mitotic cells in nuclear suspensions is relatively low (usually less than 5% of the total population), it would not be the preparation of choice for monitoring cytokinetic perturbations induced by metaphase-arrest agents (i.e., vincristine).

Data observed in our laboratory (40) and by Dethlefsen et al (14) have attempted to define the cytokinetic characteristics of cell suspensions obtained with two dispersal procedures. Dethlefsen et al. (14) used mechanical disaggregation to disperse murine mammary tumors—traditionally difficult to disperse into single-cell suspensions—and evaluated DNA distributions and phase-specific cell loss in the dispersed cells. Although DNA distributions were not of high quality and a clonogenic assay was not available for survival measurements, phase-specific selection did not occur. In subsequent studies in his laboratory (39), tetraphenylboron was used to disperse mammary tumors. Cell viability was not maintained (virtually 100% of the cells stained with trypan ·blue). However, preferential phase-specific selection did not occur, and cell yields were improved, as were the DNA distributions. As discussed earlier, dispersal of the KHT tumor, a relatively soft sarcoma, with either trypsin/DNase or neutral protease provides cells with clonogenic capability, high-quality DNA distributions, acceptable cell yields, and no phase-specific cell loss. Possible selection of other subpopulations (i.e., cell clones with different drug sensitivities and so on) were not examined.

A review of the literature describing dispersal techniques for solid tissues indicates that few of the cell suspensions used in cell cycle studies have been adequately characterized from a cytokinetic viewpoint. The choice of a specific dispersal protocol for cytokinetic studies should be done in a rational manner using criteria germane to cytokinetic analyses. Although a specific disaggregation technique may not completely fulfill the suggested set of criteria, knowledge of the characteristics of the cell suspension in view of the endpoint to be studied will contribute to accurate extrapolation of cytokinetic phenomena measured in single cells to those occuring in the tissue *in situ*.

ACKNOWLEDGMENTS

The expert assistance of Mrs. Angela Riggs and Miss Lil Mitchell in various aspects of manuscript preparation is greatly appreciated. Work was performed under the auspices of the US Department of Energy by the Lawrence Livermore National Laboratory under contract number W-7405-ENG-48 with support from USPHS Grants CA 25782 and CA 14533.

REFERENCES

1. Barlogie, B., Gohde, W., Johnston, A., Smallwood, L., Schuman, J., Drewinko, B., and Freireich, D. J. Determination of ploidy and proliferative characteristics of human solid tumors by pulse cytophotometry. Cancer Res., *38:* 3333–3339, 1978.

2. Bashor, M. M. Dispersion and disruption of tissues. *In*: (W. B. Jakoby and I. H. Pasten, eds.), Methods in Enzymology, vol. LVIII, New York: Academic, 1979.

3. Bauer, F. W., Boezeman, J., and deGrood, R. M. Cell cycle analysis in normal and pathologic skin. Pulse-Cytophotometry, *III*: 533–538, 1978.

4. Berry, M. N. and Friend, D. S. High-yield preparation of isolated rat liver parenchymal cells. J. Cell Biol., *43*: 506–520, 1969.

5. Berwick, L. and Corman, D. R. Some chemical factors in cellular adhesion and stickiness. Cancer Res., *22*: 982–986, 1962.

6. Boyse, E. A. A method for the production of viable cell suspensions from solid tumors. Transplant. Bull., *7*: 100–104, 1960.

7. Brown, J. M., Twentyman, P. R., and Zamvil, S. S. Response of the RIF-1 tumor *in vitro* and in C3H/Km mice to x-irradiation (cell survival, regrowth delay, and tumor control), chemotherapeutic agents, and activated macrophages. J. Natl. Cancer Inst., *64*: 605–611, 1980.

8. Burlington, H. and Cronkite, E. P. Characteristics of cell cultures derived from renal glomeruli. Proc. Soc. Exp. Biol. Med., *142*: 143–149, 1973.

9. Burns, E. R., Bagwell, B. C., Hinson, W. G., Pipkin, J. L., and Hudson, J. L. Preparation and stability of sixteen murine tissues and organs for flow cytometric cell cycle analysis. Cytometry, *4*: 150–160, 1983.

10. Camazine, S. M., Ryan, G. B., Unanue, E. R., and Karnovsky, N. J. Isolation of phagocytic cells from the rat renal glomerulus. Lab. Invest., *35*: 315–326, 1976.

11. Cook, G. M. W., Heard, D. H., and Seamna, G. V. F. A sialomucopeptide liberated by trypsin from the human erythrocyte. Nature, *138*: 1011–1012, 1960.

12. Darzynkiewicz, Z., Traganos, F., Sharpless, T., and Melamed, M. R. Lymphocyte stimulation: A rapid multiparameter analysis. Proc. Natl. Acad. Sci. USA, *73*: 2881–2884, 1976.

13. Darzynkiewicz, Z., Traganos, F., and Melamed, M. R. Detergent treatment as an alternative to cell fixation for flow cytometry. J. Histochem. Cytochem., *29*: 329–330, 1981.

14. Dethlefsen, L. A., Gray, J. W., George, Y. S., and Johnson, S. Flow cytometric analysis of the perturbed cellular kinetics of solid-tumors: Problems and promises. *In*: (W. Gohde, J. Schumann, and Th. Bucher, eds.), Pulse-Cytophotometry, Ghent, Belgium: European Press, 1976.

15. Dolbeare, F., Gratzner, H., Pallavicini, M. G., and Gray, J. W. Flow cytometric analysis of total DNA content and incorporated bromodeoxyuridine. Proc. Natl. Acad. Sci. USA, *80*: 5573–5577, 1983.

16. Finkelstein, J. N. and Shapiro, D. L. Isolation of Type II alveolar epithelial cells using low protease concentrations. Lung, *160*: 85–98, 1982.

17. Goodenough, D. A. and Gilula, N. B. The splitting of hepatocyte gap junctions and zonae occludentes with hypertonic disaccharide. J. Cell Biol., *61*: 575–590, 1974.

18. Gottschalk, A., Belyavin, G., and Biddle, F. *In*: (A. Gottschalk, ed.), Glycoproteins: Their Composition, Structure, and Function, 2nd Ed., part B., Amsterdam: Elsevier, 1972.

19. Hamburger, A. W., White, C. P., and Tencer, K. Effect of enzymatic disaggregation on proliferation of human tumor cells in soft agar. J. Natl. Cancer Inst., *68*: 945–949, 1982.

20. Harris, C. C. and Leone, C. A. Some effects of EDTA and tetraphenylboron on the ultrastructure of mitochondria in mouse liver cells. J. Cell Biol., *28*: 405–408, 1966.

21. Harrison, D. D. and Webster, H. L. The preparation of isolated crypt cells. Exp. Cell Res., *55*: 257–260, 1969.

22. Hill, R. P. and Pallavicini, M. G. Hypoxia and the radiation response of tumors. *In*: (H. I. Bicher and D. F. Bruley, eds.), Oxygen Transport to Tissue, vol. IV, New York: Plenum, 1983.

23. Hommes, F. A., Draisman, M. I., and Molenaar, I. Preparation and some properties of isolated rat liver cells. Biochem. Biophys. Acta., *222*: 361–371, 1970.

24. Hosick, H. L. and Strohman, R. Changes in ribosome-polyribosome balance in chick muscle cells during tissue dissociation, development in culture and exposure to simplified culture medium. J. Cell Physiol., *77*: 145–156, 1971.

25. Kikkawa, Y., Yoneda, K., Smith, F., Packard, B., and Suzuki, K. The type II epithelial cells of the lung. II. Chemical composition and phospholipid synthesis. Lab. Invest., *32*: 295–302, 1975.

26. Kleinman, H. K., Klebe, R. J., and Martin, G. R. Role of collagenous matrices in the adhesion and growth of cells. J. Cell Biol., *88*: 473–485, 1981.

27. Kreisberg, J. I. Isolation and culture of homogeneous populations of glomerular cell types. *In*: (T. G. Pretlow and T. P. Pretlaw, eds.), Cell Separation. Methods and Selected Applications, vol. I, New York: Academic, 1982.

28. Laerum, O. D. Oxygen consumption of basal and differentiating cells from hairless mouse epidermis. J. Invest. Dermatol., *52*: 204–211, 1969.

29. Lawson, A. J., Smit, R. A., Jjefers, N. A., and Osborne, J. W. Isolation of rat intestinal crypt cells. Cell Tissue Kinet., *15*: 69–80, 1982.

30. Levinson, C. and Green, J. W. Cellular injury resulting from tissue disaggregation. Exp. Cell Res., *39*: 309–317, 1965.

31. Lewin, M. J. M., Cheret, A. M., and Sachs, G. Separation of individual cells from the fundic gastric mucosa. *In*: (T. G. Pretlow and T. P. Pretlow, eds.), Cell Separation: Methods and Selected Applications, vol. I, New York: Academic, 1982.

32. Madden, R. E. and Burk, D. Production of viable cell suspensions from solid tumors. J. Natl. Cancer Inst., *27*: 841–861, 1961.

33. Magee, W. E., Sheek, N. R., and Sagik, B. P. Methods of harvesting mammalian cells grown in tissue culture. Proc. Soc. Exp. Biol. Med., *99*: 390–392, 1958.

34. McDivitt, R. W., Stone, K. R., and Meyer, J. S. A method for dissociation of viable human breast cancer cells that produces flow cytometric kinetic information similar to that obtained by thymidine labeling. Cancer Res., *44*: 2628–2633, 1984.

35. Mead, J. S., Horan, P. K., and Wheeless, L. L. Syringing as a method of tissue dispersal. I. Effect on intermediate and superficial squamous cells. Acta Cytol., *22*: 86–90, 1978.

36. Melamed, M. R., Mullaney, P. F., and Mendelsohn, M. L., eds. Flow Cytometry and Sorting. New York: John Wiley, 1979.

37. Milas, L. and Mujagic, H. The effect of spleenectomy on fibrosarcoma "metastasis" in lungs of mice. Int. J. Cancer, *11*: 186–190, 1973.

38. Moran, J. and Cohen, L. Use of papain in the preparation of adult skeletal muscle for tissue culture. In Vitro, *10*: 188–190, 1974.

39. Pallavicini, M. G., Cohen, A. M., Dethlefsen, L. A., and Gray, J. W. Dispersal of solid tumors for flow cytometric (FCM) analysis. *In*: (D. Lutz, ed.), Pulse-Cytophotometry, part III, Ghent, Belgium: European Press, 1978.

40. Pallavicini, M. G., Folstad, L. J., and Dunbar, C. Solid KHT tumor dispersal for flow cytometric cell kinetic analysis. Cytometry, *2*: 54–58, 1981.

41. Pallavicini, M. G., Gray, J. W., and Folstad, L. J. Quantitative analysis of the cytokinetic response of KHT tumors *in vivo* to 1-β-D-arabino-furanosylcytosine. Cancer Res., *42*: 3125–3131, 1982.

42. Pallavicini, M. G., Ng, C. R., and Gray, J. W. Bivariate flow cytometric analysis of murine intestinal epithelial cells for cytokinetic studies. Cytometry, 5: 55–62, 1984.

43. Penning, J. J. and Levan, J. H. A modified enzymatic technique for production of cell suspensions of a murine fibrosarcoma. J. Natl. Cancer Inst., 66: 85–87, 1981.

44. Phillips, H. J. Some metabolic changes resulting from treating kidney tissue with trypsin. Can. J. Biochem., 48: 1495–1504, 1967.

45. Piwnicka, M., Darzynkiewicz, Z., and Melamed, M. R. RNA and DNA content of isolated nuclei measured by multiparameter flow cytometry. Cytometry, 3: 269–275, 1983.

46. Pollack, A., Prudhomme, D. C., Greenstein, D. B., Irwin, G. C., Claffin, A. J., and Block, N. C. Flow cytometric analysis of RNA content in different cell populations using pyronin y and methyl green. Cytometry, 3: 28–35, 1982.

47. Rappaport, C. and Howze, G. Dissociation of adult mouse liver by sodium tetraphenylboron, a potassium complexing agent. Proc. Soc. Exp. Biol. Med., 121: 1010–1016, 1966.

48. Rasey, J. S. and Nelson, N. J. Effect of tumor disaggregation on results of *in vitro* cell survival assay after *in vivo* treatment of the EMT6 tumor: X-rays, cyclophosphamide, and bleomycin. In Vitro, 16: 547–553, 1980.

49. Reimann, J., Ehman, D., and Miller, R. G. Differential binding of lectins to lymphopoietic and myeolopoietic cells in murine marrow as revealed by flow cytometry. Cytometry, 5: 194–203, 1984.

50. Reinhold, H. S. A cell dispersal technique for use in quantitative transplantation studies with solid tumors. Eur. J. Cancer, 1: 67–71, 1965.

51. Rinaldini, L. The isolation of living cells from animal tissues. Int. Rev. Cytol., 7: 587–647, 1958.

52. Rockwell, S., Kallman, R. F., and Fajando, L. F. Characteristics of a serially transplanted mouse mammary tumor and its tissue-culture adopted derivatives. J. Natl. Cancer Inst., 49: 735–749, 1972.

53. Rockwell, S. and Kallman, R. Growth and cell population kinetics of single and multiple KHT sarcomas. Cell Tissue Kinet., 5: 449–457, 1972.

54. Roters, M., Linden, W. A., and Heienbrok, W. Comparison of three different methods for the preparation of human tumors for flow cytometry (FCM). *In*: (D. Lutz, ed.), Pulse Cytometry, part IV, Ghent, Belgium: European Press, 1978.

55. Russell, S. W., Doe, W. F., Hoskins, R. G., and Cochrane, C. G. Inflammatory cells in solid murine neoplasms. I. Tumor disaggregation and identification of constituent inflammatory cells. Int. J. Cancer, *18*: 322–330, 1976.

56. Schapp, G. H., de Josselin de Jong, J. E., and Jonkind, J. F. Fluorescence polarization of six membrane probes in embryonal carcinoma cells after differentiation as measured on a FACS II cell sorter. Cytometry, *5*: 188–193, 1984.

57. Seglen, P. O. Preparation of rat liver cells. Exp. Cell Res., *74*: 377–389, 1971.

58. Severin, E., Zold, H., and Spies, I. Proposals for isolation and further handling of liver cells in flow-cytometry. *In*: (D. Lutz, ed.), Pulse-Cytometry, part III, Belgium: Ghent, European Press, 1978.

59. Slocum, H. K., Pavelic, Z. P., Kanter, P. M., Nowak, N. J., and Rustum, Y. M. The soft agar clonogenicity and characterization of cells obtained from human solid tumors by mechanical and enzymatic means. Cancer Chemother. Pharmacol., *6*: 219–225, 1981.

60. Slocum, H. K., Pavelic, Z. P., Rustum, Y. M., Creaven, P. J., Karskousis, C., Takita, H., and Greco, W. R. Characterization of cells obtained by mechanical and enzymatic means from human melanoma, sarcoma, and lung tumors. Cancer Res., *41*: 1428–1434, 1981.

61. Snow, C. and Allen, A. The release of radioactive nucleic acids and mucoproteins by trypsin and ethylenediaminetetraacetate treatment of baby hamster cells in tissue culture. Biochem. J., *119*: 707–714, 1970.

62. Steinberg, M. S. The role of temperature in the control of aggregation of dissociated embryonic cells. Exp. Cell Res., *21*: 1–10, 1961.

63. Thomson, J. E. and Rauth, A. M. An *in vitro* assay to measure the viability of KHT tumor cells not previously exposed to culture conditions. Radiat. Res., *58*: 262–276, 1974.

64. Thornwaite, J. T., Sugarbaker, E. V., and Temple, W. J. Preparation of tissues for DNA flow cytometric analysis. Cytometry, *1*: 229–237, 1980.

65. Traganos, F., Darzynkiewicz, Z., Sharpless, T., and Melamed, M. R. Simultaneous staining of ribonucleic and deoxyribonucleic acid in unfixed cells using acridine orange in a flow cytofluorometric system. J. Histochem. Cytochem., *25*: 46–56, 1977.

66. Trotman, C. N. A. Isolation of gastrointestinal mucosa. *In*: (E. Reid, ed.), Cell Populations, Chichester, England: Horwood/Wiley, 1979.

67. Twentyman, P. R. An artefact in clonogenic assays of bleomycin cytotoxicity. Br. J. Cancer, *36*: 642–644, 1977.

68. Twentyman, R. R. and Yuhas, J. M. Use of bacterial neutral protease for disaggregation of mouse tumors and multicellular tumor spheroids. Cancer Lett., *9*: 225–228, 1980.

69. Vindelov, L. L. Flow microflurometric analysis of nuclear DNA in cells from solid tumors and cell suspensions. Virchows Arch. Cell Pathol., *24*: 227–242, 1977.

70. Vindelov, L. L., Christensen, I. J., and Nissen, N. I. A detergent-trypsin method for the preparation of nuclei for flow cytometric DNA analysis. Cytometry, *3*: 323–327, 1983.

71. Visser, J. W. M. and Bol, S. J. L. A two-step procedure for obtaining 80-fold enriched suspensions of murine pluripotent hemopoietic stem cells. Stem Cells, *1*: 240–249, 1981.

72. Wake, N., Slocum, H. R., Rustum, Y. M., Matsui, S., and Sandberg, A. A. Chromosomes and causation of human cancer and leukemia XLIV. A method for chromosome analysis of tumors. Cancer Gen. Cytogen., *3*: 1–10, 1981.

73. Waymouth, C. To disaggregate or not to disaggregate: Injury and cell disaggregation, transient or permanent? In Vitro, *39*: 97–111, 1974.

74. Waymouth, C. Obtaining cell suspensions from animal tissues. *In*: (T. G. Pretlow and T. P. Pretlow, eds.), Cell Separation: Methods and Selected Applications, vol I., New York: Academic, 1982.

75. Weiser, M. Intestinal epithelial cell surface membrane glycoprotein synthesis. I. An indicator of cellular differentiation. J. Biol. Chem., *248*: 2536–2541, 1973.

76. Weiser, R. S., Heise, E., McIvor, K., Han, S., and Granger, G. *In vitro* activities of immune macrophages. *In*: (R. T. Smith and R. A. Good, eds.), Cellular Recognition, New York: Appleton-Century-Crofts, 1969.

Chapter 7

Multivariate Cell Analysis

Techniques for Correlated Measurements of DNA and Other Cellular Constituents

Harry A. Crissman
and
John A. Steinkamp

1. INTRODUCTION

Analysis of the cell cycle was significantly advanced by the development of photometric methods for quantitative measurement of biochemical constituents in single cells. Population biochemical analysis could then be performed on a cell-by-cell basis, and distinct subpopulations discriminated and quantified. Such methods as used in early studies by Casperson and Schultz (*10*) and others [*see* review by Swift (*84*)] demonstrated the potential of ultraviolet (260 nm) absorption cytophotometry for determining cellular nucleic acids content. Interestingly, Kamentsky et al. (*51*), in one of the earliest reports on flow cytometry, utilized this analytical technique along with cellular light scattering measurement. Cytochemical analysis incorporating colorometric procedures, such as the Feulgen reaction (*33*), allowed for microspectrophotometric quantitation of DNA in single cells. Pyronin Y staining was used by Brachet (*7*) and later by Kurnick (*54*) for RNA determinations. The quantitative cytophotometric methods employed in these early studies (*9*) accurately established that both nucleic acid as well as protein contents were elevated in rapidly growing cells compared to cells in stationary phase—correlated biochemical events that are now well established for the cell cycle.

Flow cytometry (FCM) incorporates many of the advantageous principles established in these previous cytophotometric studies. In addition FCM provides ease, speed, and statistical accuracy of the measurements, as well as other desirable features. For example, when biochemical molecules are labeled stoichiometrically with fluorochromes, subsequent fluorescence measurements by flow cytometry are less sensitive than absorption cytophotometry to "distributional errors" (65) caused by the nonhomogeneous distribution of the materials within the cells. Furthermore, the loss of quantitative accuracy resulting from fluorescence fading does not appear to limit the precision of flow measurements since stained cells are exposed uniformly and only briefly (i.e., 3–5 μs) to high-intensity excitation sources. Staining with more than one dye makes it possible to analyze several biochemical constituents on a cell-to-cell basis. When such data are acquired and stored in a computer in a correlated list mode fashion (72), subsequent reprocessing of the data allows for study of the interrelationships of various biochemical moieties, including DNA. In this way, cellular features such as RNA and protein contents, for example, can be investigated for cells in all phases of the cell cycle.

The similarity of design and the widespread availability of FCM instruments, along with ease and reproducibility of staining methods for DNA and other cellular parameters, have allowed for easy comparison of data from laboratories throughout the world. For the most part, the versatility and acceptance of FCM technology have increased as a result of improvements in cell preparation and staining. Except for recent specific designs for detailed chromosome analysis (12), flow instruments have, in principle, remained much the same for the past 15 years.

In this chapter, we will summarize the cytochemical techniques and approaches for analysis of DNA content and various other cellular constituents that can be potentially useful for elucidating biochemical events involved in control of cell cycle progression and proliferation. Data presented demonstrate the feasibility of the methods and illustrate the additional information obtainable from multivariate analysis for cell cycle studies.

2. FLUORESCENCE PRINCIPLES (GENERALIZED)

Flow cytometric analysis of cellular biochemical content is achieved by labeling or staining the moiety of interest with a fluorochrome. During analysis the bound fluorochrome in stained cells is excited at an appropriate wavelength and fluorescence is emitted isotropically. The excitation wavelength and color of emitted fluorescence is characteristic of each fluorochrome. If the dye is bound specifically to the particular cellular constituent of interest, and the staining reaction is stoichiometric, then the fluorescence intensity should be proportional to the cellular constituent content. This scheme is an oversimplification; however, in principle, it provides the systematic, logical basis for development of quantitative cytochemistry for flow cytometry.

3. CHARACTERISTICS OF FLUORESCENT DYES

Excitation and emission spectra of dyes in solution are analyzed in a fluorescence spectrophotometer. However, the spectral properties of the free dye are influenced by the nature of the solvent. Also, changes in the spectral characteristics of the free dye can occur when it binds to a particular biochemical. Thus, fluorochromes should be analyzed in the solution designed for cell staining and bound either to the biochemical component of interest in solution or in cells. Upon binding to DNA many fluorochromes exhibit a significant increase in their emission intensity (i.e., increased quantum yield) accompanied by a shift in either or both the excitation and emission wavelength spectra. For example DNA complexes of ethidium bromide (EB) or propidium iodide (PI) excite at wavelengths approximately 50 nm longer than free dyes (Fig. 1). The emission peak position shifts only slightly. Preliminary fluorescence analysis is critical for determining the appropriate excitation and emission wavelengths required for staining cells with a combination of dyes. The specific aim is to achieve appropriate excitation and optimal color separation of the emitted fluorescence for

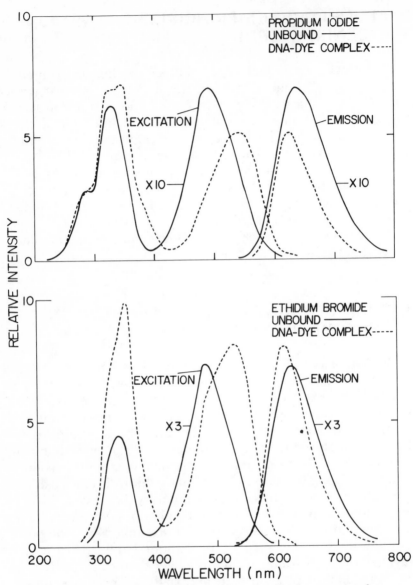

Fig. 1. Excitation and emission spectra for the fluorescent dyes propidium iodide (PI) and ethidium bromide (EB). Both dyes exhibit a dramatic shift in excitation to the longer wavelength when complexed with DNA (dotted line) compared to the unbound state (solid line). Excitation and emission peak values of PI–DNA solutions in the visible range are elevated approximately 10 nm with respect to EB–DNA.

quantitative accuracy. Spectral characteristics of dyes frequently used in FCM are shown in Fig. 2. Reference to the specific fluorochromes will be made throughout this chapter.

4. SPECIFICITY AND STOICHIOMETRY OF FLUOROCHROME REACTIONS

The degree of specificity of staining is often established on the basis of results obtained either following enzymatic degradation of the cellular substrate or by chemically blocking the molecular sites to which the dye normally binds. Subsequent cell staining should be negative or significantly diminished. Such tests are performed on ethanol-fixed cells with permeabilized membranes, since enzymes do not penetrate viable cells. Formalin fixation is avoided since some aldehyde complexes of DNA, RNA, and protein are often not readily accessible to the particular enzymes.

Stoichiometry of DNA stains can initially be established from the ratio of the mean fluorescence intensities of G2 + M cells and G1 cells in DNA distributions. Ideally this ratio should be close to a value of 2.0. Also the computer-fit of the DNA distribution should provide nearly the same percentage of S-phase cells as calculated by an independent method, such as with tritiated thymidine (29,39). Coulson et al. (11) tested the stoichiometry of various DNA-staining reactions using cells from different animal species that varied in DNA content. Of the methods tested, propidium iodide (PI) staining of ethanol-fixed cells following RNase treatment (15) gave the most linear results. The method used by Coulson et al. (11) is one of the best approaches for establishing linearity of staining with cellular content. Similar studies can be designed for establishing stoichiometry of staining reactions for other cellular constituents (i.e., RNA, protein, and so on) by the appropriate choice of several cell populations whose cellular content for a given moiety differ over a given range as determined by an alternative, independent method.

5. ENERGY TRANSFER

When two fluorochromes are bound in close proximity (i.e., generally within 50 Å) and the emission spectrum of one dye

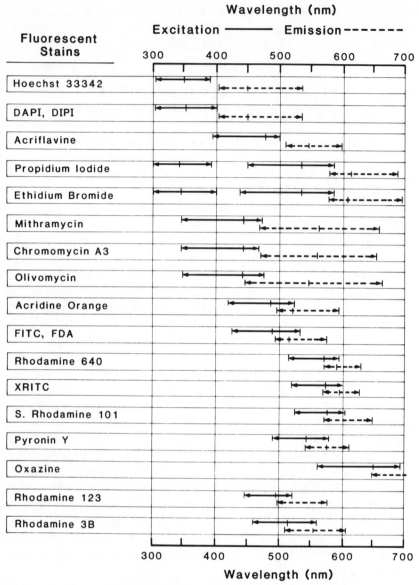

Fig. 2. Fluorescence excitation (solid arrows) and emission (dotted arrows) wavelength ranges for selected fluorescent dyes. The short vertical lines designate the position of the relative intensity peak in the spectral curves.

(donor) significantly overlaps the excitation spectrum of the second dye (acceptor), energy transfer can occur (*71*). An enhancement in fluorescence intensity of the acceptor molecule is observed coincidental to a diminution or quenching of fluorescence of the donor. The extent of energy transfer depends then on the physical orientation of the dyes and their fluorescence properties. Energy transfer can occur between a fluorescent molecule such as a Hoechst DNA reactive dye and a nonfluorescent molecule such as BrdUrd, in which case only quenching of the Hoechst fluorescence is observed. Such phenomena have been exploited in some cell cycle studies mentioned later in this chapter. This method can also provide information on the proximity of two molecular moieties within the cell.

6. INTRINSIC FLUORESCENCE OF UNSTAINED CELLS (AUTOFLUORESCENCE)

Naturally occurring fluorescent materials, generally pyrimidines and flavin nucleotides (*5*), are found in many cell types. Phagocytes such as macrophages often contain ingested fluorescent particles that can produce ''autofluorescence'' when excited at appropriate wavelengths. When levels of autofluorescence are far below that produced from cell staining, analytical problems are not encountered. However, difficulties can arise in some studies when (a) the number of bound dye molecules is small, such as for antibodies against cell surface antigens, or (b) the quantum yield of the dye is also small, and (c) the autofluorescence spectra overlap that of the labeling fluorochrome. Under these conditions, discrimination between fluorochrome labeling and autofluorescence becomes difficult and quantitative analysis is seriously impaired. Fluorescence analysis of *unstained cells* at the excitation wavelength, laser power, and electronic gain setting predetermined for the fluorochrome-labeled studies will potentially detect and determine the degree of analytical distortion caused by autofluorescence. The alternative use of a fluorochrome with spectral properties different from cellular autofluorescence can often minimize the problem.

7. CELL PREPARATION AND FIXATION

Plasma membranes of viable cells exclude most of the fluorochromes currently used in FCM. For rapid cell staining, membranes are permeabilized by brief treatment with nonionic detergents (26,36,87,94), or with hypotonic solutions (53), proteolytic enzymes (100), and/or low pH (26). Such techniques are useful for rapid nuclear staining and FCM cell cycle analysis; however, cell membrane components and cytoplasmic constituents may be lost.

Alternatively, cells can be treated with ethanol, which also perforates cell membranes, but seems to preserve most cytoplasmic and cell surface constituents. Ethanol (70%) does not appear, based on analytical results, to seriously denature molecules such as DNA, RNA, and protein. No detailed studies have been done, however, to determine the extent of preservation of these and other cellular constituents after fixation in ethanol. Such studies generally involve radioisotope labeling with precursors specific for the constituent of interest and subsequent quantitative determination of the cellular loss of radioactivity during fixation.

For ethanol fixation we routinely harvest cells from culture or from tissue dispersal solutions by centrifugation. Cells are then thoroughly resuspended in one part cold "saline GM" (g/L: glucose, 1.1; NaCl, 8.0; KCl, 0.4; $Na_2HPO_4 \cdot 12 H_2O$, 0.39; KH_2PO_4, 0.15) containing 0.5 mM EDTA for chelating free calcium and magnesium ions. Addition of three parts cold 95% nondenatured ethanol to the cell suspension with mixing makes the final ethanol concentration about 70%. It is most important that the cells are *thoroughly* resuspended before the addition of alcohol to prevent cell aggregation. Ethanol fixation is most commonly used when staining nucleic acids and other intracellular constituents for cell cycle analysis, since it does not cross-link molecules and thus does not interfere with staining. However, it is not appropriate for the Feulgen procedure or when cell membrane constituents are to be stained.

For Feulgen-DNA staining, cells are fixed for at least 12 h in saline GM containing 10% formalin (i.e., 3.7% formaldehyde). Preservatives, such as methanol, may be added to concentrated commercial formalin. Fresh formalin (pH 7.0), free of precipitate, is recommended for cell fixation. Cacodylate buffers (0.05–1.0M) containing 1.0% paraformaldehyde preserve cell membranes well.

Electronic cell volume distributions of fixed cells are nearly identical to those obtained for viable cells at the same electronic gain setting. Cells fixed in glutaraldehyde or formalin sometimes show increased autofluorescence, and unstained cells excited at 488 nm are fluorescent at elevated electronic gain setting. Stewart (personal communication) has found it useful in some studies to first react unfixed cells with appropriate fluorochromes or fluorochrome-labeled antibodies to cell surface antigens. Cells are then rinsed and fixed 2 h in paraformaldehyde, rinsed again, and stored for at least 1 mo in Ca- and Mg-free phosphate-buffered saline.

Paraformaldehyde is often used for investigation of cell membrane properties (i.e., antigenic sites). Aldehydes are not useful as fixatives for most cell cycle studies, however, since the cross-links they induce interfere with commonly used DNA stains.

Recently, Hedley et al. (42) introduced a new method for preparing nuclei from formalin-fixed, paraffin-embedded tissue. Following removal of paraffin with xyline, tissue samples are rehydrated, treated 30 min with aqueous 0.5% pepsin-HCl, and stained for DNA with DAPI. DNA distributions measured for these nuclei compare well with distributions obtained from fresh material stained with ethidium bromide–mithramycin (EB–MI). The authors suggest that pepsin digestion provides good nuclei suspensions and also apparently breaks cross-linkages in chromatin produced during initial formalin fixation. If the method continues to prove successful for many types of tissue, it may be useful for long-term storage (i.e., years) of material for subsequent FCM analysis.

8. CURRENT FLOW-SYSTEM MODIFICATIONS FOR MULTIVARIATE ANALYSIS

Simultaneous analysis of several cellular constituents may be achieved by selectively staining the constituents with fluorochromes with different spectral characteristics so that the dye contents can be determined independently during flow cytometry. Dyes that are useful for such multivariate studies typically fluoresce at different wavelengths or excite at different wavelengths. As shown in Fig. 2, however, the fluorescence emission and excitation spectra of many stains overlap to some extent. Thus, flow systems must be used that provide for selective excitation and optimal separation of emission from each dye. Dual- or triple-beam flow systems have been developed by several laboratories for this purpose (30, 74,76–79). In many of these systems, the arrangement of the excitation source and the flow chamber are similar to that graphically illustrated in Fig. 3 (75,77). An essential feature of this arrangement is the spatial separation of three laser beams, each of which is tuned to a wavelength near optimal for excitation of at least one dye (77). Appropriate color separation filters ensure fluorescence measurement over a wavelength range specifically selected to measure the emission of only one of the dyes at each point of excitation on the cell stream. Three or more fluorescence signals from

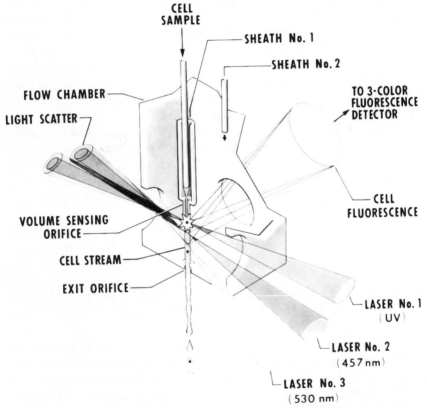

Fig. 3. Diagrammatic cutaway display of the flow chamber in the three-laser excitation FCM system. The system has capabilities for analysis of cell volume, light scattering at three wavelengths, and three colors of fluorescence. Laser beams at the indicated wavelengths are spacially separated by 250 μm for sequential excitation and selective wavelength analysis of stained cells.

each cell are collected and stored by computer in list mode fashion (72), and measurements can be correlated subsequently. The detailed description in section 10.1.5. illustrates the usefulness of the three-laser excitation flow systems in actual studies.

9. DNA CONTENT ANALYSIS BY FLOW CYTOMETRY

Most of the early studies in FCM were applied to quantitative measurement of cellular DNA content. Analysis of the DNA distributions provides data on the relative frequency of cells in various phases of the cell cycle. Comparisons of computer-fit analyses of the FCM-derived fluorescence histograms and data obtained from conventional assays of DNA replication involving autoradiographic detection of [³H]-thymidine confirmed the accuracy and precision of the FCM approach to cell cycle analysis (29,39). These early studies provided the credibility necessary to rapidly and firmly establish flow cytometry as an important analytical tool for biological analysis. After more than 15 yr, cell staining with DNA-specific fluorochromes for FCM cell cycle analysis still remains one of the most frequent applications of the technology. Details of DNA distribution analysis are given in chapter 5 (40).

The attractive features of the methodology include the rapidity and ease of both cellular DNA staining and FCM procedures, and other advantages discussed previously (88), which collectively include the ability to (a) monitor cell cycle distributions in ongoing experiments, with the added option of altering an experiment in progress in response to a population change; (b) localize cells within S-phase and distinguish between early-, mid-, and late S-phase; (c) analyze populations with radiolabeled RNA and/or protein so that [³H]-thymidine autoradiography is not necessary; (d) monitor populations composed of slowly progressing or arrested cells; (e) analyze populations devoid of cells in the S- or mitotic phases; (f) analyze populations containing cells unable to transport, incorporate, or metabolize [³H]-thymidine; (g) detect abnormalities in progression through mitosis such as nondisjunction or polyploidization (52); and (h) distinguish between intact and fragile (dying) cells in drug-treated or virus-infected cells. There is also the option of sorting cells based on DNA content for identification and further analysis. Throughout this chapter we present data that illustrate many of these features.

In spite of the above, it is important to recognize that although the technique can accurately determine cell cycle position, it has certain limitations. Flow cytometric DNA content analysis alone fails to assess the metabolic state of cells and therefore does not provide information concerning capacity and the rate at which cells will traverse the cell cycle. By analogy the cell cycle frequency distribution is equivalent to a "snapshot" photograph of individuals in a race. The position of all runners in the course of the race at any instance is evident; however, the speed and the capacity of each runner to successfully complete the race is indeterminant. Such is also the case for cells represented in a FCM cell cycle distribution. For instance, two exponentially growing cell populations may have very different doubling times, but still yield similar DNA content distributions if the durations of the individual phases of the cell cycle relative to the total cell cycle duration are similar. Also under experimental conditions it is difficult to ascertain proportions of cycling and noncycling cells within the various phases.

DNA content analysis has also been useful for detecting subclones with different DNA content within a given population. From a clinical standpoint such determinations can provide information of prognostic value (57). The terms aneuploid or heteroploid, which refer to chromosome number, have occasionally been used to specify subclones within the DNA content distribution. In a recent publication (43), guidelines are proposed to standardize nomenclature and procedural descriptions for DNA content analyses.

In recent years FCM has been used to elucidate physiological aspects, which in addition to DNA metabolism, regulate and control cell proliferation. Many current studies now include staining and subsequent measurement of other cellular constituents, such as protein and RNA, simultaneously with DNA. Cellular levels of such descriptors and others are known to be important indicators of cell cycle progression capacity.

We will present some current methods and criteria for multicolor fluorescent staining and analysis of those cellular biochemicals that, in addition to DNA content, can yield information on cell cycle traverse. In most instances the techniques involve modification of preexisting staining methods that will allow for the appropriate combination of the dyes and near-optimal quantitative analysis.

10. DNA-SPECIFIC CELL STAINING

In previous publications we (*13,14,17,18*) and others (*47,48,56, 59,62,86*) have reviewed various dyes, their mode of binding, and methods for their use as DNA stains in FCM studies. In this chapter only those aspects are addressed that are important for combining these dyes in protocols for multicolor staining and correlated analysis of several biochemical compounds and organelles in cells.

10.1. Feulgen–DNA Reaction

The Fuelgen procedure (*33*) represents one of the earliest and most well-established cytochemical procedures for labeling and quantitating cellular DNA content. Modification of the Culling and Vassar adaption (*22*) of the procedure were used in earlier FCM studies employing the fluorochromes auromine 0 (*90,92*) or acriflavine (*13,89*). In this procedure, formalin-fixed cells are treated with 4N HCl and then stained with acriflavine or auramine-*O* in a sulfite solution (0.2% dye, 0.5N HCl, and 10% $K_2S_2O_5$ in a ratio of 20:2:1) (*90*). In one recent application, Swartzendruber (*83*) used the acriflavine procedure as a control, since Feulgen–DNA analysis is not affected by BrdUrd-substituted DNA in cells. In general, however, the use of the Feulgen method is somewhat limited since it is time-consuming and produces cell loss and nonspecific staining in many cell types, such as those contained, for example, in gynecological samples. In addition, HCl hydrolysis, an essential step in the protocol, depurinates DNA, removes RNA and histones, and possibly damages other cellular components, including membranes. Therefore, the remainder of this section will address the use of more useful DNA dyes.

10.2. DNA Reactive Hoechst Dyes, DAPI (4-6-diamidino-2-phenylindole), DIPI [4-6-bis-(2-imidazolynyl-4H, 5H)-2-phenylindole], and LL 585

The Hoechst dyes are benzimidazole derivatives that emit blue fluorescence when excited by UV light at about 350 nm (Fig. 2). They have a high specificity for DNA and bind preferentially to A–T base regions, but do not intercalate (*67*). The Hoechst 33258 derivative was originally used by Hilwig and Gropp (*44*) in mouse chromosome banding studies. Latt (58) later showed that fluorescence of this dye was quenched when bound to BrdUrd-substituted DNA, and used the phenomenon in a technique for detecting regions of sister chromatid exchange in metaphase chromosomes. Ardnt-Jovin and Jovin (3), using flow cytometric analysis, first demonstrated the use of Hoechst 33342 for quantitative DNA staining and sorting of viable cells. The structural and spectral properties of DAPI and DIPI are similar, and each can be used interchangeably with the Hoechst 33258 or 33342 derivatives for staining ethanol-fixed cells.

Stock solutions of the Hoechst dyes, DAPI or DIPI, prepared in distilled water (1.0 mg/mL) and refrigerated in dark-colored containers, may be used for at least 1 mo without noticable degradation. These dyes tend to precipitate in PBS at concentrations above 50 μg/mL. Fluorescence of ethanol-fixed Chinese hamster ovary (CHO) cells stained 15–20 min at room temperature with Hoechst 33342 (0.5–1.0 μg/mL) in PBS remains stable for at least 5–6 h. Stained cells are usually analyzed in equilibrium with the dye, using a UV laser beam or mercury-arc lamp excitation source. Dye concentrations for fixed cells may vary slightly depending upon the cell type and flow instrumentation used for analysis.

Latt et al. (61) have recently examined the binding properties and cell-staining characteristics of several fluorescent compounds produced by the Eastman Kodak Co. Of three compounds studied, one, designated LL 585, appears most useful for DNA-specific staining. The spectral characteristics of LL 585 in the visible range (data not shown) are similar to propidium iodide (Fig. 2); however, as with the Hoechst dyes, it binds preferentially to A–T sites and does not seem to intercalate into double-stranded DNA. In contrast to HO 33342, LL 585 does not readily penetrate viable cells even after a 3-h staining period. Human lymphoblasts, permeabilized with 0.1% Triton X-100 and stained with LL 585 yielded the expected DNA distributions (coefficient of variation, 2.9%) when analyzed by FCM using the 514-nm line of an argon-ion laser beam. Cells treated with RNase gave essentially the same distribution. The usefulness of the other two Eastman Kodak dyes VL772 and EK4, examined by Latt et al. (61) remains to be further demonstrated.

10.3. Viable Cell Staining for DNA With Hoechst 33342

DNA-specific staining in viable cells, followed by FCM analysis and cell sorting, provides an approach for selecting and recovering living cells from various phases of the cell cycle. In experimental studies the sorted cells could be cultured and examined with regard to long-term viability, functional activity, immunological properties, and other physiological aspects. Such studies would significantly advance the understanding and interpretation of cellular aspects that regulate and control cycle progression and cell proliferation.

Unfortunately, most DNA-specific fluorochromes do not enter viable cells at concentration levels sufficient for optimal staining and analysis purposes. Dyes such as DAPI can penetrate cell membranes, but they are very cytotoxic. Of the fluorochromes utilized in FCM, HO 33342 remains the preferred dye for viable cell staining. However, experimental usage of this compound also has limitations.

Viable cell staining for DNA is usually performed by direct addition of HO 33342 (final concentration, 2.0–5.0 μg/mL) to cells in culture medium (37 °C) for incubation periods of 30–90 min, depending on the cell type. Samples stained at 4 °C yield poor DNA distributions, indicating that conditions favoring active dye transport are required. However, even at optimal conditions, dye uptake, cytotoxicity, DNA binding, and analytical resolution, as judged by coefficients of variation (CV) in the FCM–DNA histograms, are cell-type dependent. The permeability factor is extremely variable and CVs for stained cell populations can range from 4.6 (in cultured colon 26 cells) to 14.0% (in CHO cells). The accuracy of sorting cells from specific cycle phases is seriously impaired when the quality (i.e., CV value) of the DNA distribution is poor. Also the stoichiometry of the staining reaction is questionable in such cases.

If the cell membrane is perforated with nonionic detergents (i.e., NP40, Triton X-100) or ethanol, staining in most cell types is rapid (i.e., 5–10 min) even at dye concentrations of 0.5 μg/mL. Cells with damaged membranes also stain rapidly. Fixed CHO cells stained with HO 33342 can yield DNA content histograms with CVs as low as 2%. These results indicate that membrane permeability of different cell types is responsible for observed variations in staining and not DNA binding *per se.*

Heterogeneous cell populations, as contained in bone marrow samples, present a special problem for obtaining uniform population stainability because of the variability in HO 33342 uptake by the different cell types. In bivariate DNA content and cellular light scatter studies, Visser (*95*) demonstrated that many of the large myeloid progenitor cells did not stain as intensely as the smaller lymphocytes, even after several hours of staining at pH 6.7. Later Visser et al. (96) improved dye uptake in the larger cells by increasing the pH to 7.2; however, cell survival at the elevated pH, even in the absence of HO 33342, was significantly decreased (Visser, personal communication). Interestingly, Loken (64) took advantage of differences in HO 33342 stainability (i.e., membrane transport) to differentiate viable T and B cells in spleen samples (i.e., the T cells were less fluorescent). These differences were not apparent in the fixed-spleen sample. Lalande et al. (55) have demonstrated that Colcemid-resistance—a diminished drug-uptake phenomena in different clones of CHO cells—varied somewhat in proportion to the uptake of HO 33342 (i.e., resistant clones took up less dye and less Colcemid).

It is not surprising that dyes such as HO 33342 that bind to DNA can also be cytotoxic. Fried et al. (35) studied the effects of the dye on the survival of various cell types, and demonstrated that HeLa S-3 cells were highly resistant, but SK-DHL2 cells were highly sensitive to the fluorochrome. Pallavicini et al. (68) have shown that X-irradiated cells were much more sensitive to HO 33342 than nonirradiated cells, indicating a possible synergistic cytotoxic effect. Flow cytometric studies in our laboratory have shown that cell kill is also dependent on laser excitation intensity. Chinese hamster ovary cell populations stained with 2.0 μg/mL HO 33342 for 90 min (37°C) and then analyzed at 100–200 mW power had a cloning efficiency of 80–85%. Above this power, cell kill increased exponentially so that at 500 mW, population survival was about 2%. Unstained cells exposed to 500 mW had a 80–85% cloning efficiency.

EMT6 cells and macrophages were found to be more sensitive and laser power below 200 mW was necessary for optimal survival. A useful technique developed by Stohr and Vogt-Schaden (80) combines two dyes, HO 33342 and propidium iodide (PI), to differentiate viable and dead cells. Viable cells exclude PI and fluoresce blue, whereas dead cells (i.e., cells with damaged membranes) fluoresce both red and blue. Hoechst fluorescence in dead cells is significantly quenched as a result of energy transfer to PI.

In summary, the use of HO 33342 for staining and recovery of viable cells within specific cell cycle phases is useful in some studies, providing that control studies are designed to test for any potential adverse drug effects.

10.4. Cell Staining With Mithramycin, Chromomycin, or Olivomycin

Mithramycin (MI) (Pfizer, Co., Groton, CT) is a green-yellow fluorescent DNA-reactive antibiotic similar in structure and dye binding characteristics to chromomycin A3 (*47,99*) and olivomycin (*see* Fig. 2). Ward et al. (*98*) have shown that these compounds, when complexed with magnesium ions, preferentially bind to G–C base pairs in DNA by nonintercalating mechanisms. Mithramycin has been studied extensively with regard to its spectral characteristics, specificity for cellular DNA, and as a quantitative DNA stain for use in flow cytometry (*17*). Spectrofluorometric analysis of mithramycin or chromomycin A3–Mg complexes bound to DNA in PBS show two excitation peaks—a minor peak in the UV (about 320 nm) and a major peak at about 445 nm. A broad green-yellow emission spectrum is observed with a peak at 575 nm. Olivomycin by comparison has slightly lower wavelength excitation and emission characteristics (*see* Fig. 2).

We originally used MI to stain cells for DNA content (*18*). Ethanol (25%) was included in the staining solution specifically to permeabilize cell membranes. However, more recently (*16,17*) DNA distributions with improved CV's (3.0% for the new protocol vs 6.0% for the older method) were obtained when CHO cells were fixed with 70% ethanol, removed from the ethanol by centrifugation, and stained. Fixation periods as short as 30 min proved to be adequate. Also, CHO samples stored in fixative for over a year at 4°C produced MI–DNA fluorescence distributions of good quality.

Currently, ethanol-fixed CHO cells are stained 20 min at room temperature and analyzed at 457 nm in solutions of mithramycin (100 μg/mL) in 50 mM Tris-HCl (pH 7.4) containing 20 mM MgCl$_2$. As mentioned, dye uptake by viable cells is insufficient for cell staining and analysis purposes.

Experimental drugs that interact directly with DNA or interfere with DNA metabolism can have effects on subsequent staining. In in vivo studies, Alabaster et al. (*1*) found differences in MI–DNA stainability of untreated (control) L-1210 ascites cells and cells treated with a combination of cytosine arabinoside and adriamycin. An unexpected finding reported by Cunningham et al. (*23*) was more related to the conditions of ethanol fixation prior to MI-staining. They observed an aberrant peak, slightly elevated above the G1 peak position in the fluorescence distributions of human lymphocytes that had been fixed for extended periods at 4 °C. In contrast, this spurious peak was absent or diminished in samples first fixed for 7 h at room temperature or for 2 h at 37 °C prior to refrigeration. Similar reports for other cell types have not been found in the literature, indicating that human lymphocytes may pose a unique problem.

10.5. Staining With Propidium Iodide (PI) or Ethidium Bromide (EB)

Propidium iodide and ethidium bromide are red fluorescent compounds that have similar chemical structures and DNA dye binding properties. Both dyes intercalate between base pairs of both double-stranded DNA and RNA without base specificity (63). The dyes can be used as DNA-specific stains following pretreatment of fixed cells with RNase.

Ethidium bromide (EB) was used by Casperson et al. (8) to determine DNA content in single chromosomes by scanning microspectrofluorometry. Dittrich and Gohde (31) demonstrated the use of the dye for cell cycle analysis studies in flow cytometry. Propidium iodide (PI), a dye used by Hudson et al. (45) in buoyant density procedures for isolating closed circular DNA, was first introduced in FCM in a double-staining technique for analysis of DNA and protein, in which fluorescein isothiocyanate was used for protein labeling. Krishan (53), Fried et al. (36), and Vindelov et al. (94) have developed methods for rapid staining of unfixed cells with PI. Techniques vary with procedures used for permeabilizing cell membranes and/or tissue disaggregation. Vindelov et al. (93) have also proposed the use of PI-stained nucleated trout and chick red blood cells as internal standards for assessing relative DNA content. Alternatively, Frankfurt (34) and Wallen et al. (97) measured double-stranded RNA content in fixed cells stained with PI following DNase pretreatment.

Flow cytometric analysis of PI-stained cells yields DNA (fluorescence) distributions comparable to those obtained for cells stained with several other DNA stains (17). Spectral studies on PI bound to calf thymus DNA in PBS show two excitation peaks—a minor but substantial peak in the UV region (340 nm) and a major peak at about 540 nm. One emission peak was observed at 615 nm (*see* Fig. 1). Solutions of PI (5–20 μg/mL) in PBS containing 50 μg/mL RNase (Worthington, code R) are used to stain ethanol-fixed CHO cells for 30 min at 37°C prior to FCM analysis at 488 nm excitation wavelength (*15,16*). As mentioned, PI does not penetrate viable cells sufficiently for cell-staining purposes. Cells with damaged membranes stain readily with PI, so the dye is useful for differentiating and quantitating viable and dead cells in a given population.

Mazzini and Giordano (*66*) have shown that the quantum efficiency of PI and EB are significantly enhanced in solutions of deuterium oxide. Dye concentrations of 1.0 μg/mL were adequate for cell staining and FCM analysis. This technique can be used for flow measurements providing there is no significant mixing of the deuterium oxide-stain solution in the cell stream with the saline or water in the sheath fluid (Mazzini, personal communication).

10.6. Multiple Fluorochrome Labeling of DNA

Although DNA-specific stains are most often used as single dye agents, several FCM studies have demonstrated advantages in combining some of these dyes. Berkhan (*6*) used HO in combination with EB and both Zante et al. (*100*) and Barlogie et al. (*4*) used MI and EB in protocol designed to take advantage of energy transfer from HO or MI to EB. Under these conditions there was an increase in the red fluorescence signal-to-noise ratio, thereby increasing resolution (i.e., reduced CV) in the single-color fluorescence (DNA) histograms. Gray et al. (*41*) double stained chromosomes with HO 33258 and chromomycin A-3 (CA3), and using dual-beam cytometry, resolved human chromosomes having similar DNA content but different DNA base composition. These studies took advantage of the A–T and G–C binding of Hoechst 33258 and chromomycin A3, respectively. Van Dilla et al. (*91*) also used differences in DNA base composition to distinguish three bacterial strains stained with HO and CA3, and Dean et al. (*28*) used this same dye combination to directly compare cell cycle analysis results from cells stained with several DNA stains.

We have recently used a combination of HO, MI, and PI in a three-color FCM technique to detect rearrangements in chromatin structure. Such structural modifications are reflected in changes in the relative proportion of specific regions on DNA available for dye binding. The design shown in Fig. 3 allows for sequential excitation and emission analysis and alleviates difficulties in spectral resolution of HO 33342 (blue), MI (green), and PI (red) fluorescence. This approach takes advantage of the differences in excitation and/or emission characteristics of each dye (*see* spectra, Fig. 2). For example at the position of the UV beam in Fig. 3, excitation of HO in stained cells is optimal, PI near optimal, and MI suboptimal. However, appropriate filters that transmit only blue fluorescence (400–485 nm range), predominantly pass fluorescence from HO 33342. Exposure of stained cells to the second (457) nm beam excites MI optimally and PI suboptimally; however, only green-yellow fluorescence over the 510–575 nm range, predominantly from MI is analyzed. At 530 nm (beam 3), only PI is excited and red fluorescence (580–800 nm) is analyzed. Ratio analysis of HO/MI (blue/green) or PI/MI (red/green) fluorescence (*15,72*) allows for direct comparison of different regions available for dye binding in DNA on a cell-to-cell basis. Detailed descriptions of experimental design for cell staining, sequential excitation, and fluorescence analysis will be described in a separate publication (manuscript in preparation).

Near optimal, simultaneous three-color staining and analytical conditions for exponentially growing cells ideally should provide independent fluorescence distributions, one for each dye with relatively low coefficients of variation (CV). Also, all distributions should yield comparable estimates of the fractions of cells in the G1-, S-, and G2M-phases. Such results have been obtained from populations of CHO cells (Fig. 4 and Table 1), as well as HL-60 and Friend erythroleukemia cells (data not shown). For these cells, the percentages of cells in the G1-, S-, and G2 + M-phases (Table 1) are in good agreement and the CVs are acceptable for accurate computer analysis. Compared to cells stained separately with each dye, the cells stained with three dyes exhibit a 75–80% decrease in HO fluorescence, a 35–40% decrease in MI, and a decrease of 20–25% in PI fluorescence. This could be a result of either energy transfer and/or nonspecific quenching by unbound dye in stain solution surrounding cells during analysis.

The correlation of fluorescence intensities from the various dyes (same data as for Fig. 4) is illustrated in the computer-generated bivariate displays shown in Fig. 5A–E. In addition, Fig. 6 shows a bivariate PI/MI fluorescence ratio vs HO fluorescence for a population of CHO cells treated for 4 h with Colcemid to accumulate a subpopulation of cells in mitosis. The PI/MI ratio for the exponentially growing untreated cells (Fig. 5E) is higher than for cells in

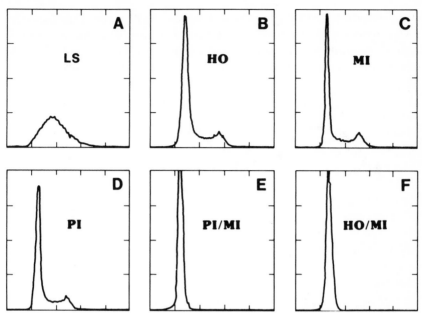

Fig. 4. Univariate frequency distributions of light scatter, LS (A); Hoechst 33342, HO (B); mithramycin, MI (C); propidium iodide, PI (D); fluorescence and fluorescence ratios of PI/MI (E); and HO/MI (F), obtained by analysis of stained CHO cells. Exponentially growing CHO cells were fixed in 70% ethanol and subsequently stained with a solution containing all three DNA stains. Sequential excitation and analysis were performed as described in the text. Coefficients of variation and cell cycle frequency distributions are shown in Table 1.

TABLE 1
Comparison of Computer-Fit Analysis of DNA Histograms
Shown in Fig. 4[a]

	CV,	Percentage of cells			Ratio
	%	G1	S	G2 + M	G2 + M to G1
HO	8.5	60.2	32.1	7.7	2.02
MI	6.6	59.2	34.0	6.8	1.98
PI	5.9	60.4	33.8	5.8	1.98

[a]*See* chapter 8 for details of DNA distribution analysis.

mitosis (Fig. 6B) with highly condensed chromatin. This subpopulation of mitotic cells is also observed in bivariate MI vs PI distributions (Fig. 6A). The single-variable PI-fluorescence distribution showed a broadening in the G2 + M-phase and the expected increase in the fraction of cells in the G2 + M region. Details of chromatin changes detected in differentiating HL-60 and FL cells will be presented elsewhere (manuscript in preparation).

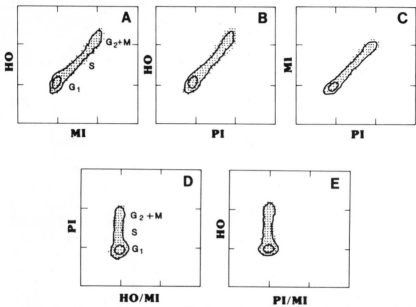

Fig. 5. Bivariate contour displays showing the correlation of the various DNA dyes (A–C) in three-color stained CHO cells and ratio for HO/MI (D) and PI/MI (E) throughout the cell cycle as derived by the HO (D) and PI (E) fluorescence. Data are the same as used for Fig. 3.

11. TWO-COLOR DNA AND PROTEIN STAINING

Simultaneous FCM analysis of DNA and protein allows for direct assessment of the cell growth at various phases of the cell cycle. Within heterogeneous populations, cells with distinct protein content variations can also be identified. Such determinations have been used in a number of studies involving a wide variety of cell types. For most studies, two-color staining of DNA and protein for cells in suspension has been accomplished using a method that combines ethidium bromide (EB) and fluorescein isothiocyanate (FITC) (37) or propidium iodide (PI) and FITC (15). Stohr et al. (81), however, have also evaluated several other dye combinations. Of

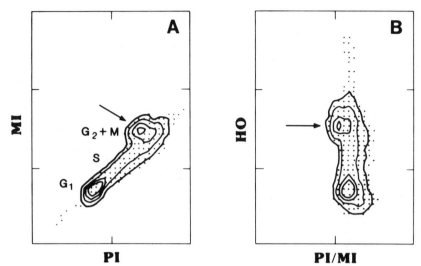

Fig. 6. Detection of mitotic cells in populations stained with three DNA dyes. Chinese hamster ovary cells were cultured 4 h in medium containing 0.1 μg/mL Colcemid to accumulate cells in mitosis. Mitotic cells (indicated by arrow) show a decrease in PI stainability compared to MI, and are thus detected by bivariate analysis.

these, DAPI and SR101 was the most rapid combination and gave the best results. Recently, Roti-Roti et al. (*70*) have also used a PI–FITC combination to examine and directly compare nuclear protein and DNA contents of cells in various phases of the cell cycle. Recently we described rapid, one-step staining procedures for simultaneous cellular DNA-protein analysis using single- or dual-laser beam FCM (*16*). RNase-treated cells were stained with PI and FITC for single-laser FCM analysis. Alternatively, the DNA stains, MI or HO 33342, were applied in combination with either of the red protein stains, X-rhodamine isothiocyanate (X-RITC) or rhodamine 640 (R640) without RNase, and the cells were analyzed by dual-laser FCM. Rhodamine 640 and HO 33342 were also used to stain unfixed cells. In all cases the dye concentrations were adjusted to allow for staining without the usual washing and centrifugation steps. Cells were then analyzed in the stain solution. For comparative purposes the cell concentration in the staining solution was kept constant from sample to sample. Spectral properties of these protein dyes are shown in Fig. 2.

We have previously shown the advantage of simultaneous DNA–protein analysis for detecting myeloma cells in human bone marrow (21). Recently we used sequential dual-laser excitation (i.e., UV and 488 nm) and HO 33342–FITC staining for DNA and protein analysis. Using this method, we resolved subpopulations of cells in samples analyzed from non-Hodgkins lymphoma patients, such as shown in Fig. 7 (univariate) and Fig. 8 (bivariate). The total DNA distribution pattern in Fig. 7A provides no evidence of the two subpopulations indicated in the LS, protein, and protein-to-DNA ratios. Conversely, in the absence of DNA content analysis, the cycling activity of these subpopulations cannot be determined. However, gated analysis (Figs. 7B and C) or bivariate analysis (Fig. 8) reveal at least two populations of cells, both having cells in G1, S, and G2 + M. Tritiated-thymidine studies, FCM analyses, cell sorting, and autoradiography could be used to determine the actual cyling capacity, as well as the sensitivity of the two populations to various chemotherapeutic agents.

12. CORRELATED ANALYSIS OF DNA, RNA, AND PROTEIN

In chapter 9, Darzynkiewicz (24) describes the use of acridine orange (AO) for staining and FCM analysis of DNA vs RNA. Data presented show the interaction of the nucleic acids in cell-cycle-related events. There also appears to be a close coupling of the metabolic patterns of protein content with RNA and DNA content in maintenance of the state of balanced growth. However, under experimental conditions such as drug treatment, the DNA synthetic and cell division patterns are often grossly perturbed, resulting in an uncoupling of transcriptional and translational activity (19). Correlated studies on cellular DNA, RNA, and protein are extremely useful in detecting such abnormalities in the cell cycle, as well as for elucidating normal cell growth and cycle progression patterns.

We have recently developed a technique for simultaneous analyses of DNA, RNA, and protein (20) based on modifications of FCM methods described by Arndt-Jovin and Jovin (3) for HO 33342–DNA staining; and by Tanke et al. (85), Shapiro (73), and Pollack et al. (69) for pyronin Y (PY)–RNA staining, and by Gohde et al. (37) and us (15,16) for FITC protein staining. Sequential excitation of stained cells (UV, 458 and 530 nm) is followed by fluorescence analysis of HO–DNA (blue), FITC–protein (green), and PY–RNA (red), respectively (see spectra in Fig. 9). Fluorescence ratio

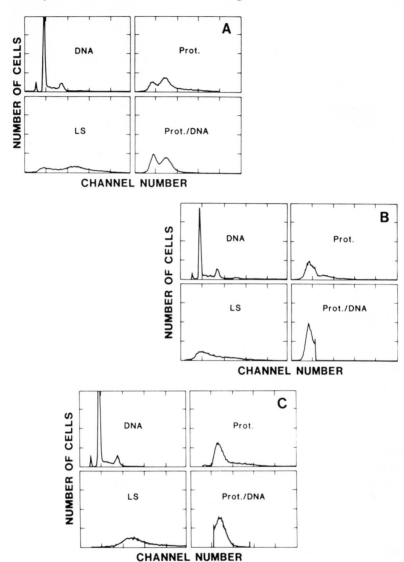

Fig. 7. Univariate distribution of DNA content, protein, light scatter (LS), and the protein/DNA ratio obtained for stained cells from non-Hodgkins lymphoma sample. Ethanol-fixed cells were stained simultaneously with HO 33342 (0.5 μg/mL) and FITC (0.08 μg/mL) for 20 min prior to analysis using the UV and 488 nm (dual-laser) sequential excitation. Results for the total population are shown in (A). Gated analysis results for subpopulations with a low (B) or high (C) protein-to-DNA ratio are also provided.

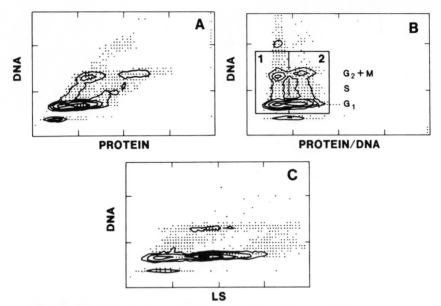

Fig. 8. Bivariate contour displays showing the correlation of protein and DNA contents (A), protein/DNA ratio vs DNA (B), or light scatter and DNA (C) from same data used for Fig. 6. The analysis in (B) detects at least two subpopulations with relatively low or high protein/DNA ratios, and with cells in G1, S, and G2 + M as well.

analysis allows for direct correlation of RNA/protein (red/green) and RNA/DNA contents (red/blue) throughout the cell cycle.

Comparisons of results obtained from analyses of CHO cells stained with each dye alone vs those with all three dyes simultaneously demonstrate the good resolution that can be obtained with the staining protocol combining these three dyes (*see* Fig. 10A–D). The Hoechst-DNA fluorescence distribution is essentially unchanged (compare Figs. 10A and D). A slight decrease in FITC fluorescence occurs in the multiply-stained cells (Fig. 10B and D); probably because of energy transfer to PY. Comparison of Figs. 10C and D show a slight increase in PY fluorescence in the triply-stained cell population. In general, however, the shape of the distributions is similar, demonstrating adequate fluorescence resolution to quantify DNA vs RNA vs protein.

This staining and analysis technique was used to characterize and compare cycling and noncycling (G1 arrested) CHO cells, and

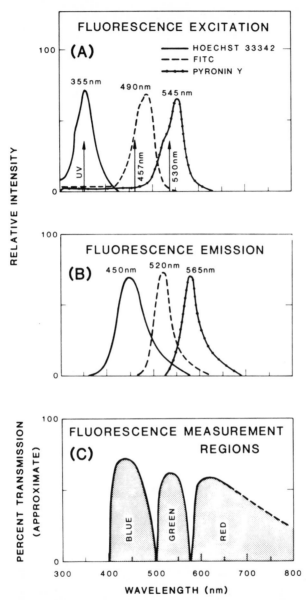

Fig. 9. Excitation (A) and emission (B) spectra for Hoechst 33342, fluorescein isothiocyanate (FITC), and pyronin Y. Selected fluorescence measurement regions (C) are used for analysis of the respective dyes.

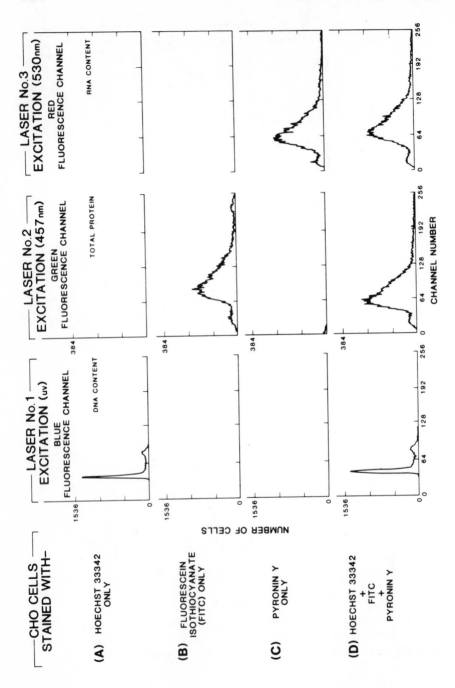

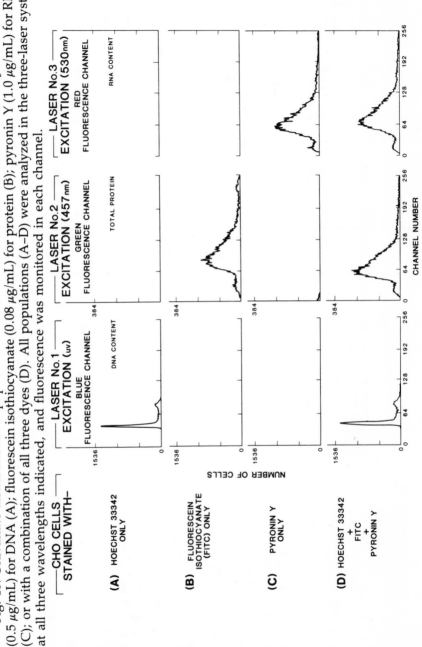

Fig. 10. Univariate distributions for populations of ethanol-fixed CHO cells stained with *only* Hoechst 33342 (0.5 µg/mL) for DNA (A); fluorescein isothiocyanate (0.08 µg/mL) for protein (B); pyronin Y (1.0 µg/mL) for RNA (C); or with a combination of all three dyes (D). All populations (A–D) were analyzed in the three-laser system at all three wavelengths indicated, and fluorescence was monitored in each channel.

to correlate cellular DNA, RNA, and protein, as well as the ratios of RNA/DNA and RNA/protein throughout the cell cycle for both populations (*see* Fig. 11). It is evident that G1 cells with a low RNA

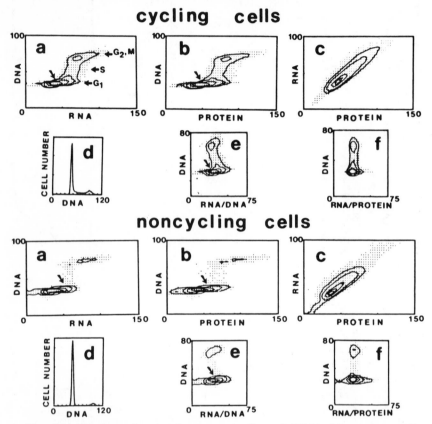

Fig. 11. Contour isometric maps (a–f) and DNA histograms (d) representing the distribution of cells with respect to DNA, RNA, and protein content of CHO cells from exponentially growing cultures ("cycling cells") and from cultures deprived of isoleucine for 36 h ("noncyling cells"). The dotted areas (at least 5 cells/dot) enclosed in the sequential contours represent increasing isometric levels equivalent to 10, 50, 250, 500, and 1500 cells, respectively; 40,000 cells were measured per sample. There were 57% cells in G1, 24% in S, and 19% in G2 + M in the exponentially growing culture, as compared with 93% in G1, 2% in S, and 5% in G2 + M in the isoleucine-deprived population. The arrows indicate the threshold RNA or protein content of G1 cells; the cells with RNA or protein below the threshold values do not immediately enter S-phase. The experiment was repeated three times with essentially similar results.

or protein content do not enter the S-phase. Detailed characterization of the critical RNA and protein thresholds for entry of G1 cells into S-phase are presented by Darzynkiewicz in chapter 9. The G1-arrested (noncyling) cells show a much greater heterogeneity in RNA and protein compared to the cycling population; however, in both populations the RNA-to-protein contents are well correlated.

Analysis of the cellular RNA/DNA ratio in relation to DNA content reveals a characteristic pattern reflecting changing rates of DNA replication and transcription during the cell cycle. Thus, during G1 when DNA content is stable, cells accumulate increased quantities of RNA, but at different rates so that the G1 phase is quite heterogeneous with respect to RNA. However, during progression through S-phase, the rate of DNA replication exceeds RNA accumulation, giving rise to a nonvertical, negative slope of the S-phase cell cluster. Cells in G2 + M have RNA/DNA ratios in the same range as the majority of the G1 cells.

The RNA/protein ratio throughout the cell cycle remains quite constant, as would be expected for cell populations in a balanced growth state. However, CHO cells arrested in mitosis with Colcemid do show a lower RNA/protein ratio compared to G2 cells (Fig. 12B). Ratios for untreated populations are shown in Fig. 12A. The method represents a very powerful approach for studying mechanisms and control of the cell cycle. In addition, we have used this technique to study tumor cell growth and cellular differentiation.

13. CORRELATED ANALYSIS OF DNA, RNA, OR PROTEIN AND MITOCHONDRIA

Johnson et al. (50) first reported the use of rhodamine 123 (R123) as a fluorescent probe for selectively staining mitochondria in viable cells. In subsequent studies, Johnson et al. (49) further characterized mitochondrial specificity of R123 and several other rhodamine and cyanine dyes. These cationic dyes accumulate as a result of the electronegativity of the inner mitochondrial membranes. Loss of the negative charge or interference with electron transport results in a loss or decrease in fluorescence intensity. These findings show that these cationic dyes do not attach to specific molecules, as in the case of DNA, RNA, or protein stains. Therefore an increase in fluorescence of stained cells is a function of the increase in mitochondrial number and/or electronegativity.

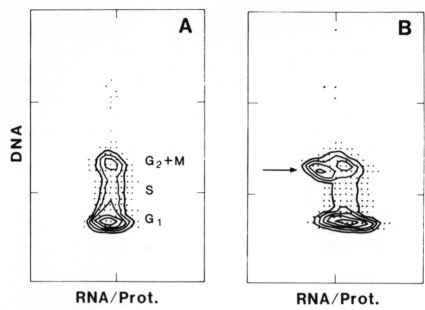

Fig. 12. RNA/protein ratios vs DNA for untreated CHO cells (A) and cells treated 4.5 h with 0.1 μg/mL Colcemid (B). The subpopulation of accumulated mitotic cells (B, arrow) shows a decreased RNA/protein ratio as well as decreased fluorescence of DNA–Hoechst compared to G2 cells.

Darzynkiewicz et al. (25), in FCM studies, showed a dramatic increase in R123 fluorescence of PHA-stimulated lymphocytes concominant to lymphoblast activation. Based on these studies, they suggested the use of R123 as a probe for distinguishing between cycling and quiescent cells. James and Bohman (46) also used FCM techniques to study proliferation of mitochondria during the cell cycle of HL-60 cells. Darzynkiewicz et al. (27) later showed decreased mitochondrial (R123) fluorescence in differentiating FL cells, as well as in stationary L-1210 and CHO cells compared to the respective exponentially growing populations.

In an effort to correlate DNA, RNA, or protein and mitochondrial activity, we incorporated the green fluorescent dye R123 into the protocol described by Shapiro (73) for viable cell-staining of DNA and RNA, respectively, with HO 33342 and PY. Alternatively, for protein analysis the red fluorescent dye rhodamine 640 (R640) was applied to the viable cells in place of PY. The three-laser excitation was employed as described using the 457 nm laser line for

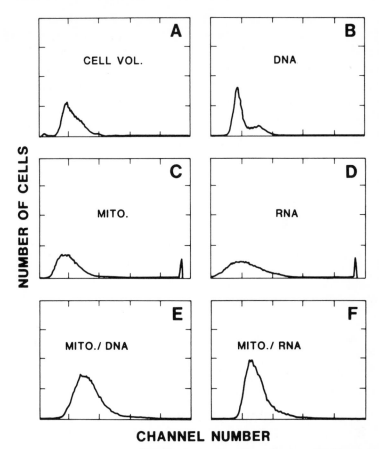

CHANNEL NUMBER

Fig. 13. Univariate distribution of cell volume, DNA content, mitochondrial (R123) fluorescence, and ratios of mitochondrial (green) to stained DNA (HO) fluorescence and mitochondrial to pyronin Y-RNA (red) fluorescence in viable CHO cells. Growing cells in suspension were stained in culture for 1 h with HO 33342 (0.5 μg/mL) and PY (2 μg/mL). Rhodamine 123 (R123) was then added for 30 additional min prior to analysis in the dye-containing medium. Cells were analyzed in the three-laser system set at the excitation wavelengths shown in Fig. 3.

R123 excitation and the 528 line for PY or the 568 line for R640 excitation (*see* fluorescence spectra, Fig. 2). Frequency distributions for volume, DNA, and RNA contents, mitochondrial (R123) fluorescence, and ratios of mitochondrial/DNA (green/blue) and mitochondria/RNA (green/red) for CHO cells are shown in Fig. 13.

196

For cells in balanced growth, the protein content (R640 analysis) did not significantly differ from the PY–RNA content distributions shown here. Mitochondria fluorescence correlated well, but not directly with cell volume and RNA content. However, it was weakly correlated with DNA content. Bivariate analysis (Fig. 14) allows for direct comparison of the mitochondrial activity as a function of the cell cycle (Figs. 14A and C), or as related to cell volume (Figs. 14B

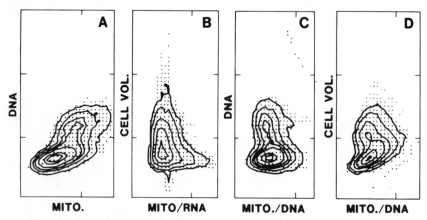

Fig. 14. Bivariate distributions graphically illustrating the correlation of DNA, RNA, and cell volume to mitochondrial (MITO) fluorescence. Data are the same as used for Fig. 13.

and D). The mitochondria/RNA ratio appears to remain constant with increasing cell size (Fig. 14B). However, the mitochondria/ DNA ratio shows two slopes with increasing cell size (Fig. 14D). The negative slope in larger cells is presumably caused, in part, by the more rapid increase in DNA content during S-phase. Extension of these preliminary studies will prove useful in examining populations of drug-treated and irradiated populations. Care should be taken in interpretation of mitochondrial fluorescence, since agents that are potentially cytotoxic may induce alterations in membrane potential that change mitochondrial fluorescence without any real fluctuation in mitochondrial number and/or size.

14. CELL CYCLE ANALYSES USING BROMODEOXYURIDINE (BrdUrd)

Cells cultured in medium containing BrdUrd will incorporate the analog in place of thymidine during the DNA synthetic (S)-

phase. Latt (58) showed that the fluorescence of HO 33258 was quenched when bound to BrdUrd-substituted DNA in chromosomes. Later, Latt et al. (60) used FCM analysis and demonstrated the detection of cycling BrdUrd-containing cells as based on BrdUrd quenching of HO 33342. Swartzendruber (82,83) observed that CHO cells cultured in 30 μM BrdUrd for one- and two-cell cycle division periods showed an increase in MI fluorescence intensity of 25 and 40%, respectively. These results suggest that BrdUrd substitution for thymidine in DNA induces structural rearrangement in chromatin that allows for increased binding by MI. Based on the spectral properties of BrdUrd and MI, and the distant proximity of G–C-bound-MI to BrdUrd sites on DNA, it is difficult to attribute increased MI fluorescence to energy transfer from BrdUrd. Latt (59) has previously suggested that BrdUrd substitution in DNA may promote conformational changes in chromatin, thus favoring the hypothesis mentioned above.

Recently Dolbeare et al. (32) developed a sensitive method for simultaneous staining for cellular BrdUrd and DNA content using FITC-conjugated to a monclonal antibody to BrdUrd (anti-BrdUrd) and PI. The method is based on the development by Gratzner (38) of an immunologic procedure to stain for BrdUrd incorporated into cellular DNA. PI–DNA staining and bivariate analysis (488 nm excitation) showed the antibody was bound to cells in S-phase. Dolbeare et al. (32) used only short, 10–30 min periods for exposure of cells in culture to BrdUrd (10 μM concentrations). This illustrates the very good sensitivity of the method [*see* chapter 5 (40) for further details and examples of this technique].

Alternatively, we have simultaneously used the three DNA stains and analytical methods described in section 10.1.5 to directly detect quenching of HO by BrdUrd-substituted DNA. The HO/MI fluorescence ratio is sensitive for detecting HO quenching in cells, since MI–DNA fluorescence is not significantly affected by short-term (i.e., 1 h) exposure of cultured cells to BrdUrd (82). Thus, on a cell-to-cell basis, the HO fluorescence intensity is normalized to the MI intensity, which remains proportional to DNA content. Unaffected PI fluorescence stochiometrically related to DNA content accurately reflects the distribution of cells in various phases of the cell cycle. Bivariate analysis of CHO cells exposed to 10, 30, or 50 μM BrdUrd for 1 or 2 h, as indicated, are shown in Fig. 15. A decrease in HO/MI fluorescence ratio, compared to the ratios for untreated cells shown in Fig. 5D, is noted for cells in S-phase as defined by the PI-DNA distribution (Y axis). Currently, we are using this technique in studies to examine cell growth in tumors and bone marrow, as well as in cell populations exposed to drugs and X-irradiation.

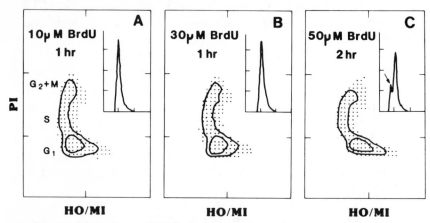

Fig. 15. Detection of BrdUrd-substituted DNA quenching of Hoechst 33342 fluorescence, in S-phase CHO cells. Cells were cultured for 1 or 2 h in medium containing 10, 30 or 50 μM BrdUrd prior to harvest, ethanol-fixation and simultaneous staining with the three DNA dyes. Univariate distributions of HO/MI fluorescence ratios are inserted in (A), (B), and (C). Fluorescence of MI and PI were not significantly affected when bound to the BrdUrd-substituted DNA.

15. CONCLUSIONS

In this chapter we have discussed some current staining methods used for analysis of cellular DNA, RNA, protein, and mitochondria.

The data presented serve only as illustrations of approaches that might provide new and useful information for cell cycle related studies. Although some analyses involved a three-laser system, the techniques are applicable for univariate or bivariate analysis.

ACKNOWLEDGMENTS

The authors thank T. Robinson, C. Goolsby, and C. Stewart for their comments and helpful suggestions on this chapter. The work was performed under the auspices of the Los Alamos National Flow Cytometry and Sorting Research Resource, funded by the Division of Research Resources of NIH (Grant P41-RR01315-02) and the US Department of Energy.

REFERENCES

1. Alabaster, O., Tannenbaum, E., Habbersett, M. C., Magrath, I., and Herman, C. Drug-induced changes in DNA fluorescence intensity detected by flow microfluorometry and their implications for analysis of DNA content distributions. Cancer Res., *38*: 1031–1035, 1978.

2. Arndt-Jovin, D. J., Grimwade, B. G., and Jovin, T. M. A dual laser flow sorter utilizing a CW pumped dye laser. Cytometry, *1*: 127–131, 1980.

3. Arndt-Jovin, D. J. and Jovin, T. M. Analysis and sorting of living cells according to deoxyribonucleic acid content. J. Histochem. Cytochem., *25*: 585–589, 1977.

4. Barlogie, B., Spitzer, G., Hart, J. S., Johnston, D., Buchner, T., and Schumann, J. DNA histogram analysis of human hemopoietic cells. Blood, *48*: 245–258, 1976.

5. Bensen, R. C., Meyer, R. A., Zaruba, M. E., and McKhann, G. M. Cellular autofluorescence—Is it due to flavins? J. Histochem. Cytochem., *27*: 44–48, 1979.

6. Berkhan, E. Pulse-cytophotometry as a method for rapid photometric analysis of cells. *In*: (C. A. M. Haanen, H. F. P. Hillen, and J. M. C. Wessels, eds.), Pulse Cytophotometry. Ghent: European Press, 1975.

7. Brachet, J. La detection histochimique des acides pintose nucleiques. Comptes Rendus Des Seances De La Societe De Biologie, *133*: 89–90, 1940.

8. Casperson, T., Zech, L., Modest, J., Foley, G. E., Wagh, U., and Simonsson, E. DNA-binding fluorochromes for the study of the organization of the metaphase nucleus. Exp. Cell. Res., *58*: 141–152, 1969.

9. Casperson, T. and Santesson, L. Studies on protein metabolism in the cells of epithelial tumors. Acta Radiol. (suppl.), *46*: 1–105, 1942.

10. Casperson, T. and Schultz, J. Nucleic acid metabolism of the chromosomes in relation to gene reproduction. Nature, *142*: 294–295, 1938.

11. Coulson, P. A., Bishop, A. O., and Lenarduzzi, R. Quantitation of cellular deoxyribonucleic acid by flow cytometry. J. Histochem. Cytochem., *25*: 1147–1153, 1977.

12. Cram, L. S., Bartholdi, M. F., Wheeless, L. L., and Gray, J. W. Morphological analysis by scanning flow cytometry. *In*: (M. A. Van Dilla, P. N. Dean, O. D. Laerum, and M. R. Melamed, eds.), Flow Cytometry Instrumentation and Data Analysis, New York: Academic, 1985.

13. Crissman, H. A., Mullaney, P. F., and Steinkamp, J. A. Methods and applications of flow systems for analysis and sorting of mammalian cells. *In*: (D. M. Prescott, ed.), Methods in Cell Biology, vol. 9, New York: Academic, 1975.

14. Crissman, H. A., Orlicky, D. A., and Kissane, R. J. Fluorescent DNA probes for flow cytometry. J. Histochem. Cytochem., *27*: 1652–1654, 1979.

15. Crissman, H. A. and Steinkamp, J. A. Rapid simultaneous measurement of DNA, protein and cell volume in single cells from large mammalian cell populations. J. Cell Biol., *59*: 766–771, 1973.

16. Crissman, H. A. and Steinkamp, J. A. Rapid, one step staining procedures for analysis of cellular DNA and protein by single and dual laser flow cytometry. Cytometry, *3*: 84–90, 1982.

17. Crissman, H. A., Stevenson, A. P., Kissane, R. J., and Tobey, R. A. Techniques for quantitative staining of cellular DNA for flow cytometric analysis. *In*: (M. R. Melamed, P. F. Mullaney, and M. L. Mendelsohn, eds.), Flow Cytometry and Sorting. New York: John Wiley, 1979.

18. Crissman, H. A. and Tobey, R. A. Cell cycle analysis in 20 minutes. Science, *184*: 1297–1298, 1974.

19. Crissman, H. A., Darzynkiewicz, Z., Tobey, R. A., and Steinkamp, J. A. Normal and perturbed CHO cells: Correlation of DNA, RNA and protein by flow cytometry. J. Cell. Biol., *101*: 141–147, 1985a.

20. Crissman, H. A., Darzynkiewicz, Z., Tobey, R. A., and Steinkamp, J. A. Correlated measurements of DNA, RNA and protein in individual cells by flow cytometry. Science, *228*: 1321–1324, 1985b.

21. Crissman, H. A., Von Egmond, J. V., Holdrinet, R. G., Pennings, A., and Hannen, C. Simplified method for DNA and protein staining of human hematopoietic cell samples. Cytometry, *2*: 59–62, 1981.

22. Culling, C. and Vassar, P. Desoxyribose nucleic acid. A fluorescent histochemical technique. Arch. Pathol., *71*: 88–92, 1961.

23. Cunningham, R. E., Skramstad, K. S., Newburger, A. E., and Schackney, S. E. Artifacts associated with mithramycin fluorescence in the clinical detection and quantitation of aneuploidy by flow cytometry. J. Histochem. Cytochem., *30*: 317–322, 1982.

24. Darzynkiewicz, Z. Cytochemical probes of cycling and quiescent cells applicable for flow cytometry. *In*: (J. W. Gray and Z. Darzynkiewicz, eds.), Techniques in Cell Cycle Analysis. New Jersey: Humana, 1986.

25. Darzynkiewicz, Z., Staiano-Coico, L., and Melamed, M. R. Increased mitochondrial uptake of rhodamine 123 during lymphocyte stimulation. Proc. Natl. Acad. Sci. USA, *78*: 2383–2387, 1981.

26. Darzynkiewicz, Z., Traganos, F., Sharpless, T., and Melamed, M. R. Lymphocyte stimulation: A rapid, multiparameter analysis. Proc. Natl. Acad. Sci. USA, *76*: 358–362, 1976.

27. Darzynkiewicz, Z., Traganos, F., Staiano-Coico, L., Kapuscinski, J., and Melamed, M. R. Interaction of rhodamine 123 with living cells studied by flow cytometry. Cancer Res., *42*: 799–806, 1982.

28. Dean, P. N., Gray, J. W., and Dolbeare, F. A. The analysis and interpretation of DNA distributions measured by flow cytometry. Cytometry, *3*: 188–195, 1982.

29. Dean, P. N. and Jett, J. H. Mathematical analysis of DNA distributions derived from flow microfluorometry. J. Cell Biol., *60*: 523–527, 1974.

30. Dean, P. N. and Pinkel, D. High resolution dual laser flow cytometry. J. Histochem. Cytochem., *26*: 622, 1978.

31. Dittrich, W. and Gohde, W. Impulse fluorometry with single cells in suspension. Z. Naturforsch, *24B*: 360, 1969.

32. Dolbeare, F., Gratzner, H. G., Pallavicini, M. G., and Gray, J. W. Flow cytometric measurement of total DNA content and incorporated bromodeoxyuridine. Proc. Natl. Acad. Sci. USA, *80*: 5573–5577, 1983.

33. Feulgen, R. and Rossenbeck, H. Mikroskopisch-chemischer nachweis einer nucleinsaure von typus der thymonucleinsaure und die darauf beruhende elektive farbung von zellkerzen in mikroskopischen praparaten. Z. Physiol. Chem., *135*: 203–244, 1924.

34. Frankfurt, O. S. Flow cytometric analysis of double-stranded RNA content distributions. J. Histochem. Cytochem., *28*: 663–669, 1980.

35. Fried, J., Doblin, J., Takamoto, S., Perez, A., Hansen, H., and Clarkson, B. Effect of Hoechst 33342 on survival and growth of two tumor cell lines and on hematopoietically normal bone marrow cells. Cytometry, *3*: 42–47, 1982.

36. Fried, J., Perez, A. G., and Clarkson, B. D. Flow cytofluorometric analysis of cell cycle distributions using propidium iodide. Properties of the method and mathematical analysis of the data. J. Cell Biol., *71*: 174–181, 1976.

37. Gohde, W., Spies, I., Schumann, J., Buchner, T., and Klein-Dopke, G. Two parameter analysis of DNA and protein content of tumor cells. *In*: (T. Buchner, W. Gohde, and J. Schumann, eds.), Pulse-Cytophotometry, Gent, Belgium: European Press, 1976.

38. Gratzner, H. G. Monoclonal antibody to 5-bromo- and 5-iododeoxy-uridine: A new reagent for detection of DNA replication. Science, 218: 474–475, 1982.

39. Gray, J. W. Cell cycle analysis of perturbed cell populations: Computer stimulation of sequential DNA distributions. Cell Tissue Kinet., 9: 499, 1976.

40. Gray, J. W., Dolbeare, F., Pallavicini, M., and Vanderlaan, M. Flow cytokinetics. In: (J. W. Gray and Z. Darzynkiewicz, eds.), Techniques in Cell Cycle Analysis, New Jersey: Humana, 1986.

41. Gray, J. W., Langlois, R. G., Carrano, A. V., Burkhart-Schultz, K., and Van Dilla, M. A. High resolution chromosome analysis: One and two parameter flow cytometry. Chromosoma, 73: 9–27, 1979.

42. Hedley, D. W., Friedlander, M. L., Taylor, I. W., Rugg, C. A., and Musgrove, E. A. Method for analysis of cellular DNA content of paraffin embedded pathological material using flow cytometry. J. Histochem. Cytochem., 31: 1333–1335, 1983.

43. Hiddemann, W., Schumann, J., Andreeff, M., Barlogie, B., Herman, C. J., Jeff, R. C., Mayall, B., Murphy, R. F., and Sandberg, A. Convention on nomenclature for DNA cytometry. Cytometry, 5: 445–446, 1984.

44. Hilwig, I. and Gropp, A. Decondensation of constitutive heterochromatin in L cell chromosomes by a benzimidazole compound ("33258 Hoechst"). Exp. Cell Res., 81: 474–477, 1973.

45. Hudson, B., Upholt, W. B., Deninny, J., and Vinograd, J. The use of an ethidium bromide analogue in the dye-bouyant density procedure for the isolation of closed circular DNA. The variation of the superhelix density of mitochondrial DNA. Proc. Natl. Acad. Sci. USA, 62: 813–820, 1969.

46. James, T. W. and Bohman, R. Proliferation of mitochondria during the cell cycle of the human cell line (HL-60). J. Cell Biol., 89: 256–260, 1981.

47. Jensen. R. H. Chromomycin A_3 as a fluorescent probe for flow cytometry of human gynecological samples. J. Histochem. Cytochem., 25: 573–579, 1977.

48. Jensen, R. H., Langlois, R. G., and Mayall, B. H. Strategies for choosing a DNA stain for flow cytometry of metaphase chromosomes. J. Histochem. Cytochem., 25: 954–964, 1977.

49. Johnson, L. V., Walsh, M. L., Bockus, B. J., and Chen, L. B. Monitoring of relative mitochondrial membrane potential in living cells by fluorescence microscopy. J. Cell Biol., 88: 526–535, 1981.

50. Johnson, L. V., Walsh, M. L., and Chen, L. B. Localization of mitochondria in living cells with rhodamine 123. Proc. Natl. Acad. Sci. USA, 77: 990–994, 1980.

51. Kamentsky, L. A., Melamed, M. R., and Derman, H. Spectrophotometer: New instrument for ultrarapid cell analysis. Science, *150*: 630–631, 1965.

52. Kraemer, P., Deaven, L., Crissman, H., and Van Dilla, M. DNA constancy despite variability in chromosome number. *In*: (E. J. DuPraw, ed.), Advances in Cell and Molecular Biology, vol. 2, New York: Academic, 1972.

53. Krishan, A. Rapid flow cytophotometric analysis of mammalian cell cycle by propidium iodide staining. J. Cell Biol., *66*: 188–193, 1975.

54. Kurnick, N. B. Histological staining with methylgreen-pyronin. Stain Technol., *27*: 233–242, 1952.

55. Lalande, M. E., Ling, V., and Miller, R. G. Hoechst 33342 dye uptake as a probe of membrane permeability changes in mammalian cells. Proc. Natl. Acad. Sci. USA, *78*: 363–367, 1981.

56. Langlois, R. G. and Jensen, R. H. Interaction of DNA specific fluorescent stains bound to mammalian cells. J. Histochem. Cytochem., *27*: 72–79, 1979.

57. Latreille, J., Barlogie, B., Johnston, D., Drewinko, B., and Alexanian, R. Ploidy and proliferative characteristics in monoclonal gammopathies. Blood, *59*: 43–51, 1982.

58. Latt, S. A. Microfluorometric detection of deoxyribonucleic acid replication in human metaphase chromosomes. Proc. Natl. Acad. Sci. USA, *70*: 3395–3399, 1973.

59. Latt, S. A. Fluorescent probes of DNA microstructure and synthesis. *In*: (M. R. Melamed, P. F. Mullaney, and M. L. Mendelsohn, eds.), Flow Cytometry and Sorting, New York: John Wiley, 1979.

60. Latt, S. A., George, Y. S., and Gray, J. W. Flow cytometric analysis of bromodeoxyuridine-substituted cells stained with 33258 Hoechst. J. Histochem. Cytochem., *25*: 927–934, 1977.

61. Latt, S. A., Marino, M., and Lalande, M. New fluorochromes compatible with high wavelength excitation for flow cytometric analysis of cellular nucleic acids. Cytometry, *5*: 339–347, 1984.

62. Latt, S. A. and Stetten, G. Spectral studies on 33258 Hoechst and related bisbenzimidazole dyes for fluorescent detection of deoxyribonucleic acid synthesis. J. Histochem. Cytochem., *24*: 24–33, 1976.

63. LePecq, J. B. and Paoletti, C. A fluorescent complex between ethidium bromide and nucleic acids. J. Mol. Biol., *27*: 87–106, 1967.

64. Loken, M. R. Separation of viable T and B lymphocytes using a cytochemical stain, Hoechst 33342. J. Histochem. Cytochem., *28*: 36–39, 1980.

65. Mayall, B. H. and Mendelsohn, M. L. Errors in absorption cytophotometry: Some theoretical and practical considerations. *In*: (G. L.

Wied and G. G. Bahr, eds.), Introduction to Quantitative Cytochemistry II, New York: Academic, 1970.

66. Mazzini, G. and Giordano, P. Effects of some solvents on fluorescence intensity of phenoantridinic derivatives-DNA complexes: Flow cytometric applications. *In*: (O. D. Laerum, T. Lindmo, and E. Thorud, eds.), Flow Cytometry IV, Bergen, Norway: Universitetsforlaget, 1979.

67. Mueller, W. and Gautier, F. Interaction of heteroaromatic compounds with nucleic acids. A-T specific nonintercalating DNA ligands. Eur. J. Biochem., *54*: 385–394, 1975.

68. Pallavicini, M. G., Lalande, M. E., Miller, R. G., and Hill, R. P. Cell cycle distribution of chronically hypoxic cells and determination of the clonogenic potential of cells accumulated in G2 + M phases after irradiation of a solid tumor in vivo. Cancer Res., *39*: 1891–1897, 1979.

69. Pollack, A., Prudhomme, D. L., Greenstein, D. B., Irvin III, G. L., Claflin, A. J., and Block, N. L. Flow cytometric analysis of RNA content in different cell populations using pyronin Y and methyl green. Cytometry, *3*: 28–35, 1982.

70. Roti-Roti, J. L., Higashikubo, R., Blair, O. C., and Vygur, H. Cell cycle position and nuclear protein content. Cytometry, *3*: 91–96, 1982.

71. Sahar, E. and Latt, S. A. Enhancement of banding patterns in human metaphase chromosomes by energy transfer. Proc. Natl. Acad. Sci. USA, *75*: 5650–5654, 1978.

72. Salzman, G. C., Wilkins, S. F., and Whitfill, J. A. Modular computer programs for flow cytometry and sorting: the LACEL system. Cytometry, *1*: 325–336, 1981.

73. Shapiro, H. M. Flow cytometric estimation of DNA and RNA content in intact cells stained with Hoechst 33342 and pyronin Y. Cytometry, *2*: 143–150, 1981.

74. Shapiro, H., Schildkraut, R., Curbelo, R., Brough-Turner, R., Webb, R., Brown, D., and Block, M. Cytomat R: A computer-controlled multiple laser source multiparameter flow cytophotometer system. J. Histochem. Cytochem., *25*: 836–844, 1977.

75. Steinkamp, J. A. Flow cytometry. Rev. Sci. Instrum., *55*: 1375–1400, 1984.

76. Steinkamp, J. A., Orlicky, D. A., and Crissman, H. A. Dual-laser flow cytometry of single mammalian cells. J. Histochem. Cytochem., *27*: 273, 1979.

77. Steinkamp, J. A., Stewart, C. C., and Crissman, H. A. Three-color fluorescence measurements on single cells excited at three laser wavelengths. Cytometry, *2*: 226–231, 1982.

78. Stohr, M. Double beam application in flow techniques and recent results, *In*: (W. Gohde, J. Schumann, and Th. Buchner, eds.), Pulse Cytophotometry, Ghent: European Press, 1976.

79. Stohr, M., Eipel, H., Goerttler, K., and Vogt-Schaden, M. Extended application of flow microfluorometry by means of dual laser excitation. Histochemistry, *51*: 305–313, 1977.

80. Stohr, M. and Vogt-Schaden, M. A new dual staining technique for simultaneous flow cytometric DNA analysis of living and dead cells. *In*: (O. D. Laerum, T. Lindmo, and E. Thorud, eds.), Flow Cytometry IV. Bergen, Norway: Universitetsforlaget, 1979.

81. Stohr, M., Vogt-Schaden, M., Knobloch, M., Vogel, R., and Futterman, G. Evaluation of eight fluorochrome combinations for simultaneous DNA-protein flow analyses. Stain Technol., *53*: 205–215, 1978.

82. Swartzendruber, D. E. A bromodeoxyuridine-mithramycin technique for detecting cycling and noncycling cells by flow microfluorometry. Exp. Cell Res., *109*: 439–443, 1977a.

83. Swartzendruber, D. E. Microfluorometric analysis of cellular DNA following incorporation of bromodeoxyuridine. J. Cell Physiol., *90*: 445–454, 1977b.

84. Swift, H. Analytical microscopy of biological materials. *In*: (G. L. Weid, ed.), Introduction to Quantitative Cytometry, New York: Academic, 1966.

85. Tanke, H. J., Nieuwenhuis, A. B., Koper, G. J. M., Slats, J. C. M., and Ploem, J. S. Flow cytometry of human reticulocytes based on RNA fluorescence. Cytometry, *1*: 313–320, 1981.

86. Taylor, I. W. and Milthorpe, B. K. An evaluation of DNA fluorochromes, staining techniques and analysis for flow cytometry. J. Histochem. Cytochem., *28*: 1224–1232, 1980.

87. Thornthwaite, J. T., Sugarbaker, E. V., and Temple, W. J. Preparation of tissues for DNA flow cytometric analysis. Cytometry, *1*: 229–237, 1980.

88. Tobey, R. A. and Crissman, H. A. Unique techniques for cell cycle analysis utilizing mithramycin and flow microfluorometry. Exp. Cell Res., *93*: 235–239, 1975.

89. Tobey, R. A., Crissman, H. A., and Kraemer, P. M. A method for comparing effects of different synchronizing protocols on mammalian cell cycle traverse. J. Cell Biol., *54*: 638–645, 1972.

90. Trujillo, T. T. and Van Dilla, M. A. Adaptation of the fluorescent Feulgen reaction to cells in suspension for flow microfluorometry. Acta Cytol., *16*: 26–30, 1972.

91. Van Dilla, M. A., Langlois, R. G., and Pinkel, D. Bacterial characterization by flow cytometry. Science, 220: 620–622, 1983.

92. Van Dilla, M. A., Trujillo, T. T., Mullaney, P. F., and Coulter, J. R. Cell microfluorometry: A method for rapid fluorescence measurement. Science, 163: 1213–1214, 1969.

93. Vindelov, L. L., Christensen, I. J., Jensen, G., and Nissen, N. I. Standardization of high resolution flow cytometric DNA analysis by the simultaneous use of chicken and trout red blood cells as internal reference standards. Cytometry, 3: 328–331, 1983.

94. Vindelov, L. L., Christensen, I. J., and Nissen, N. I. A detergent-trypsin method for the preparation of nuclei for flow cytometric DNA analysis. Cytometry, 3: 323–327, 1983.

95. Visser, J. W. M. Vital staining of hemopoietic cells with the fluorescent bis-benzimidazole derivatives Hoechst 33342 and 33258. In: (O. D. Laerum, T. Lindmo, and E. Thorud, eds.), Flow Cytometry IV, Bergen, Norway: Universitetsforlaget, 1979.

96. Visser, J. W. M., Bol, S. J. L., and Van den Engh, G. J. Characterization and enrichment of murine hemopoietic stem cells by fluorescence activated cell sorting. Exp. Hematol., 9: 644–655, 1981.

97. Wallen, C. A., Higashikubo, R., and Dethlefsen, L. A. Comparison of two flow cytometric assays for cellular RNA-acridine orange and propidium iodide. Cytometry, 3: 155–160, 1982.

98. Ward, D. C., Reich, E., and Goldberg, I. H. Base specificity in the interaction of polynucleotides with antibiotic drugs. Science, 149: 1259–1263, 1965.

99. Waring, M. Variation of the supercoils in closed circular DNA by binding of antibiotics and drugs: Evidence for molecular models involving intercalation. J. Mol. Biol., 54: 247–279, 1970.

100. Zante, J., Schumann, J., Barlogie, B., Gohde, W., and Buchner, Th. Preparation and staining procedures for specific and rapid analysis of DNA distributions, In: (W. Gohde, J. Schumann, and Th. Bucher, eds.), Pulse Cytophotometry, Ghent: European Press, 1976.

Chapter 8

Data Analysis in Cell Kinetics Research

Phillip N. Dean

1. INTRODUCTION

Cell kinetics is a science that has as its objective an understanding of how cells mature with time and how they respond to outside influences (e.g., exposure to radiation and drugs). Emphasis is placed on the rates at which these events occur. Consider, for example, the application of cell cycle kinetics in the treatment of cancer. Therapeutic agents currently in use often are not cell-type-specific; they act on normal and abnormal cells indiscriminately. They are, however, often cell-phase-specific, preferentially killing cells that are synthesizing DNA (i.e., in S-phase). For example, the application of the agent (e.g., cytosine arabinoside, ara-C) will kill cells in S-phase, both normal and abnormal. Most of the cells in the body are in a quiescent state, not cycling, and will not be affected by the drug. However, in addition to the cancer cells, a large number of normal cells are in cycle, notably in the gut and bone marrow, and it is these cells that are at risk in therapy. Therapy schedules must be devised that will spare these cells. One method of treatment might utilize the drug ara-C to "synchronize" the cycling cells and to selectively kill cells in S-phase. As a synchronizing agent, ara-C blocks cells at the end of G1-phase by inhibiting the synthesis of DNA. All cells in G1- and G2M-phases will continue to progress through the cell cycle until they reach the end of G1-phase, at which time their progress stops. Then, as the effect of the drug wears off, the cells are released from the block and enter S-phase. Use might then be made of the fact that normal and cancer cells grow at different rates, the latter usually more slowly. As the population of normal cells passes through S-phase and enters G2-phase, the cancer cells enter and proceed through S-phase. Ara-C is administered as a killing agent when the fraction of normal cells in S-phase is low and the fraction of tumor cells is high. This cycle is repeated to reduce the number of cancer cells, hopefully to the point where the body can eliminate them altogether. To be able to decide when and for how long to administer the drug, it is apparent that a detailed knowledge of the growth kinetics of the normal and cancer cells, and their response to the therapeutic agent, is required. To be able to predict the effect of modifications in the therapy protocol, a model of the cell cycle is needed. To generate the model, the procedure outlined below is followed. First a biological model is constructed, using experimentally derived data, preferably from the kind of cells to be treated. Then a mathematical model has to be developed, from which optimized therapy protocols can be obtained. There is a large body of pharmacological data on drugs and cell kinetic data on cells grown in vitro that can be used in developing the parameters of the model. If the models are comprehensive and very general, it may be possible to predict the response of drugs on cells on which the drugs have not previously been used.

The development of a mathematical analysis procedure is a three-step process: (1) A conceptual model of the biological process being studied must be established. It would include all that is known and observable about the process. (2) A mathematical formalism is developed to describe the process. This would typically consist of one or more equations with several parameters. The independent variable would be the observable quantity; in cell kinetics this might be the DNA content of the cells. The parameters could be the rate constants for passage between phases of the cell cycle. (3) The parameters of the model are adjusted until the data computed from the model match the experimentally obtained data. If the computed and experimental data match, then the parameters are the correct rate constants that describe the biological process. If the two sets of data do not match, then the model is incorrect and a new one must be formulated.

Mathematical models can be very simple, providing only the fraction of cells in each phase, or complex, providing phase durations and dispersions in the durations, and cell loss. For example, the fraction of a cell population in each phase of the life cycle is often used to monitor the effects of a particular drug therapy schedule. This can be done using flow cytometry (FCM) to measure the DNA content distribution and a relatively simple mathematical model to resolve the distribution into its components. Although this method of obtaining cell kinetic information is simple and rapid, it is very limited in its application, since it provides so little information, and this information can be misleading. For example, a cell can have an S-phase DNA content, but be reproductively dead (nonproliferating). Thus the method yields only a rough measure of the proliferative activity of the population under study.

There has long been a method for obtaining cell cycle traverse rates for asynchronously growing cells, called the fraction labeled mitosis (FLM) method. In this method, described in detail in chapter 2 (29,30), the cell population is exposed to tritiated thymidine ([³H]-TdR), a DNA precursor. The [³H]-TdR is incorporated into cells in S-phase—cells that are actively synthesizing DNA. Time sequential samples of the cells are then taken and covered with a photographic emulsion. When the emulsion is developed, the cells are examined microscopically and those cells in mitosis that also contain tritium (show grains in the emulsion) are scored. The fraction of the mitotic cells that are labeled is then plotted as a function of time in culture. These data, when evaluated with the proper model, yield the phase durations. The mathematical models involved can be complex, especially when compared to the models used in the analysis of single DNA distributions.

Flow cytometric measurements have been used to provide comparable data to the FLM method, with improved precision and additional capability. In these studies, the cell population is also exposed to [³H]-TdR. However, instead of scoring cells in mitosis, a procedure that is very time-consuming and can be subjective, cells with a specific DNA content (mid-S-phase) are physically sorted and measured for radioactivity per cell using liquid scintillation counting (RCS_i) (19). These data utilize the same mathematical models as the FLM data, with some changes, and yield the same kind of information.

As the amount of data obtained increases, and our knowledge of the cell cycle improves, the mathematical models increase in complexity. The advent of the bromodeoxyuridine (BrdUrd) method, presented in chapter 5 (20), has provided this increase in information, while decreasing the amount of experimental work required. This method utilizes a monoclonal antibody to BrdUrd (*10,12,16*). In this method the cells under investigation are exposed for a short period of time in vitro or in vivo to BrdUrd, an analog of thymidine. The BrdUrd is incorporated into the DNA of cells in S-phase. A fluorescently labeled monoclonal antibody to the BrdUrd is added to the cells, along with a DNA-specific dye to measure total DNA content. The bivariate measurement of BrdUrd and DNA content shows the cells in S-phase to be well discriminated from both G1- and G2M-phase cells. For phase fraction analysis, only the most simple mathematical model is required; the summing of specific areas of the bivariate distribution. However, as will be shown later, the BrdUrd/DNA method provides additional information at little cost in time and effort. Models have been developed to utilize these data, as will be illustrated.

2. METHODS

Each method of analysis will be dealt with in turn. First the biological model will be presented, followed by the mathematical model. Examples will then be used to illustrate the methods. Since the analysis of single DNA distributions is currently the most widely used cytokinetic analysis procedure, it will be treated first, in some detail. Detailed applications of various methods of DNA distribution analysis are presented in chapters 5 (*20*), 10 (*5*), and 12 (*22*) of this book.

2.1. DNA Distribution Analysis

The general biological model used in all of the methods of analysis is that introduced by Howard and Pelc (*24*) and presented in Fig. 1A. This model, based on DNA content, divides the cell cycle into four phases, G1, S, G2, and M. As cells progress through G1-phase, they maintain the same DNA content, although their age (maturity) increases. As the cells proceed through S-phase, they copy their DNA such that at the end of S-phase they have doubled their DNA content. This amount of DNA is maintained through G2- and M-phases until the cells divide, with each daughter cell receiving equal amounts of DNA and proceeding into G1-phase. This process is illustrated in Fig. 1B, resulting in the DNA distribution shown.

(A)   (B)

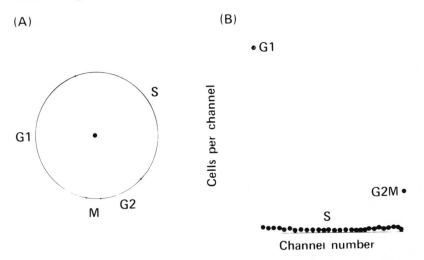

Fig. 1. (A). The DNA model of the cell cycle, as introduced by Howard and Pelc (24). The life cycle is divided into four phases, G1, S, G2, and M. (B). The DNA distribution of a population of cells.

Unfortunately, FCM measurements of DNA content of a cell are not perfect, with DNA content measurement coefficients of variation (CV = standard deviation divided by the mean) ranging from 1 to 10%. A typical distribution is shown in Fig. 2. In this distribution the CV is about 3%. The objective of all methods of analyzing this type of data is to resolve the distribution into three different phases to obtain the fraction of cells in each phase. The G2- and M-phases are combined into a G2M-phase, since they cannot be resolved based on DNA content alone.

The broadening effect is illustrated in Fig. 3. The unbroadened distribution is given by the open circles; only a few of the S-phase compartments (channels) are plotted. Each S-phase point is also represented by a Gaussian function, with the CV of the G1 peak, as shown by the solid lines. The G1-phase Gaussian is also shown. Adding together all of the S-phase Gaussians produces the dashed line, which would be the distribution of a pure S-phase population. When all Gaussians are added together, the solid circles representing the measured distribution, are obtained. Note the overlap between G1- and S-phase. Some of the S-phase cells have a measured DNA content of less than the mean G1-phase content, and half of the G1-phase cells have the measured DNA content of S-phase cells. Any mathematical model used to analyze DNA distributions must have some method of resolving this overlap.

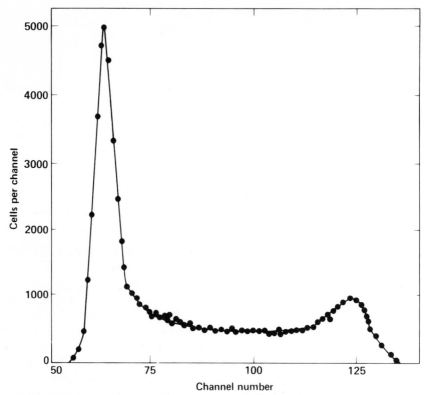

Fig. 2. Frequency distribution of Chinese hamster ovary (CHO) cells stained with the DNA-specific dye, chromomycin A3.

Errors in phase fraction estimates will be sensitive to the CV (degree of overlap between the populations) and to the magnitude of the phase fractions. In many cell kinetics studies, the S-phase fraction is the most important. If it is small, then the accuracy of results will be very sensitive to the CV. The following paragraphs present several popular models used to analyze DNA distributions, differing primarily in the way they represent S-phase. Since the true S-phase distribution is unknown, none of the methods can model it exactly and, consequently, all make errors. Following the description of the methods is a comparison of results obtained using all of the methods on the same data sets.

2.1.1. Graphical Methods

Two graphical methods of analysis were proposed early in the development of FCM. These methods are illustrated in Figs. 4A

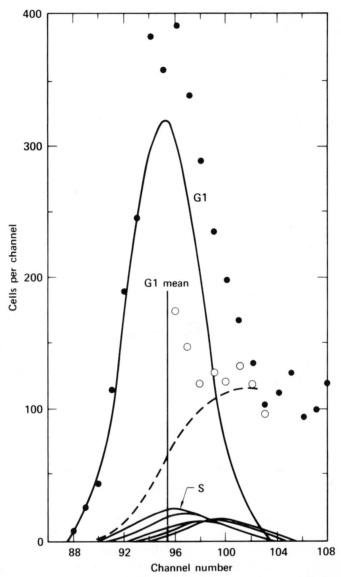

Fig. 3. An illustration of the overlap between G1- and S-phase cells. S-phase begins at the mean of the G1 peak. Variability in the measurement of the DNA content of a cell causes the overlap, as explained in the text.

and B for L-1210 cells in exponential growth. The first method, Fig. 4A, called here the ''peak reflect'' method (3), makes the assump-

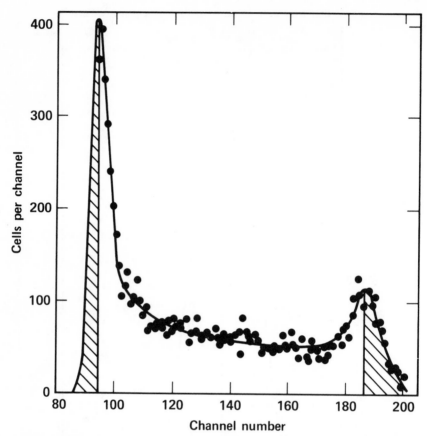

Fig. 4(A). The "peak reflect" method of analysis in which the area of the left side of the G1 peak is doubled to obtain the G1 population and the right side of the G2M peak is doubled to obtain the G2M population. The S fraction is obtained by subtraction.

tion that the left side of the G1 peak contains no S-phase cells, and that the G1 peak mode and mean are the same. The procedure followed is to sum the counts to the left of the G1 peak model channel and double the number to obtain the population of G1-phase cells. This same procedure is used on the right side of the G2M peak. Since this method makes no allowance for overlap of the phases, it will almost certainly overestimate the G1- and G2M-phase fractions and underestimate S-phase. The magnitude of the error will increase with the CV and with the size of the S-phase fraction. A simple improvement in this model is to compute the G1

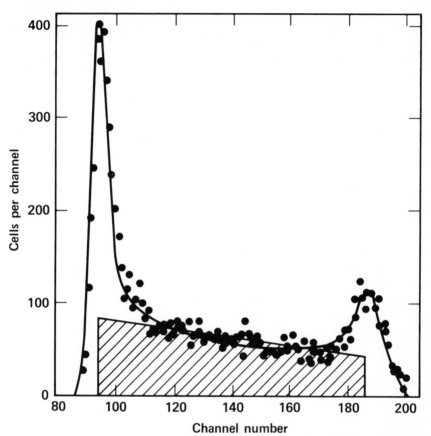

Fig. 4(B). The "rectangle" method of analysis in which S-phase is represented by a rectangle placed between the G1 and G2M means. The data are from L-1210 cells grown in suspension culture.

peak mean numerically, using the left side and one half of the right side of the G1 peak. A further improvement can be made by estimating the degree of overlap, considering the magnitude of mid-S-phase. This method can yield adequate estimates of the G1 fraction, and of S-phase fractions in some cases, especially if the S-phase fraction is high and/or the CV is low. G2M-phase fraction estimates using the method are not reliable.

The second graphical method of analysis, called here the "rectangle" method (2), consists of drawing a straight horizontal line through the S-phase data and between the G1 and G2M peak modes. The area under the line defines the S-phase population, and the G1 and G2M populations are obtained by subtraction. Although this method compensates to some extent for the phase overlap, it will also tend to overestimate the G1- and G2M-phase fractions and underestimate the S-phase fraction. It has the same dependence on CV and S-phase fraction as the peak reflect method. Its accuracy is also improved by calculating the G1 peak mean, and by allowing the S-phase line to have nonzero slope, as illustrated in Fig. 4B.

There are several variants of the rectangle method that provide better compensation for the phase overlap. For very low S-phase fractions (10% and less), and for low CVs (2% or less), graphical methods produce results for G1- and S-phases that are accurate compared to measurement and sampling errors, but are not dependable for G2M-phase.

All of these graphical methods should be used only with cells in exponential growth, where the S-phase distribution has low curvature.

More recently, graphical analysis has been applied to BrdUrd/DNA distributions (*see* chapter 5 for a description of the BrdUrd method). Figure 5 shows data for Chinese hamster ovary (CHO) cells elutriated on the basis of sedimentation velocity (*10*). The number of cells in each phase is obtained by establishing a region around each subpopulation, as shown in the figure. Figure 5A is for cells from fraction 14, where 17% of the cells were in G1-phase, 82% in S-phase, and 1% in G2-phase. Figure 5B is for cells from fraction 16, where 2% of the cells were in G1-phase, 90% in S-phase, and 8% in G2M-phase. This method is remarkably accurate since cells incorporating BrdUrd for only a few seconds are well resolved from G1-phase cells (*see* chapter 5). Therefore, the boundaries between the phases can be easily established. Errors of as much as 10% in location of the regions result in errors of less than 5% in the phase fractions.

The boxing technique can be used to further parameterize the data. Figure 6 shows how the rate of DNA synthesis can be obtained from such analyses. Using tritiated BrdUrd Dolbeare et al. (*13*) have shown a linear correlation between radioactivity per cell and amount of fluoresceinated antibody per cell. Therefore, assuming that the rate of BrdUrd incorporation is a good estimate of the rate of DNA synthesis, the fluorscence intensity of a cell (antibody content) is directly proportional to its rate of DNA synthesis. The higher the amount of antibody bound to the cells, the greater the

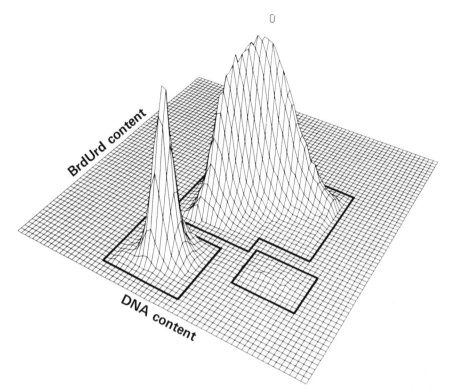

Fig. 5. Bivariate data for CHO cells measured simultaneously for DNA content with propidium iodide and for incorporated BrdUrd with a fluoresceinated monoclonal antibody. The G1, S, and G2M subpopulations are clearly separated, as shown by the solid lines. Phase fractions are obtained by summing the numbers of cells in each region. Data are from elutriated cells. (A) A fraction where most cells are in early S-phase. (B) A fraction where most of the population is in mid S-phase.

rate of DNA synthesis. To obtain the rate of synthesis across S-phase, rectangular regions are drawn in S-phase, with different DNA content boundaries, illustrated by region 1 in Fig. 6A. Region 2 includes all of S-phase. The BrdUrd content distribution for cells within the selected narrow range of DNA ccontent is shown in Fig. 6B. The mean of this distribution represents the mean rate of BrdUrd incorporation at the selected DNA content. By selecting regions throughout S-phase and computing the mean rate for each region, the data of Fig. 6C are obtained. These data are comparable to those obtained for similar cells in other experiments (21).

0

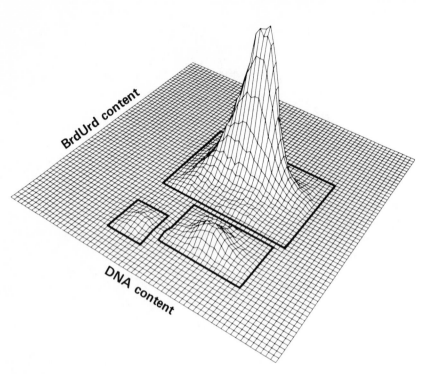

Fig. 5B.

The windowing technique, using polygons to define regions of interest in the distribution and summing the numbers of cells in the regions, can be used to obtain detailed information about the various subpopulations of the total population. For a sequence of distributions from perturbed cells, the same regions can be used in each distribution and the sums plotted vs time. These data can then be input to the mathematical models described later in this chapter.

2.1.2. Fitting Methods

Fitting methods are those that attempt to describe experimental data with a mathematical function, the parameters of which are adjusted iteratively until the function matches the data as well as possible. As discussed in the Introduction, the function is ideally written to model the biological process being measured. If the fitting

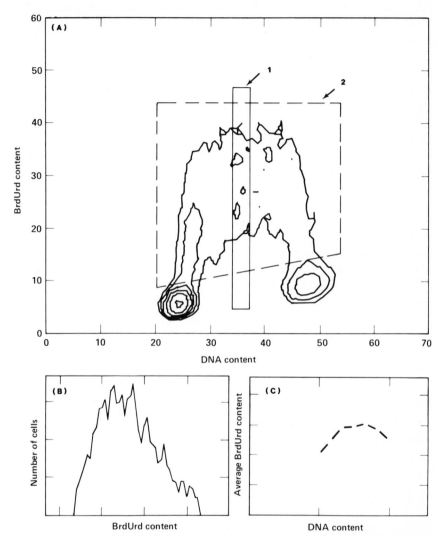

Fig. 6. Bivariate data illustrating the use of BrdUrd method to obtain the rate of DNA synthesis in S-phase. (A) is a contour plot of asynchronously growing cells. (B) is the BrdUrd content distribution of the cells in region 1 of (A), and (C) is the rate of DNA synthesis as described in the text.

procedure results in a good match of the function to the data, then the model is probably correct and the parameters yield useful information. If the fit is poor, then the model is incorrect. It is also possible that the data are poor; they do not represent the biology—a result of instrumental or other errors. The values of the parameters of the equation that best describes the data are usually obtained by the method of least squares. This is an important procedure in virtually every method of analysis. The concept will be illustrated by applying it to a simple linear regression. Suppose a set of data obtained by repeated measurements of some variable could be described rather well with a straight line. How can we determine the slope and intercept of the line that best "fits" the data? The equation of a straight line is:

$$Y = A + BX \qquad [1]$$

where Y is the dependent variable (e.g., number of cells), X is the independent variable (e.g., size), and A and B are the intercept and slope of the line (i.e., the parameters), respectively. Suppose N data points have been measured at a variety of sizes (X values). The method of least squares finds the best estimates of A and B by minimizing the sum of the squares of the differences between the measured values, Y_i, and the values computed from the equation. The difference for data point i is D_i,

$$D_i = Y_i - A - BX_i \qquad [2]$$

where i varies from 1 to N. If there are N data points, then there are N of these equations. Each equation must be squared, since we will be minimizing the sum of the squares of the differences (also called the deviations). The summation can be expressed as:

$$S = \Sigma \, [Y_i - A - BX_i]^2 \qquad [3]$$

From calculus we know that at the minimum of a function, its slope (i.e., derivative) is zero. Therefore, to find the minimum of S for Eq. [3] with respect to a parameter, we take the partial derivative of the function with respect to the parameter and set it equal to zero.

$$\frac{\partial S}{\partial A} = \Sigma_i \, Y_i - N - B \, \Sigma_i \, X_i = 0 \qquad [4]$$

$$\frac{\partial S}{\partial B} = \Sigma_i \, X_i \, Y_i - A \, \Sigma_i \, X_i - B \, \Sigma_i \, X_i^2 = 0 \qquad [5]$$

Since Y_i and X_i are measured quantities, we now have two equations in two unknowns (A and B) and can solve for A and B with determinants. These values will be the least squares best estimates of the two parameters of the equation. Note that even though the curve defined by A and B is the best fit of a straight line to the data, there is no guarantee that a straight line provides the best fit to the data. This is the problem of model selection. Essentially any function can be fit to any set of data. Whether or not the parameters have any meaning relative to the biology is another question, which must be determined by the experimenter, not the mathematics.

The foregoing is a very simple example of how least squares solutions are obtained. As the number of parameters increases, the time required for analysis increases as well. Also, note that the equation is linear in the parameters. This means that as the value of the parameter changes, the value of the function changes linearly with it. This is true for both A and B, and indeed for each parameter in a polynomial function. However, consider an exponential function, such as can be used to describe the number of cells as a function of time, for cells in exponential growth:

$$Y = Ae^{BX} \tag{6}$$

This function is linear in A, but not in B. To perform a least squares fit of this equation to a set of data, the procedure illustrated cannot be used and nonlinear fitting methods are required. Although the principles are the same, these methods are much more complex than that just described and will not be presented here. The interested reader is referred to an excellent review of this topic by Bevington (4).

From a practical standpoint, linear methods of analysis are preferred. This is because the result is unique. Iterations are not required and the calculations are fast. This type of analysis can be performed on microcomputers. In nonlinear methods of analysis, multiple iterations are required in which the parameters are changed by small amounts each time. There are many computations per iteration and these methods should be used on midsize or larger computers. As will be seen, both methods are required for different kinds of data in different situations.

The simplest fitting method used in the analysis of single DNA distributions, called SFIT, is illustrated in Fig. 7 for the same data as used in Fig. 4. The mathematical model uses a second degree polynomial to fit that part of S-phase that is not overlapped by either G1- or G2M-phase (6). With a CV of 3%, this region amounts to 75% of S-phase. The polynomial (concave upward) provides for a low rate of DNA synthesis in early and late S-phase, which has been observed experimentally. The parameters of the polynomial are derived using the linear least squares method described above. To obtain the S-phase population, the polynomial is integrated between the G1 and G2M peak means. Thus the assumption is made that the shape of mid-S-phase is representative of all of S-phase, and a polynomial fit to mid-S-phase accurately represents all of S-phase. The G1 peak mean is estimated by computing the arithmetic mean of the G1 peak, using all of its left side and two thirds of its right side. This estimate is usually within 1 or 2% of the true mean, dependent to some extent on the CV and S-phase fraction. The user must provide either the G2M mean or the G2M/G1 peak mean ratio. Since a polynomial is linear in its parameters, linear fitting methods can be used, which makes the method very fast. This method yields good accuracy (10% or better) when the CV is 6% or less. As the CV increases the amount of S-phase not overlapped by G1- or G2M-phase decreases and the accuracy becomes very sensitive to counting statistics and to the degree to which the polynomial adequately describes S-phase. This method is applicable only to those distributions in which S-phase can be described by a second degree polynomial (usually asynchronously growing cells).

A more complete (and complex) mathematical model is illustrated in Fig. 8, for the L-1210 cells of Figs. 4 and 7. In this model, the G1 and G2M peaks are represented by Gaussian (normal distribution) functions, and S-phase is represented by a second degree polynomial, broadened with a Gaussian function to account for measurement variability (9). The complete equation used to fit the distribution is:

$$Y = \frac{A_1}{\sqrt{2\pi}\,\sigma_1}\, e^{-\frac{1}{2}\left(\frac{x-\mu_1}{\sigma_1}\right)^2} + \frac{A_2}{\sqrt{2\pi}\,\sigma_2}\, e^{\frac{1}{2}\left(\frac{x-\mu_2}{\sigma_2}\right)^2} + P(x) \qquad [7]$$

where

$$P(x) = \sum_{j=1}^{N} \frac{1}{\sqrt{2\pi}\,\sigma_j}\, e^{-\frac{1}{2}\left(\frac{x-\mu_j}{\sigma_j}\right)^2} (a + bx + cx^2) \qquad [8]$$

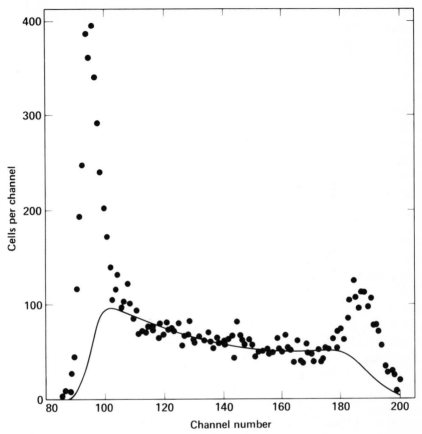

Fig. 7. The "SFIT" method of analysis in which that part of S-phase that is not overlapped by G1- or G2M-phase cells is fit with a second degree polynomial, which is then extended to the G1 and G2M peak means. The integral of the polynomial yields the S population and the G1 and G2M populations are obtained by differences.

Distributions are usually accumulated in pulse height analyzers, where the fluorescence intensity is converted to channel number. Therefore Y is the number of cells per channel and X is the channel number. A_1 and A_2 are the areas, σ_1 and σ_2 are the standard deviations, and μ_1 and μ_2 are the means of the G1 and G2M peaks, respectively. The polynomial parameters are a, b, and c, and N is typically 50. The polynomial begins at the G1 mean, μ_1, and ends at the G2M mean, μ_2. This procedure amounts to fitting S-phase with a series of Gaussians, where the areas of the Gaussians are constrained to lie along a second degree polynomial. All nine of

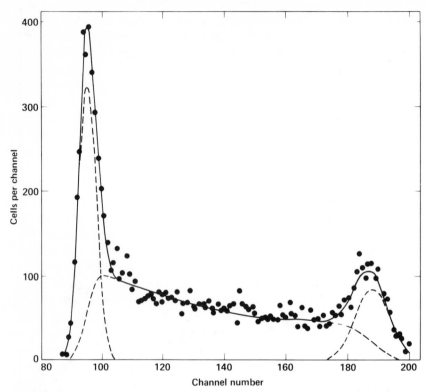

Fig. 8. The "broadened polynomial" method of analysis applied to the same L-1210 cells of Figs. 4 and 5. The solid circles are the data points, the solid line is the overall fit to the data, and the dotted lines are the individual components.

the parameters shown are allowed to vary during the fitting procedure. The equation just presented is fit to the data by the method of least squares, using nonlinear curve fitting techniques (4). In nonlinear curve fitting it is very important to have good initial estimates of the parameters. In the method just described the initial estimates are made using the first parametric method described (SFIT). As with the graphical and SFIT methods, the broadened polynomial model is appropriate only for asynchronously growing cells, or when there is no apparent structure in S-phase; i.e. a second degree polynomial is adequate to describe the S-phase distribution. This method has been found to be quite accurate for a wide variety of cell types and phase fractions. In those distributions in which the peak means or CVs are not well defined, they can be introduced into the model as constants; i.e., entered by the user and not allowed to change during the analysis. In addition, the CVs of the two peaks can be determined independently or required to be the same. A background function such as a straight line or an exponential can also be included in the fitting procedure.

Many variations of this model have been proposed in an attempt to fit data that do not comply with the initial conditions, i.e., asynchronously growing cells of a single population. One must recognize that poor sample preparation, nonuniform staining, complex aneuploid populations, or poor instrument performance may produce data that cannot be analyzed by any method with any degree of confidence.

For perturbed cells, a wide variety of different mathematical models have been developed. In all cases, S-phase is divided into a set of compartments. The methods differ in the number, spacing, and widths of the compartments, and in the functions used to represent them. An early model proposed using a series of evenly spaced Gaussian functions to represent S-phase (*15*). The number of Gaussians depends upon the CV of the measurement; the better the resolution the more Gaussians can and should be used. An example of the use of this model is shown in Fig. 9, for the same data used in Fig. 8. For this method to work consistently, the CV and the means of the S-phase Gaussians must be held constant during the fitting procedure. Often, the CV is made the same for all the Gaussians, including the G1 and G2M peaks. In some cases the standard deviations of the G1 and G2M peaks are allowed to change and the standard deviations of the S-phase peaks are scaled between them. The function used to fit the data is:

$$Y = \sum_{i=1}^{N} \frac{A_i}{\sqrt{2\pi}\,\sigma_i}\, e^{-\frac{1}{2}\left(\frac{x - \mu_i}{\sigma_i}\right)^2} \qquad [9]$$

where $i = 1$ is the G1 peak and $i = N$ is the G2M peak. For a CV of 4%, I recommend an N on the order of 8–10. The primary advantage of this model is speed. If a constant CV and fixed S-phase peak means are selected, there are a minimum number of parameters to be derived. The principal disadvantage is that the first and last S-phase Gaussians cannot be positioned closely enough to the G1 and G2M peaks to correctly model S-phase. This is illustrated

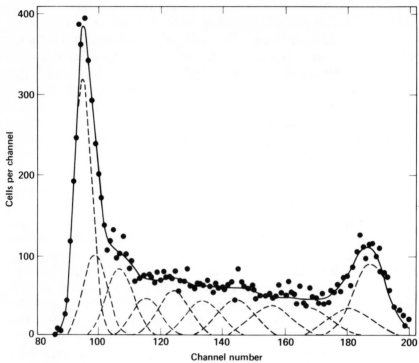

Fig. 9. The "sum of Gaussians" method of analysis as applied to the same data of Fig. 8. The symbols are also the same as in Fig. 8.

in Fig. 10. This figure shows the G1 peak and the first part of S-phase. These are data synthesized with a cell cycle model, where the rate of DNA synthesis is assumed to have a Gaussian shape, as has been measured experimentally. The data points are the distribution that would be measured, including statistical and measurement errors. The solid line is the Gaussian curve used to represent G1-phase. The dashed line is the S-phase data broadened with a Gaussian function to represent measurement error. The dashed lines show the calculated data for Gaussian functions used to represent the first few S-phase compartments. The mean of the first S-phase peak is separated from the G1 peak mean by 2.5 standard deviations, a distance determined experimentally to be the minimum that can be reliably used. It is clear that the S-phase distribution composed of the sum of the S-phase Gaussian functions does not well represent the true S-phase distribution. In particular, the

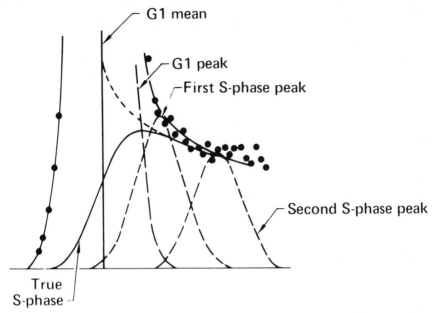

Fig. 10. The G1 peak and early S-phase of a DNA distribution. The solid circles are the data points. One solid line is drawn through the points; another shows the shape of the true S-phase. Dashed lines show the Gaussian function used for G1 peak, and the first two Gaussian functions used in an attempt to fit S-phase.

S-phase contribution to the left side of the G1 peak is too small. Consequently the S-phase fraction will be underestimated and the G1-phase fraction overestimated. The magnitude of the error increases with increased CV.

Another and somewhat more dependable and accurate model represents S-phase with a series of rectangles (*1*). The widths of the rectangles can be uniform, although a logarithmic width distribution (i.e., width increasing with DNA content) is preferred. The rectangles are broadened to account for measurement error by convolving them with a Gaussian function, whose CV is scaled between the G1 and G2M peak CVs. An example of the use of this model is presented in Fig. 11, for the same L-1210 cell data used in Figs. 8 and 9. Since the first S-phase rectangle begins at the G1 peak mean, this model does a better job of accounting for the overlap between G1- and early S-phase. The same is true for late S- and G2M-phases. An application of this model to data from perturbed cells is presented in Fig. 12. These data are for KHT cells 6 h after treatment with cytosine arabinoside (ara-C), a chemotherapeutic drug. A newer version of this method replaces the rectangles with trapezoids whose corners are connected. This feature prevents such things as negative peak areas and fits the S-phase overlap region better than a rectangle.

For DNA distributions of many perturbed cell populations, there is no clear G1 or G2M peak or indication of the CV of the measurement. In such cases control (untreated) populations are

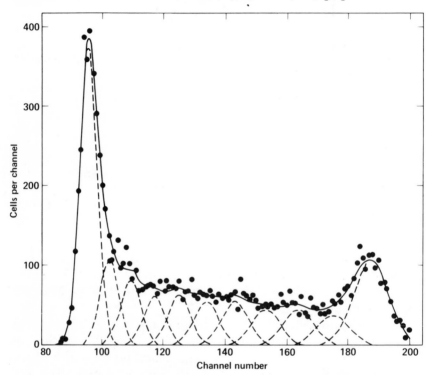

Fig. 11. The "sum of broadened rectangles" method of analysis as applied to the same data of Fig. 8. The symbols are also the same.

used to obtain the peak means and CVs. This approach is fraught with danger since it is not even clear whether drug-treated and normal cells bind DNA dyes with the same efficiency. However, it is the best method available for the analysis of some DNA distributions from pertrubed cells. Since the means and CVs of the peaks are specified, not free, parameters, the function becomes a linear one and the analysis can be performed rapidly with quite small computers.

All of the methods described in this section were used to analyze the same data, L-1210 cells, as shown in Figs. 4, 7, 8, 9, and 11. The results of these analyses are presented in Table 1. They divide roughly into two groups; (1) the POLY and PEAKSR methods, and (2) the others. Other studies to be presented later show the first two methods to be the most accurate for all three phases. As predicted in the descriptions of the other methods, they underestimate the S-phase fraction by 10–15% and overestimate G1-phase by 10–30%. For this set of data, however, none of the results are remarkably bad, compared to sampling and other errors (*11*).

TABLE 1
DNA Histogram Analysis for L-1210 Cells
in Asynchronous Growth[a]

Method	G1	S	G2M
POLY[b]	0.24	0.63	0.13
PEAKSR[c]	0.23	0.64	0.13
PEAKSG[d]	0.28	0.56	0.16
SFIT[e]	0.27	0.62	0.11
Peak reflect[f]	0.25	0.58	0.16
Rectangle[g]	0.32	0.56	0.10
Trapezoid[h]	0.28	0.59	0.13

[a]In the first three methods G1 and G2M peaks are fit with a Gaussian function.
[b]POLY, S-phase is fit with a second degree polynomial.
[c]PEAKSR, S-phase is fit with a series of broadened rectangles.
[d]PEAKSG, S-phase is fit with a series of Gaussian functions.
[e]SFIT, mid-S-phase is fit with a second degree polynomial.
[f]Peak reflect, described in section 2.1.1.
[g]Rectangle, S-phase represented by a rectangle.
[h]Trapezoid, S-phase represented by a trapezoid.

A more complete study of the accuracy of these various models has been performed (*7*) using DNA distributiions synthesized with the use of a cell cycle model developed by Gray (*17*). This model uses as input parameters the cell transit time through each phase, dispersions in these times, a variable rate of DNA synthesis through S-phase, and instrumental measurement error. The histograms used to test the models had a wide range of CVs and phase fractions, and included perturbed as well as asynchronously growing populations of cells.

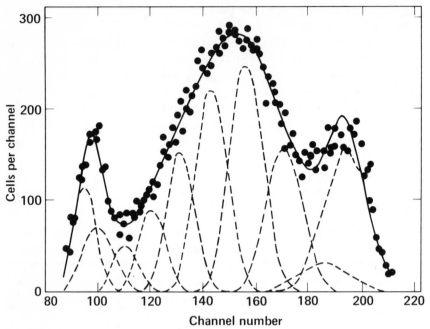

Fig. 12. The sum of broadened rectangles method of analysis as applied to KHT cells 6 h after treatment with cytosine arabinoside.

The results were the following:

1. Rectangle method: G1 and G2M fractions were overestimated, with the error increasing as the fraction decreased. Results were mostly independent of CV, although errors were relatively large; +15% for G1, −20% for S, and +20% for G2M.
2. Peak reflect method: For high G1 fractions (30%) with CV's of 4% or less, errors were less than 10%. Errors in G2M-phase estimates were always above 25% and errors in S-phase estimates were above 20%.
3. SFIT method: For G1- and S-phases, errors were less than 10% in all cases. For G2M, errors exceeded 10% for CV's of 6% and above.
4. POLY method: For G1 and S estimates, errors were less than 5% for CVs up to 8% and phase fractions as low as 10%. For G2M, errors were as large as 20% for CVs of 8 and 10% and phase fractions below about 8%.

5. Broadened rectangles method: Errors were only slightly larger than for the POLY method. For perturbed populations, errors in phase fraction estimates were very sensitive to the magnitude of the G1 and G2M populations, particularly to the presence of distinct peaks. Where there is a large early or late S-phase population, errors in G1- and G2M-phase fraction estimates could easily reach 100%, for any method.

2.2 Kinetic Models

2.2.1. Asynchronous Populations

To permit a detailed analysis of the kinetics of normal and tumor cells and their response to therapeutic procedures, mathematical models of the cell cycle are required. These models have a very wide range of complexity according to the kind of data being studied. The models must be biologically reasonable. Since no mathematical function or functions can exactly mimic the biology, the models usually contain constraints on their parameters. Models have been developed that involve only the number of cells vs time and the growth curve, and have only two parameters. Other models attempt to describe the complete cell cycle, accounting for cycling and quiescent cells in each phase, variable rates of DNA synthesis in S-phase, cell-dependent durations of each phase, and even a capability to block cells in various parts of the cell cycle. The latter is useful in modeling the effects of various chemotherapeutic drugs. These models can have 15 or more parameters. To use these models effectively and efficiently, manual interaction with the analysis procedure is usually required. Representative models will be presented, along with examples of their use.

The simplest model of cell growth assumes that when a cell divides, two viable cells result. Thus the number of cells increases exponentially with time according to the relationship:

$$N(t) = N(0)\, e^{kt} \qquad [10]$$

where $N(t)$ is the number of cells at time t, $N(0)$ is the number at time $t = 0$, and k is ln(2) divided by the doubling time of the cell population. These kind of data are presented in Fig. 13, with the number of cells plotted on a logarithmic scale. The slope of a line

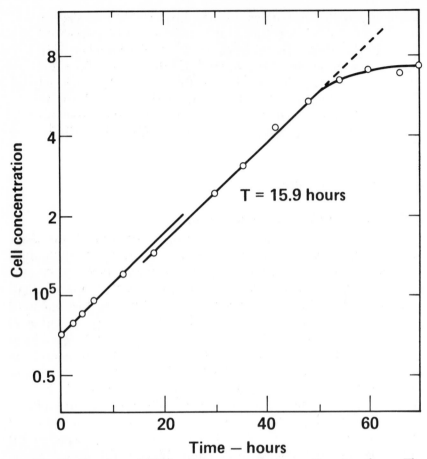

Fig. 13. Number of CHO cells in a population vs time in culture. The central portion of the data can be described by a straight line, yielding a doubling time of 15.9 h.

fit to the data yields the doubling time of the cells. Also note that the data deviate from the exponential at late times after the cell culture was begun. This is because the large number of cells in culture begin to deplete essential nutrients in the culture medium and many leave the growth cycle or slow down, usually while in G1-phase. Some cells also seem to be affected by cell density in the culture, with the same response. Similar growth characteristics have been observed for tumors in vivo.

Much of the data on cell cycle traverse rates that have been reported in the literature have been for the fraction labeled mitosis

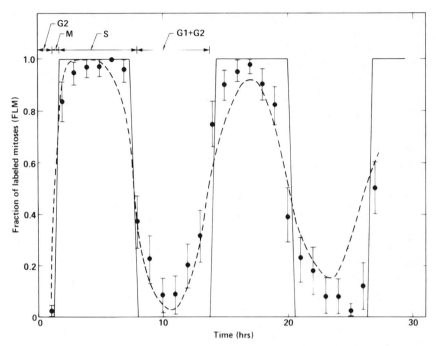

Fig. 14. Fraction labeled mitoses as a function of time after exposure to tritiated thymidine. The data points are for CHO cells in suspension culture. The solid line represents the ideal curve that would be obtained if all cells had the same cycle time, and the dashed line shows the result of fitting a mathematical model to the data. The model derived parameters are given in the text.

(FLM) method of analysis (27). The method of obtaining these data was described in the Introduction. For a hypothetical population in which all cells cycle at the same rate, the solid curve of Fig. 14 would be obtained. The first labeled cells to appear in mitosis are those that were in late S-phase when the tritiated thymidine was added and the delay in time between adding the [³H]-TdR and detecting the labeled mitotic cells is the minimal duration of G2-phase. The fraction of the mitotic population that is labeled would then increase to a maximum of 1.0, which would occur at a time equal to the duration of G2- and M-phases. The fraction would then stay at 1.0 for a time equal to the length of S-phase, minus the length of M-phase (assuming all S-phase cells are unambiguously labeled). Then no labeled mitotic cells will appear until the cells have gone through a complete cycle and once again undergo mitosis. This type of analysis allows for the direct measurement of the length of all four phases of the life cycle. However, the actual situation is not as simple as the hypothetical curve shown. The points plotted in Fig. 14 are for Chinese hamster cells grown in suspension culture. Several factors combine to produce the less-than-ideal data shown. Most importantly, all cells do not traverse the cell cycle at the same rate. In addition, the FLM values are highly dependent on the criteria selected to define a labeled cell. For the method to work well, all cells would have to become unambiguously labeled at the moment the tritiated thymidine is introduced. Since the rate of DNA synthesis is not constant throughout S-phase (8,21), cells in different parts of S-phase will take up different amounts of the [^{3}H]-TdR. This produces S-phase cells with a wide range of grain counts and some cells can be misassigned to G1- and G2-phases. It is well known that the scoring of labeled mitotic cells can vary considerably among individuals (30), being very sensitive to the definition of a labeled cell (number of grains above background). All of these factors combine to make it difficult to graphically estimate the phase times from such data.

To improve the reliability of results from graphical analysis of FLM data, and to allow for corrections for such things as cells leaving the cycle or dying, mathematical analysis procedures were developed (23,25,31). A multicompartment model (31) will be used to illustrate the method. In this model, the life cycle is divided into a number of compartments (states), k. The total cycle time is T_C and individual phase times are T_{G1}, T_S, T_{G2}, and T_M. Since all cells do not move through the life cycle at the same rate, the model includes coefficients of variation in the phase transit times; CV_{G1}, CV_S, CV_{G2}, and CV_M. The compartments are numbered from 1 to k, where the first compartment in the cycle is the first compartment in G1-phase and the last compartment is the one in M-phase in which the cells divide. Cells move from one compartment to the next with a probability density, λ, where the probability density is the same throughout a phase. That is, there are four probability densities, λ_{G1}, λ_S, λ_{G2}, and λ_M. In this example all cells are assumed to be in cycle and stay in cycle. This leads to the relations, for S-phase:

$$T_S = N_S/\lambda_S \qquad [11]$$

and

$$CV_S = \frac{1}{\sqrt{N_S}} \qquad [12]$$

where N_S is the number of compartments. Similar relations hold for G1- and G2M-phase.

We can now derive a function that describes the way cells progress through the cycle, and that can be used to analyze the FLM data. The efflux of cells from the i'th compartment in an infinitesimal time interval Δt is:

$$\text{EFFLUX} = \lambda_i A_i(t)\Delta t \qquad [13]$$

where $A_i(t)$ is the number of cells in compartment i at time t. Influx into the compartment is equal to the efflux from the previous compartment.

$$\text{INFLUX} = \lambda_{i-1}A_{i-1}(t)\Delta t \qquad [14]$$

Therefore, the increment of cells in compartment i is:

$$\Delta A_i(t) = \text{INFLUX} - \text{EFFLUX} \qquad [15]$$

$$= \lambda_{i-1}A_{i-1}(t)\Delta t - \lambda_i A_i(t)\Delta t \qquad [16]$$

The influx of compartment 1 is a special case, since cell division occurs in the last compartment of the cell cycle, yielding two cells for the first compartment of G1-phase.

$$\Delta A_1(t) = 2\lambda_k A_k(t)\Delta t - \lambda_1 A_1(t)\Delta t \qquad [17]$$

By taking the limits of these equations, and for similar equations for each compartment, we obtain:

$$\frac{dA_1(t)}{dt} = 2\lambda_k A_k(t) - \lambda_1 A_1(t) \qquad [18]$$

$$\frac{dA_2(t)}{dt} = \lambda_1 A_1(t) - \lambda_2 A_2(t) \qquad [19]$$

$$\vdots$$

$$\frac{dA_k(t)}{dt} = \lambda_{k-1}A_{k-1}(t) - \lambda_k A_k(t) \qquad [20]$$

The number of equations is equal to k, the number of states, and is therefore controlled by the CVs. For example, if the CV of each phase is the same, 0.2, the number of equations is $N = (1/CV)^2$ = 25 per phase, for a total of 100.

There are a total of eight independent parameters to be derived from the data, four transit times and four coefficients of variation. This is often reduced to seven by setting CV_{G2} equal to CV_M. This set of differential equations can be solved for the number of cells in each compartment, $A_i(t)$, which then yields the number of cells in each phase. This method is easily applied to the analysis of FLM data, as follows. The number of cells in mitosis, M-phase, is the sum of the cells in all M-phase compartments. It changes with time as:

$$A_M(t) = A_M(0)e^{Et} \qquad [21]$$

where E is the exponential growth rate, and $A_M(0)$ is the number of cells in M-phase at time $t = 0$. If we denote the number of labeled cells in each compartment with an asterisk, then the FLM value at time t is:

$$FLM(t) = A_M^*(t)/A_M(t) \qquad [22]$$

where:

$$A_M^*(t) = \sum_{i=j}^{k} A_i(t) \qquad [23]$$

where j is $N_{G1} + N_S + N_{G2} + 1$, the beginning of M-phase. Given the initial values $A_i^*(0)$, we can solve the differential equations for the times at which we have measured data points of FLM, giving us the computed FLM (designated SFLM). Initial conditions are derived by assuming that only cells in S-phase are labeled at $t + 0$. Thus,

$$A_i^*(0) = 0 \text{ for } i + 1, N_{G1} \text{ and } i = N_s + 1, k$$

$$A_i^*(0) = A_i(0) \text{ for } i = N_{G1}, N_s$$

The computed FLM curve that best fits the measured points is determined by the method of least squares, as discussed earlier. The function being minimized is the mean squared deviation:

$$R = \frac{N}{N-6} \sum_{j=1}^{N} \frac{m_j}{M} [SFLM(t_j) - EFLM(t_j)]^2 \qquad [24]$$

where N is the number of data points, $N - 6$ is the number of degrees of freedom, m_j is the number of mitoses counted at the

j'th data point, M is the total number of mitoses observed, and $SFLM(t_j)$ and $EFLM(t_j)$ are the calculated and experimental FLM values, respectively, at t_j. There are many methods of minimizing R with respect to the parameters, as discussed earlier in this chapter.

The dashed line of Fig. 14 shows the results of applying this type of analysis. The derived parameters were: G1-phase, 4.5 (0.20) h; S-phase, 6.30 (0.20) h; G2 + M-phases, 1.69 (0.19) h; total cycle time, 12.54 (0.12) h. The coefficients of variation are shown in parentheses.

A modification of this technique that has been introduced by Gray et al. (*18,19*) utilizes cell sorting technology. As with the FLM method, in this procedure the cells are pulse labeled with tritiated thymidine. Then samples of the cell population are removed, stained with a DNA-specific dye, sorted from mid-S-phase and G1-phase, and the radioactivity per cell obtained by liquid scintillation counting. These variables are called $RCS_i(t)$ and $RCG1(t)$. Data similar to that seen for the FLM method are obtained, as shown in Fig. 15. The same models used for FLM analysis can be used, with some modification. A similar formalism to that presented earlier is used to describe the passage of radioactivity around the cycle. The radioactivity per state, R_i, is assumed to be proportional to the number of labeled cells in that state at time $t = 0$. The change in radioactivity in state i can be written as:

$$\Delta R_i(t + dt) = \lambda_{i-1} R_{i-1}(t) - \lambda_i R_i(t) \qquad [25]$$

where λ is the probability density for transition from one state to the next, in analogy to Eq. 15, and $R_i(t)$ is the radioactivity in state i at time t. Similar equations to Eqs. 17–19 can be written for each state, except that for state 1 there is no factor of 2 since the total radioactivity per state will not change at cell division. Carrying the analogy of the similarity to Eq. 20 further, we can write the time variation of the radioactivity per cell in a window in S-phase, $RCS_i(t)$, as:

$$RCS_i(t) = \sum_{j=m}^{n} R_j(t) / \sum_{j=m}^{n} A_j(t) \qquad [26]$$

where states m and n are the first and last states in the S-phase window. The numerator is the total radioactivity in the window and the denominator is the total number of cells in the window. This is the quantity plotted in Fig. 15. This method has several advantages over the FLM method: (1) autoradiography is not required,

Dean

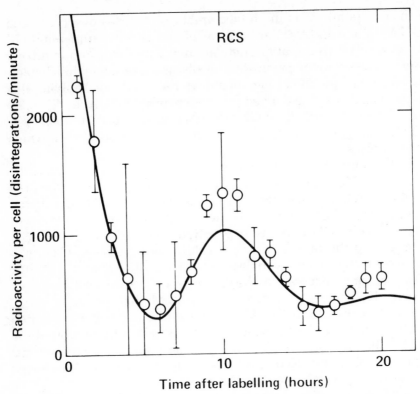

Fig. 15. Radioactivity per cell in mid-S-phase (RCS), for CHO cells in suspension culture; data obtained by sorting cells with DNA content from 1.4 to 1.5 times the G1 DNA content. The circles are the data points, the solid line is a computer fit to the data using a mathematical model. Phase durations obtained from the analysis were 2.4, 6.5, and 1.6 h for G1-, S-, and G2M-phases, respectively.

(2) it is faster and less subjective since microscopic examination is not required, and (3) more information can be obtained, e.g., rate of DNA synthesis through S-phase. Disadvantages include the necessity to disperse and stain the cells for processing and the need for a relatively expensive instrument, the cell sorter, for sorting the cells based on DNA content.

2.2.2. Perturbed Populations

As described in the Introduction, a major objective in cell kinetics is to be able to predict the effects of therapeutic agents on cell growth. To achieve this, more extensive data are needed to

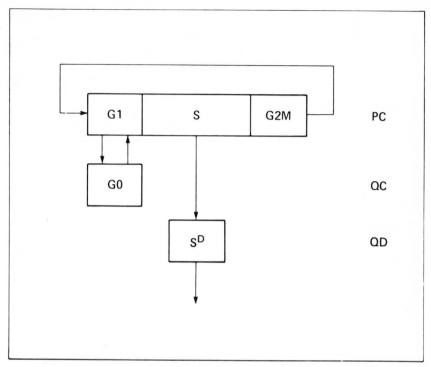

Fig. 16. A relatively simple model of the cell cycle, in which all cycling cells are assumed to be clonogenic (PC). Noncycling cells with G1 DNA content are in G0 phase and clonogenic (QC). Noncycling cells in S-phase are assumed to be destined to die (doomed, QD). Cell loss occurs only from S-phase.

fully define the effects produced by the agents. These data can be in the form of time sequences of DNA distributions, measurements of DNA precursor uptake, clonogenicity, and so on. The models must describe all of the effects known to be caused by the action of the agents, such as the killing of subpopulations and the decycling and recruitment of cells, on the cells under study. A typical cell-drug interaction is modeled conceptually in Fig. 16. In this model, describing the response of KHT cells to ara-C, there are three individual cell classes—proliferating, quiescent, and doomed. The latter class includes cells that are destined to die, but are still in the population and will be included in a measurement. Quiescent cells are those that are still viable, but have temporarily stopped cycling (maturing) or have slowed down considerably. Figure 17

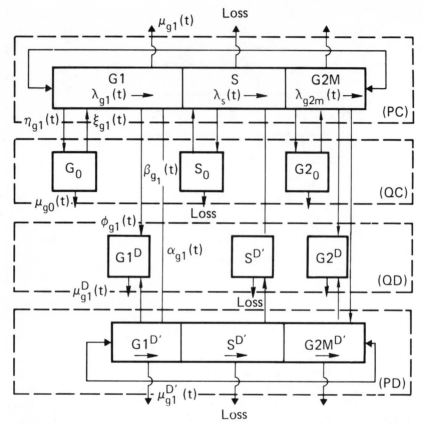

Fig. 17. A mathematical model of the life cycle, for cells undergoing therapy. Cells may be in four classes, proliferating and clonogenic (PC), quiescent and clonogenic (QC), quiescent and doomed (QD), and proliferating and doomed (PD). Quiescent and the nonproliferating are synonyms in this context.

shows a more general model, for cells in the midst of therapy (*18*). In this model there are four classes of cells; those that are proliferating and clonogenic (PC), proliferating but doomed (PD), quiescent and clonogenic (QC), and quiescent and doomed (QD). The arrows show how the cells might move between the different classes. The Greek characters are the rate constants for movement between the compartments. The model illustrated in Fig. 16 is really a subset of that illustrated in Fig. 17, and can be achieved by constraining the more general model. Many mathematical formulations have been derived to describe this type of model (*14,28,32,33*). Equation 27 shows one model in which the cell cycle is divided into a series of compartments. This formalism is an extension of the one used to describe FLM and RCS_i data in the previous section. With reference to Fig. 16, the number of cells in the i'th compartment of the PC class at time t is defined as $PC_i(t)$. The quantity that is of most interest in describing the kinetics of a population of cells is the rate of movement of the cells from one compartment to another. This can be described by the function:

$$\frac{dPC_i(t)}{dt} = \lambda_{i-1}(t)PC_{i-1}(t) + \zeta_i(t)QC_i(t) - [\lambda_i(t) + \eta_i(t)]PC_i(t)$$

[27]

where $QC_i(t)$ is the number of cells in the i'th compartment of class QC at time t; $\lambda_i(t)$ is the rate parameter for transition from the PC_i class to the PC_{i+i} class (continued proliferation); $\zeta_i(t)$ is the rate parameter for transition from the QC_i class to the PC_i class (recruitment); and $\eta_i(t)$ is the rate parameter for transition from the PC_i class to the QC_i class (cell cycle traverse inhibition). This equation describes transitions between the PC and QC classes for cells in G1-phase. Similar equations are used to describe movement of cells between all compartments of the model. Given the initial distribution of cells among the compartments, and the rate parameters, the series of equations can be solved for the number of cells in each compartment at any time. In practice, the rate parameters are adjusted until, by visual examination, the calculated data closely match the experimental data. One can also use these equations to test the effects of cell cycle-specific therapeutic agents. Figure 18 illustrates an application of this model to KHT tumor-bearing animals after treatment with 100 mg/kg ara-C in vivo (26,18). The harvested cells were stained with propidium iodide for FCM measurement. The solid circles are the measured data points. The solid line was obtained by varying the transition rates in the model depicted in Fig. 16 and computing the distribution at the different time points after treatment. The transition rates that were determined to fit the data the best are presented in Table 2. These transition rates are then used to predict the response of the tumor to different treatment schedules.

This type of analysis, although powerful, still has limitations. Acquisition of the data is still laborious and subject to errors, such as the determination of phase fractions from DNA histograms. The

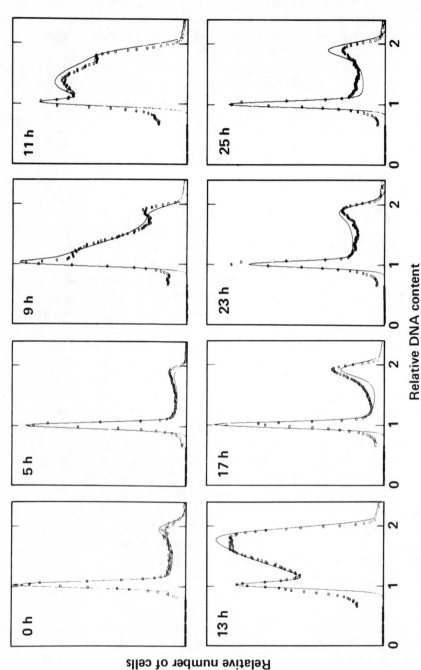

Relative DNA content

Fig. 18. DNA histograms for KHT cells from tumor-bearing animals treated with 100 mg/kg ara-C in vivo. The experimental data are shown as circles, the computer simulations are solid lines. The parameters used to obtain the curves are shown in Table 2.

TABLE 2
Rate Parameters Derived From Mathematical Analysis
of Data Obtained From ara-C-Treated KHT Tumor Cells[a]

$t < 5.5$ h		$t \geqslant 13.5$ h
λ_{G1}	$= 6.0$	12.0
λ_s	$= 1.0$	2.1
λ_{G2M}	$= 6.0$	12.0
$t < 5.5$ h		$t \geqslant 5.5$ h
$\zeta_{G0} = 0.0$		1.0
$\eta_s = 100$		0.0
$\zeta_{G1} = 100$		0.0
$t < 10$ h		$t \geqslant 10$ h
$\mu_s = 0.0$		0.5

[a]Rate parameters are explained in the text and by referring to Fig. 16.

BrdUrd method described earlier in section 1 can also be used to obtain kinetic data for a wide range of experimental conditions for asynchronously growing and perturbed cells. The method was also applied to C3H mice bearing KHT sarcomas and treated in vivo with 100 mg/kg ara-C. At several times after administration of the ara-C, the animals received 100 mg/kg BrdUrd. Thirty minutes after administration of the BrdUrd, the tumors were harvested, dispersed, and prepared for BrdUrd/DNA analysis as previously described. Figure 19 shows a bivariate distribution for the cells before exposure to the ara-C. Both the normal and tumor cell populations are readily apparent, as are the G1, S, and G2M cells of the tumor. Figure 20 shows contour plots of the data obtained at various times after exposure to the ara-C. As illustrated in the distribution of the control population, boxes are used to define three regions of interest: G1-phase cells; S-phase cells with high BrdUrd content; and S-phase cells with low BrdUrd content. The two S-phase boxes distinguish S-phase cells that can incorporate BrdUrd (i.e., that are probably capable of continued proliferation) from those that cannot. For example, at 2 h after administration of the ara-C, none of the cells are capable of incorporating the BrdUrd, whereas at 6 h, cells have begun to move from G1- to S-phase. The fractions of the population in each of the boxes are plotted vs time after ara-C, as shown in Fig. 21. The models described earlier are then used to analyze the data, with the following results:

KHT TUMOR PULSE LABELED IN VIVO

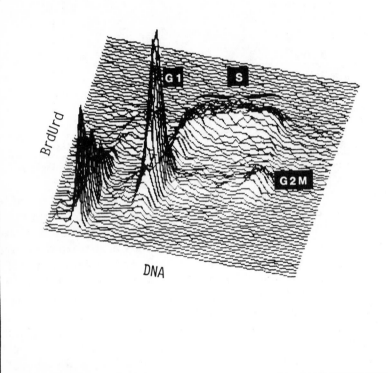

Fig. 19. Bivariate distribution of BrdUrd content vs DNA content for KHT cells, before treatment with ara-C.

1. A three-fold increase in cell cycle traverse rate occurs at 10 h post-ara-C. The altered phase durations of the perturbed population are:

    TG1   = 1.2 h
    TS   = 4.1 h
    TG2M = 1.4 h

2. Cell loss begins 8 h post-ara-C, at 5%/h.
3. G1/S block released at 4.5 h post-ara-C.

KHT SARCOMA RESPONSE TO ARA-C IN VIVO

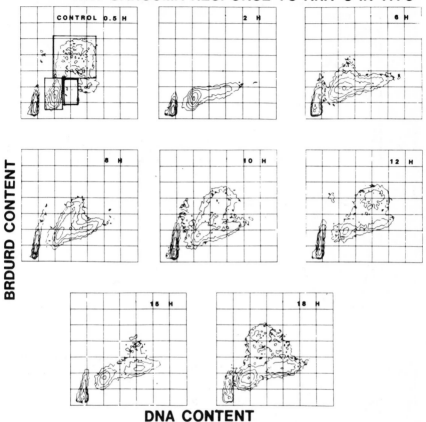

Fig. 20. Contour plots of the KHT cells from Fig. 19, at various times after treatment with ara-C.

These cytokinetic estimates agree well with those obtained using the radioactive tracer technique, RCS_i, as described in the previous paragraph.

Bivariate distributions of incorporated BrdUrd vs DNA content also provide information about the phase durations and dispersions of asynchronously growing cells. To accomplish this a mathematical procedure has been developed that includes a model of cell cycle traverse and a model for incorporation of BrdUrd (34). With information from these models, distributions of BrdUrd content vs DNA content are generated and compared with the experimental data. The cell cycle is modeled as being composed of a series of in-

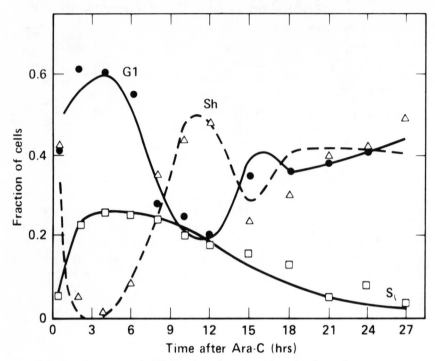

Fig. 21. Fraction of cells in G1-phase, and in S-phase for two groups of cells; those with high and with low BrdUrd contents. Boxes showing these two S-phase populations are shown in Fig. 20. The solid lines are fits to the data using a cell cycle model.

finitesimally narrow compartments. Cells mature by progressing through the compartments according to an exponential probability function. This procedure results in a series of differential equations, as described earlier for other multicompartment models. Incorporation of BrdUrd into cells assumes that the amount incorporated is directly proportional to the amount of DNA synthesized. As cells in S-phase mature, they synthesize DNA and incorporate BrdUrd. When a cell divides, the amount of BrdUrd in the cell is halved, as is the DNA content. Both continuous exposure and short time exposures to the BrdUrd can be modeled. The free parameters of this model are the mean total cell cycle duration, the duration of each phase, with the coefficient of variation of each time (time dispersion), and the DNA synthesis rate in S-phase.

Starting with initial estimates of the free parameters, distributions of DNA content vs BrdUrd content are generated. DNA content comes from evaluating the maturity distribution; the rate of DNA synthesis determines the amount of BrdUrd incorporated. The simulation takes measurement errors into account by applying a two-dimensional Gaussian broadening function to the calculated distributions. The simulated distributions are then compared to the experimental data, and the parameters adjusted until there is good agreement between them. At this time, the quality of fit is determined only by visual examination. Figure 22 shows data for Chinese hamster ovary (CHO) cells measured at various times after exposure to a 30-min pulse of BrdUrd. Figure 23 shows the final simulations for the same time plots. Although there are some differences between the sets of distributions, the agreement is quite good. The major disagreement is probably a result of errors in the assumed rate of DNA synthesis in S-phase, and its effect on incorporated BrdUrd. The phase durations (and their coefficients of variation) that resulted from the calculations were 5.6 h (0.08), 7.0 h (0.07), 7.4 h (0.16), and 14.0 h (0.05), for G1, S, G2M and total cell cycle duration, respectively. These values agree well with other published values.

3. CONCLUSIONS

Many parameters can be used to characterize a cell population: e.g., maturity; size; DNA, RNA, and protein content; the presence of specific cell surface receptors. This chapter has dealt only with the study of cell kinetics through DNA content and incorporation of BrdUrd.

Only a few of the many models proposed for the analysis of DNA distributions have been discussed. However, the models presented are representative and in the authors opinion are those most in use today. Results using graphical methods for the analysis of DNA distributions are reliable only for the G1-phase fractions and generally only for data with low CVs. The G1-phase fraction will be overestimated, but for low CVs the calculation error will probably be lower than the experimental error. The methods are not recommended for S-phase and particularly not for G2M-phase. With the advent of microcomputers and the constantly decreasing costs of minicomputers, more suitable and accurate mathematical models should be utilized. Graphical methods work well, however,

Experimental Data

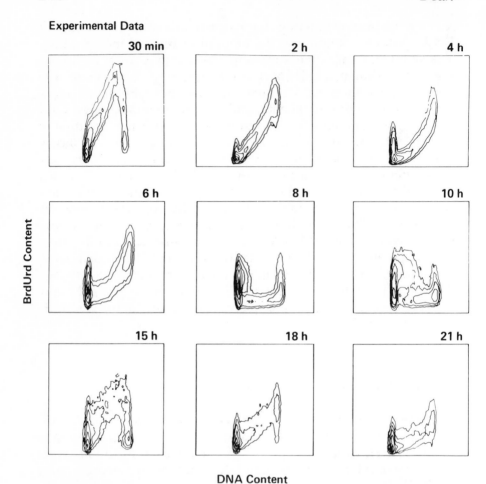

DNA Content

Fig. 22. Bivariate contour plots of BrdUrd content vs DNA for CHO cells, at various times after a 30-min exposure to the BrdUrd. The data show the labeled cohort of cells progressing through the cell cycle.

for BrdUrd/DNA distributions, where the G1, S, and G2M subpopulations are more clearly resolved.

The SFIT model is indicated for use with DNA distributions from asynchronously growing cells where the peaks are truly Gaussian shaped (symmetrical) and the CV is low (5% or less). This is a simple model suitable for use on microcomputers. For data where the G1 and G2M peak means are not so well defined, and where the CV is greater, the broadened polynomial method is recom-

Simulations

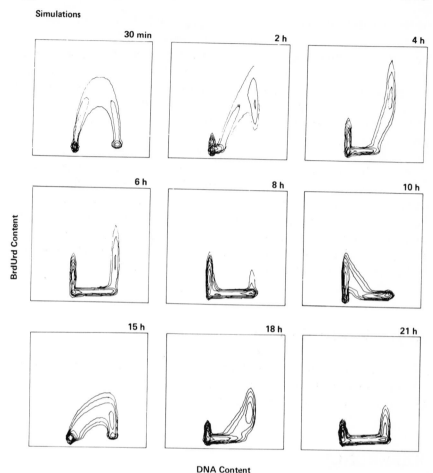

DNA Content

Fig. 23. Bivariate contour plots as in Fig. 22, simulated with a mathematical model of the cell cycle, as explained in the text. The parameters of the model are adjusted until the two sets of histograms are as alike as possible, yielding estimates of phase times and dispersions.

mended. Computation times will be longer, on the order of 2 min on a minicomputer (e.g., PDP-11, Digital Equipment Corp.). This model is not suitable for microcomputers. For perturbed cells, the broadened rectangle method is recommended as the more acurate model, although there is some penalty in computation time relative to the multiple Gaussian model. For some data it may be necessary for the user to specify the G1 and G2M peak means, and in some circumstances even the CVs. In any event, errors in phase fraction estimates are likely to be on the order of 5–10% for G1, 10–20% for S, and 20–100% for G2M phases, resulting primarily from experimental errors (e.g., not obtaining a representative sample of the cells), not from the mathematical analysis procedure. Finally, any good mathematical analysis will not only provide estimates of the phase fractions, peak means, and CVs, but in addition will provide estimates of the errors associated with these estimates. Please take note of them.

There are many mathematical models that are used to make a comprehensive analysis of cell kinetic data, with a wide range of complexity. Some of these models have been described. With the advent of a monoclonal antibody to incorporated BrdUrd, the FLM method of analysis, popular for many years, may soon be displaced. There are still some problems, but a large effort is now being expended to solve them, and the BrdUrd method is expected to become the standard method in the near future.

ACKNOWLEDGMENT

Work was performed under the auspices of the US Department of Energy by the Lawrence Livermore National Laboratory, under contract number W-7405-ENG-48.

REFERENCES

1. Bagwell, C. B. Theory and application of DNA histogram analysis. PhD thesis. Florida: Univ. of Miami, 1979.
2. Baisch, H., Gohde, W., and Linden, W. Analysis of PCP-data to determine the fraction of cells in the various phases of the cell cycle. Rad. Environ. Biophys., 12: 31–39, 1975.
3. Barlogie, B., Drewinko, B., Johnston, D. A., Buchner, T., Hauss, W. H., and Freidreich, E. J. Pulse cytophotometric analysis of synchronized cells *in vitro*. Cancer Res., 36: 1176–1181, 1976.
4. Bevington, P. Data Reduction and Error Analysis for the Physical Sciences. New York: McGraw Hill, 1969.
5. Darzynkiewicz, Z., Traganos, F., and Kimmel, M. Assay of cell cycle kinetics by multicompartment flow cytometry using the principle of

stathmokinetics. *In:* (J. W. Gray and Z. Darzynkiewicz, eds.), Techniques in Cell Cycle Analysis. New Jersey: Humana, 1986.

6. Dean, P. N. A simplified method of DNA distribution analysis. Cell Tissue Kinet., *13:* 299–308, 1980.

7. Dean, P. N. Methods of data analysis in flow cytometry. *In:* (M. Van Dilla, P. Dean, O. Laerum, M. Melamed, eds.), Progress in Flow Cytometry, San Francisco: Academic, 1985.

8. Dean, P. N. and Anderson, E. C. The rate of DNA synthesis during S phase by mammalian cells in vitro. *In:* (C. A. Haanen, H. F. P. Hillen, and J. M. Wessels, eds.), Pulse-Cytophotometry, Ghent, Belgium: European Press, 1975.

9. Dean, P. N. and Jett, J. H. Mathematical analysis of DNA distributions derived from flow microfluorometry. J. Cell Biol., *60:* 523–527, 1974.

10. Dean, P. N., Dolbeare, F. A., Gratzner, H. G., Rice, G. C., and Gray, J. W. Cell cycle analysis using a monoclonal antibody to BrdUrd. Cell Tissue Kinet., *17:* 427–436, 1984.

11. Dean, P. N., Gray, J. W., and Dolbeare, F. A. The analysis and interpretation of DNA distributions measured by flow cytometry. Cytometry, *3:* 188–195, 1982.

12. Dolbeare, F. A., Gratzner, H. G., Pallavicini, M. G., and Gray, J. W. Flow cytometric measurement of total DNA content and incorporated bromodeoxyuridine. Proceed. Natl. Acad. Sci. USA, *80:* 5573–5577, 1983.

13. Dolbeare, F. A., Beisker, W., Pallavicini, M. G., and Gray, J. W. Cytochemistry for bromodeoxyuridine/DNA analysis: Stoichiometry and sensitivity. Cytometry, *6:* 521–530, 1985.

14. Eisen, M. Mathematical Models in Cell Biology, Lecture Notes in Biomathematics, New York: Springer-Verlag, 1979.

15. Fried, J. Method for the quantitative evaluation of data from flow microfluorometry. Comp. Biomed. Res., *9:* 263–276, 1976.

16. Gratzner, H. G. Monoclonal antibody to 5-bromo- and 5-iododeoxyuridine: A new reagent for detection of DNA replication. Science, *107:* 474–475, 1982.

17. Gray, J. W. Cell-cycle analysis of perturbed cell populations: Computer simulation of sequential DNA distributions. Cell Tissue Kinet., *9:* 499–516, 1976.

18. Gray, J. W. Quantitative cytokinetics: Cellular response to cell cycle specific agents. Pharmac. Ther., *22:* 163–197, 1983.

19. Gray, J. W., Carver, J. H., George, Y. S., and Mendelsohn, M. L. Rapid cell cycle analysis by measurement of the radioactivity per cell

in a narrow window in S phase (RCSi). Cell Tissue Kinet., *10*: 97, 1977.

20. Gray, J. W., Dolbeare, F., Pallavicini, M., and Vanderlaan, M. Flow cytokinetics. *In*: (J. W. Gray and Z. Darzynkiewicz, eds.), Techniques in Cell Cycle Analysis. New Jersey: Humana, 1986.

21. Gray, J. W., Pallavicini, M. G., George, Y. S., Groppi, V., Look, M., and Dean, P. N. Rates of incorporation of radioactive molecules during the cell cycle. J. Cell. Phys., *108*: 144, 1981.

22. Grdina, D., Meistrich, M., Meyn, R., Johnson, T., and White, A. Cell synchrony techniques: A comparison of methods. *In*: (J. W. Gray and Z. Darzynkiewicz, eds.), Techniques in Cell Cycle Analysis. New Jersey: Humana, 1986.

23. Hartmann, N. R., Gilbert, C. W., Jansson, B., MacDonald, P. D. M., Steel, G. G., and Valleron, A. J. A comparison of computer methods for the analysis of fraction labeled mitosis curves. Cell Tissue Kinet., *8*: 119, 1975.

24. Howard, A. and Pelc, S. R. Synthesis of deoxyribonucleic acid in normal and irradiated cells and its relation to chromosome breakage. Heredity, *6* (suppl): 261–273, 1953.

25. Mendelsohn, M. L. Computer analyzed transit times for vertebrate cell cycle and phases. *In*: (P. A. Altman and D. D. Katz, eds.), Cell Biology, Bethesda, Maryland: Federation of American Societies for Experimental Biology, 1976.

26. Pallavicini, M. G., Gray, J. W., and Folstad, L. Quantitative analysis of the cytokinetic response of KHT tumors in vivo to cytosine arabinoside. Cancer Res., *42*: 3125–3131, 1982.

27. Quastler, H. and Sherman, F. G. Cell population kinetics in the intestinal epithelium of the mouse. Exp. Cell Res., *17*: 420–438, 1959.

28. Rotenberg, M., ed. Biomathematics and Cell Kinetics, Developments in Cell Biology, vol. 8, New York: Elsevier/North-Holland, 1981.

29. Shackney, S. and Ritch, P. Percent labeled mitosis curve analysis. *In*: (J. W. Gray and Z. Darzynkiewicz, eds.), Techniques in Cell Cycle Analysis. New Jersey: Humana, 1986.

30. Shackney, S. S., Ford, S. S., and Wittig, A. B. The effects of counting threshold and emulsion exposure duration on the percent-labeled mitosis curve and their implications for cell cycle analysis. Cancer Res., *33*: 2726–2731, 1973.

31. Takahashi, M., Hogg, S. D., and Mendelsohn, M. L. The automatic analysis of FLM curves. Cell Tissue Kinet., *4*: 505–518, 1971.

32. Valleron, A. J., ed. Mathematical Models in Cell Kinetics. Medikon, Ghent, Belgium: European Press, 1975.

33. Valleron, A. J. and MacDonald, P. D. M., eds. Biomathematics and Cell Kinetics, vol. 2, New York: Elsevier/North Holland, 1978.

34. Yanagisawa, M., Dolbeare, F., Todoroki, T., and Gray, J. W. Cell cycle analysis using numerical simulation of bivariate DNA/bromo-deoxyuridine distributions. Cytometry, in press, 1985.

Chapter 9

Cytochemical Probes of Cycling and Quiescent Cells Applicable to Flow Cytometry

Zbigniew Darzynkiewicz

1. INTRODUCTION

1.1. Definition of Quiescence/Classification of Quiescent States

For convenience the term quiescent cells will be used in this chapter to define normal or tumor cells, untreated by any specific drugs, that do not progress through the mitotic cycle for an extended period of time. The definition is purposefully wide to cover a variety of situations, often with different metabolic causes, associated with irreversible or transient cell withdrawal from the cell cycle. The even wider term, noncycling cells, will be reserved to describe, in addition to quiescent cells, cells arrested in the cycle by drugs (Table 1). (*See* chapter 3 for additional discussion of cellular quiescence.)

A variety of terms is used in the literature to describe cells that do not progress through the mitotic cycle. "Dormant," "resting," or "G0" cells are terms most often given to cells that are potentially able to proliferate (e.g., stem cells), but that have left the cell cycle temporarily and remain in the dormant state until an appropriate stimulus triggers their reentrance to the mitotic cycle (6–8,34, 35,52,57,61,69,94,103,110,115). The classic examples of such cells are nonstimulated lymphocytes normally found in the peripheral

TABLE 1
Categories of the Noncycling Cells

Noncyling cells	Characteristics	Examples
Quiescent cells	"G0," "dormant," or "resting" cells. Stem cells temporarily withdrawn from the cell cycle	Nonstimulated lymphocytes, stem hepatocytes, density-inhibited normal diploid fibroblasts
	Differentiated or differentiating cells	Granulocytes, mast cells, plasma cells, nerve cells, adipocytes, avian erythrocytes
	Nutrient- or oxygen-deprived cells	Cells in serum-isoleucine deprived or unfed cultures, anoxic cells in tumors, inner cells in spheroids
Cells arrested by drugs	Negative unbalanced growth (drugs preferentially affecting cell growth)	Actinomycin D- or cycloheximide-treated cells
	Positive unbalanced growth (drugs primarily affecting DNA division cycle)	Ara-C-, methotrexate-, or hydroxyurea-treated cells

blood. The lymphocytes can be triggered by antigens or nonspecific mitogens to enter a prereplicative phase (~ 24 h), and then to progress through S and G2 and divide (*89*). After several rounds of division, the lymphocytes return to quiescence and remain in that state until triggered again. Stem hepatocytes represent another type of dormant cells, and partial hepatectomy provides a signal for their proliferation. Normal, diploid fibroblasts that cease proliferation in crowded cultures resemble in many respects dormant lymphocytes or hepatocytes. Following trypsinization and reseeding at low density, these cells also enter a long prereplicative phase, the duration of which is proportional to the "depth" of their prior quiescence (*8*).

Another category of quiescent cells includes cells that undergo differentiation. This process consists of several stages. Depending on the cell type, the cells' ability to proliferate is lost at different stages of the differentiation process. Ordinarily, differentiation is terminal and only in exceptional situations (e.g., after cell fusion) can differentiated cells undergo dedifferentiation and reenter the mitotic cycle.

Cells that cease proliferation because of deprivation of nutrients or oxygen can be grouped in still another category. This is a very broad group and its members differ considerably depending on the cell type, paucity of the particular nutrient, and length and severity of deprivation. A more detailed description of some of these cell systems will be given later in this chapter.

The noncycling cells arrested in the cell cycle by drugs are not a subject of this review. Their general characteristics, however, should be mentioned because these cells sometimes exhibit properties similar to those of quiescent cells; thus their distinction may be of importance. The drug-arrested cells can be divided into two broad categories. In the first category are cells arrested by the drugs whose target is cell growth rather than the DNA division cycle. Thus, for instance, at low concentrations of actinomycin D, the synthesis of rRNA and subsequently proteins (i.e., cell growth) is suppressed preferentially; with time the actinomycin-D-treated cells cease progression through the cell cycle. Likewise, direct inhibition of protein synthesis primarily affects cell growth and subsequently the DNA division cycle. The second group contains cells whose DNA division cycle (DNA replication-division cycle) is the primary target of a given drug. Mitchison's concept (94), separating cell activities on the basis of growth cycle and DNA division cycle, used in the above classification, is convenient because the cell metabolic parameters that can be measured by flow cytometry are different in these two groups.

The classification introduced above (Table 1), as well as the classification based on DNA content and metabolic markers (Table 2), will both be referred to later in the chapter for discussion of particular metabolic parameters that can be measured by flow cytometry.

1.2. Metabolic Properties of Quiescent Cells

Different classes of quiescent cells exhibit different metabolic features. The metabolic properties of quiescent cells will be discussed in general before analyzing in detail particular cell features that can be measured by flow cytometry, which may in turn serve

Table 2
Noncycling Cell Populations with Different Kinetic and Metabolic Properties
That Can be Distinguished by Multiparameter Flow Cytometry

Cells[a]	Characteristics	Examples
G1Q	Quiescent cells with a G1 DNA content and very low metabolic activity. The metabolic parameters (e.g., RNA or mitochondrial contents, DNA denaturability) of these cells are distinct (no overlap) from those of G1 cells of exponentially growing populations	Peripheral blood lymphocytes (*42,43,47*), 3T3 cells in contact-inhibited cultures (*41*)
SQ, G2Q	Quiescent cells with an S or G2 DNA content, respectively, and very low metabolic activity. The metabolic parameters are distinct from those of S or G2 cells in exponentially growing populations	Chronic myeloid leukemia-blastic stage (*33,43*)
G1D	Differentiated or differentiating cells with a G1 DNA content. The "differentiated cell" phenotype	DMSO-treated-Friend leukemia, plasma, or mast cells (*45,142*)
G1A (subthreshold population)	Cells present in cultures during exponential growth, characterized by a G1 DNA content and significantly different (no overlap) RNA content or denaturability of DNA in comparison with early-S cells. The cell residence times in G1A are widely distributed with the exponential	Cells present in exponentially growing cultures (*39,41,45*)

G1A-arrested	or pseudoexponential component. Kinetic features resemble the indeterminate "A" state (123)	
G1A-arrested	G1-quiescent cells that have metabolic features similar to G1A cells from exponentially growing populations, but do not enter S	3T3 cells in serum-deprived cultures (41), CHO cells in isoleucine-deprived cultures, n-butyrate-treated cells (50)
G1-, S-, or G2-arrested (b−)	Noncycling cells, arrested in G1, S, or G2 by the drugs, which exhibit negative unbalanced growth (b−). The growth parameters (RNA, protein content) are diminished in comparison with exponentially growing cells	Cells treated with actinomycin D or dihydroxy-5-azacytidine (140)
G1-, S-, or G2-arrested (b+)	Noncycling cells arrested in G1, S, or G2 by the drugs, characterized by positive unbalanced growth. Increased RNA, protein, or mitochondrial content	Cells treated with hydroxyurea, ara-C, or ellipticine (140)
G1-, S-, or G2-arrested (b)	Noncycling cells arrested in G1, S, or G2 that have metabolic features similar to cycling cells from exponentially growing populations	Anoxic cells from solid tumors (17)

[a] Abbreviations: b, balanced growth; b+, positive unbalance; b−, negative unbalance.

as markers characterizing cycling vs noncycling populations. Table 2 presents differences between noncycling cells observed in various cell systems regardless of the cause. This is in contrast to Table 1, which lists different categories of quiescent cells based on causes or mechanisms inducing quiescence. The differences reflecting metabolic heterogeneity presented in Table 2 were detected by multiparameter flow cytometry and provide a basis for subdividing the noncycling cells into several distinct groups (34,35,45). This classification represents an attempt to analyze cell populations using objective, measurable parameters.

1.2.1. G1Q Cells

Cells in the G1Q compartment in general are characterized by a G1 DNA content and very low metabolic activity. As a criterion for recognition of G1Q cells, it was proposed that these cells have significantly lower RNA content compared to their cycling counterparts (34,35,45,47). Thus, they do not overlap at all in RNA values with exponentially growing populations of cells of the same type. G1Q cells also have a distinctly lower (no overlap) stainability of mitochondria with the fluorescent dye rhodamine 123 (R123) in comparison with cycling cells (42). These cells can also be totally separated from their cycling counterparts based on the increased denaturability of DNA *in situ* (as detected following staining with acridine orange), a parameter reflecting the high degree of condensation of their nuclear chromatin (43). In summary, G1Q cells are in deep quiescence, have low metabolic activity, and in most cell systems are equivalent to G0 or dormant cells, as originally defined by Lajtha (84). Nonstimulated lymphocytes or 3T3 cells maintained in cultures at high cell density for an extended period of time exhibit characteristics of G1Q cells (41).

1.2.2. SQ and G2Q Cells

SQ or G2Q cells are quiescent cells with a DNA content equal to that of cells in S- or G2-phase, respectively. These cells are characterized by low RNA content and condensed chromatin, at least in certain cell systems (34,43,45). By flow cytometric measurements, the SQ or G2Q cells are totally separated from cycling cells in S- or G2-phase based on differences in RNA content or chromatin structure. Quiescent G2 cells were first detected by Gelfant in epidermis; after stimulation, those cells were able to rapidly enter the cell cycle (60,61). Quiescent S and G2 cells were observed in the blood of patients with chronic myeloid leukemia during blastic crisis (*33,43*). When transferred to tissue culture, those cells rapidly accumulated RNA, then resumed progression through the cycle. Cells with this characteristic were also seen in stationary (unfed) cultures of HeLa cells (*10*) and in multicellular tumor spheroids (*2,12*).

1.2.3. G1D Cells

Concomitant with differentiation, cells lose their proliferative ability. Such cells can then be distinguished by the presence of the differentiation-specific products such as mRNA, enzyme(s), or other protein(s) (e.g., hemoglobin). At the final stages of differentiation the cells are characterized by a G1-DNA content (G1D cells). Differentiation may involve either loss or gain of a particular metabolic marker, depending on the cell type. For instance, during erythroid or myeloid differentiation, a marked decrease in RNA content and increased denaturability of DNA *in situ* parallel a decrease in cell proliferation (*142*). A decrease in cell stainability with the mitochondrial probe R123 is also observed (*26,49*). On the other hand, compared to their stem cells, fully differentiated plasma cells have increased RNA content (*45*). Thus, the metabolic features of the quiescent differentiated (G1D) cell are specific for the particular differentiated phenotype, rather than the cells possess a common pattern related to quiescence. As proposed before (*45*), based on observations of Friend erythroleukemia (FL) cells diffferentiating in the presence of dimethylsulfoxide (DMSO), the term G1D should be restricted to cells that are withdrawn from the cell cycle and exhibit the differentiated phenotype, whereas the term G1Q should be used to describe quiescent cells that do not show any differentiation-specific features.

1.2.4. G1A Cells

Another category listed in Table 2 contains G1A cells. It has been shown (*34,39,41,45*) that during exponential growth, G1-phase cells must attain a critical RNA content before entering S-phase. The cells with RNA values below this threshold were classified as G1A. A critical threshold in G1 related to chromatin structure was also detected (*34,35*). Based on this threshold, the G1 cells can also be subdivided into G1A and G1B, with G1A cells having chromatin significantly more sensitive to denaturation than early-S cells (*22,31*). Although there are minor differences in the number of G1A cells discriminated based on RNA content vs chromatin structure, the kinetic properties of these cells are similar (34). A detailed description of G1A and G1B cells and a discussion on possibly different roles of these compartments during the cells' lifetime are given elsewhere (34,35,39). Briefly, the distinction between the G1A and G1B compartments fits neatly into the concept of Mitchison (94), discriminating two parallel cell activities; the growth and the DNA division cycle. The G1A compartment appears to be the growth or equalization subphase of G1, during which cells attain a critical RNA content or decondense chromatin to a critical level that allows them to enter S-phase (34,35,39). According to this concept, the G1A represents a gap or interruption of the DNA division cycle; during the interruption, only cell growth takes place. In contrast, during G1B growth continues, but the DNA division cycle is now resumed and both activities progress in parallel. The threshold dividing G1A and G1B represents, therefore, a metabolic landmark discriminating between the growth subphase and onset of the DNA division cycle. The G1A compartment may be equivalent to the portion of G1 during which cells acquire ''competence'' to proliferation (106,132). Recent data of Seuwen et al. (118) and Adam et al. (1) indicate that the competence signal is indeed associated with accumulation of RNA, and the progression depends on achievement of a threshold number of ribosomes.

The characteristic feature of G1A cells is a wide distribution of their residence times in this compartment with a prominent exponential or quasi-exponential ''tail'' (34,35,39). Evidence for this comes from the exponentially declining slope of the G1A exit curve during stathmokinetic experiments (e.g., *see* chapter 10 in this monograph). Thus, even in exponentially growing populations there are cells that exhibit unusually long residence times in G1A prior to entrance into S. Kinetically, such cells can be considered noncycling. The kinetics of cell transit through G1A therefore justifies listing G1A cells together with other types of noncycling cells (Table 2). Although most likely a consequence of the metabolic differences between the postmitotic cells (39), the kinetics of cell transit through G1A resemble those proposed by Smith and Martin (123) for the ''indeterminate-A state'' of the cell cycle.

1.2.5. G1A-Arrested Cells

There are situations in which cells do not progress through G1 phase at all, and while remaining in G1, exhibit metabolic features of G1A cells. These G1A-quiescent, or G1A-arrested, cells have RNA content significantly lower than cells entering S-phase, but still higher than cells in G1Q (*41,50*). Likewise, the degree of chromatin condensation of these cells resembles that of G1A rather than G1Q cells. When stimulated, the G1A-quiescent cells enter S-phase, but after a considerable delay. Cells with the above characteristics were observed in 3T3 cell cultures maintained at low (0.5%) serum concentration (*41*) in tumor-origin cells growing at high density, treated with *n*-butyrate (*50*) or the mitochondrial antimetabolite R123 (*49*). Prolonged arrest in G1A under these conditions resulted in cell death rather than in a further decrease in RNA content to the level of G1Q cells. Human leukemic cells separated by elutriation into a population recognized by autoradiography as quiescent contained increased fractions of G1A cells (*109*). It is possible that normal cells under "normal" conditions facilitating their transition to quiescence (chalones, lack of mitogens) enter G1Q, whereas tumor cells remain arrested in G1A. Indeed, biochemical studies indicate that the quiescence of normal cells is associated with a marked loss of ribosomes in contrast to quiescent tumor cells, which maintain a rather high RNA content (*125*).

1.2.6. Drug-Arrested Cells

Drug treatments often produce unbalanced cell growth. The unbalance may be either negative or positive (*107*). The negative unbalance is observed when the drugs affect anabolic processes of the cell more severely than the DNA division cycle. The cells could then progress through the cell cycle, at least through a portion of it, but exhibit diminished RNA/DNA or protein/DNA ratio in comparison with cells that grow exponentially (*140*). With time the cells either die or recover when the drug is removed. No quantitative data exist as to what extent and for how long cells can tolerate negative unbalance (e.g., the degree of RNA/DNA decrease) without losing their proliferative potential. Negative unbalanced growth has been observed in cultures treated with dihydro-5-azacytidine (*140*) or 0.05 μg/mL of actinomycin D (data unpublished). L1210 cells in cultures containing agents such as *n*-butyrate, retinoic acid, or R123 arrest in G1A, i.e., as a population they have somewhat lowered mean values of RNA or protein in comparison with cells growing exponentially.

Positive unbalanced growth occurs when the DNA division cycle is specifically inhibited but growth continues. In this situation individual cells may increase in size and accumulate increasing quantities of RNA and proteins so that the RNA/DNA or protein/ DNA ratios increase. At a certain point during this positive unbalanced growth, the cells lose viability. Prior to that, if the inhibitors are removed, the cells can recover and progress through the cell cycle or even through the next several cycles (until they assume "normal" or "balanced" proportions of RNA/DNA or size) at markedly accelerated rates (59). This rapid proliferation is, perhaps, an expression of the regulatory mechanisms compensating for the unbalance. Positive unbalanced growth has been observed in cultures treated with hydroxyurea (59), methotrexate (138), or dihydroxyanthraquinone (140).

As briefly outlined above, there are different causes and different metabolic features of the cell that are associated with suppression of cell progression through the cell cycle. The classification introduced by us (41,45) is the first attempt to subdivide quiescent cells based on objective parameters that can be measured by flow cytometry. Because the "depth" of quiescence, cell survivability, sensitivity to drugs, ability to reenter into cell cycle, kinetics of the reentrance into the cell cycle, all appear to be different for each of these groups, such a subdivision could be of clinical relevance. In situations in which analysis of the patient's tumor cell population is possible (e.g., obtained by the biopsy), the knowledge of the kinetic and metabolic properties of tumor cells may be helpful in designing the treatment strategy.

In the remaining portion of this chapter, different cell parameters related to the quiescence or cycling state that can be measured by flow cytometry will be described and discussed in light of the classification scheme introduced above (Table 2).

2. CELL SIZE

2.1. Dry Weight; Image Analysis

It is generally accepted that G0 cells, classified here as G1Q, are smaller than their counterparts that are progressing through the cell cycle. For instance, the diameter of nonstimulated lymphocytes (on smears) is between 8 and 12 μm, whereas the diameter of phytohemoglutinin-stimulated lymphoblasts is between 20 and 30 μm (89). The dry weight of the stimulated lymphocytes (measured by interference microscopy) is between five and seven times higher than the weight of lymphocytes prior to stimulation (*40,129*). Both the increase in size of nucleus and cytoplasm contribute to cell enlargement. The enlargement can be conveniently measured on smears by image-analyzing instruments such as the Quantimet (*18*).

Extensive literature exists on the cell size distribution among cycling cells, especially in relation to attainment of the "critical" size or mass prior to entrance into S-phase (*7,9,34,35,94,110*). Despite the vast amount of data there is no general agreement as to whether or not a critical threshold mass is a prerequisite for initiation of DNA replication. In what are perhaps the most exhaustive studies on this subject, Killander and Zetterberg investigated L-929 cells by combined interferometry, microspectrophotometry, and time-lapse cinematography; they concluded that DNA synthesis is initiated only when cells do attain a critical mass (*79,80*). This subject has been reviewed recently by Baserga (*9*).

It should be emphasized, however, that the threshold levels of proteins ("critical mass") or RNA in G1 cells are characteristic for exponentially growing, unperturbed populations (*34,35,39*). In perturbed systems, e.g., after partial suppression of protein synthesis, cells may enter S-phase with a subthreshold amount of protein (*9,114,155*). Likewise, adenovirus-2-infected cells enter S-phase with low RNA content (*107*).

2.2. Coulter Volume Measurements

The principle and applications of the Coulter-volume measurements were extensively reviewed by Kachel (*75*). The volume of G1Q cells is clearly lower than the volume of proliferating cells of the same type. Lymphocytes were the most extensively studied and the measurements of their volume increase were repeatedly proposed as a method to assay mitogenic stimulation (*63, 126,128*). The earliest measured increase was observed between 8 and 12 h after stimulation (by concanavalin A), i.e., well before onset of DNA synthesis (*128*). The average volume of the mitogen-responding cells prior to their entrance to mitosis was four times higher than of the nonstimulated G1Q cells. However, the problem of cell aggregation induced by most mitogens (e.g., by phytohemagglutinin) complicates the interpretation of the assay because cell doublets or higher aggregates can be erroneously classified as single cells with increased volume.

Changes in cell volume during the cell cycle were measured by numerous investigators. The kinetics of volume increase during progression through G1, S, and G2 appear to be different in different cell types (e.g., 3,127). Simultaneous measurements of DNA and cell volume of cells from multicellular spheroids (73) revealed a large heterogeneity of the G1 populations, but there was no evidence that quiescent cells in the spheroids had significantly decreased volume in comparison with cells in the cycle.

2.3. Light Scatter

Cells or particles passing through a focused light beam scatter light. Measurements of the scattered light within a narrow angle in the forward direction of the laser beam provide an estimate of cell size. The attractiveness of the light scatter as a probe stems from the fact that the measurement can be made on viable, unstained cells. The probe thus introduces minimal perturbation of cell growth. Theoretical bases and practical applications of light scatter measurements are discussed extensively by Salzman (117). It should be mentioned, however, that the intensity of the light scatter signal is also influenced by (a) the refractive index of cell constituents, (b) light reflection from cell surface and internal structures, or (c) cell shape. It thus cannot be considered a direct measure of cell size.

Light scatter of nonstimulated lymphocytes was shown to be markedly lower than that of mitogen-stimulated cells (42,56). Measurements of light scatter, like volume measurements, were proposed as a quantitative assay of stimulation (56). Forward light scatter of leukemic cells in suspension cultures in the plateau phase of cell growth (induced by high cell density or nutrient deprivation) was found to be significantly lower than during their exponential growth (152). Nuclei isolated from 3T3 cells arrested in the cycle by serum deprivation also exhibited diminished light scattering in comparison with the nuclei of exponentially growing 3T3 cells (153). Initiation of DNA replication was observed to be correlated with attainment of a threshold value of nuclear size (light scatter) (153).

Light scatter measured simultaneously at several different angles provides more information on cell morphology than the single measurement in the forward direction (117). Using multiangle measurements, Visser and colleagues (144,145) were able to partially distinguish the pluripotent stem cells in the bone marrow and subdivide the differentiating cells of the marrow into several groups.

Multiangle light scatter analysis of teratocarcinoma cells, however, failed to recognize the pluripotent stem cells in these tumors (*135*).

2.4. Cell Morphology by Flow Cytometry

Analysis of the pulse-shape of the cell passing through a narrowly focused light beam provides information on cell morphology. From the measurement of the pulse width (time of flight), it is possible to estimate the cell or nuclear diameter (*121,130,151*), as well as distinguish nuclei of different shapes (*27*). This approach has been used by Johnson et al. (*74*) to analyze nuclear size during the transition of cells from G1- to S-phase; in contrast to the data of Yen and Pardee (*153*), no "critical" nuclear size was observed during the transition. In studies on lymphocyte stimulation, the pulse width of the G1Q cells was seen to be distinctly lower than that of G1 cells responding to the mitogen (*43,48*). This feature has been used by Monroe and Cambier to recognize G1Q vs G1A vs G1B cells in mitogen-stimulated B-cell cultures, which enabled them to measure expression of the surface-antigen in these respective cell subpopulations (*95*).

3. PROTEIN CONTENT

A variety of techniques has been proposed to stain cellular or nuclear proteins (e.g., *29,30,131*; also *see* chapter 7 of this monograph). As expected, cellular protein content distributions were similar to the cell volume (*29*) or RNA content (*39*) distributions, reflecting a correlation between cell size and RNA and protein content. The literature on changes in protein content during the cell cycle or during transition from quiescence to the cycle as measured by flow cytometry is scarce. Rønning and Lindmo (*114*) have observed in exponentially growing cultures of NHIK 3025 cells that a threshold protein content is required for G1 cells to enter S-phase. The cells, however, can bypass the threshold requirements when treated with a low concentration of cycloheximide.

A threshold protein content of isolated nuclei prior to entrance to S-phase of HeLa cells was reported by Roti Roti and Higashikubo (*116*). Recently, Pollack et al. (*108*) observed that quiescent cells have significantly lower nuclear protein content than cells progressing through the cycle. This confirms earlier reports of Auer and Zetterberg (*4*), who measured nuclear mass by microinterferometry.

4. RNA CONTENT

Biochemical measurements indicate that noncycling cells have lower RNA content than cycling cells (125). Simultaneous staining of DNA and RNA can be accomplished either by using a metachromatic fluorochrome acridine orange (47,141) or a combination of dyes, e.g., Hoechst 33342 and pyronin Y (119,137). RNA content was studied in a variety of cell systems during growth and quiescence and in most studies was found to be significantly lower in quiescent cells. There was no, or minimal, overlap in RNA values of the quiescent cells with the corresponding populations of cycling cells in the following cell types: (a) nonstimulated lymphocytes (42,47,113); (b) 3T3 cells induced to quiescence by growth at high cell density (41,45); (c) nonstimulated macrophages (124); (d) nonclonogenic EMT6 cells from the unfed cultures (149); (e) noncycling HeLa-S3 cells from the unfed cultures (10,53); (f) unfed mammary carcinoma cells in culture (147); (g) quiescent cells from multicellular tumor spheroids (12); and stem hepatocytes (68). It should be mentioned, however, that in the case of cells of tumor origin or virus-transformed cells, the long-term viability of cells characterized by low RNA content is in doubt (125). Thus, in all the studies listed above in which the quiescent tumor cells were shown to have diminished RNA, those cells could be nonclonogenic. In contrast, normal cells that enter quiescence can remain viable with low RNA content for a considerable period of time (125).

There are situations in which cells do not progress through the cycle, yet their RNA content is not markedly diminished. For example, 3T3 cells in serum-deprived cultures remain quiescent with RNA values typical of G1A cells (41). Likewise, quiescent Chinese hamster ovary (CHO) cells from cultures deprived of isoleucine (139) and blast lymphocytes from the peripheral blood of leukemic patients (109) remain quiescent with a G1A, rather than G1Q, RNA content, as do the cells inhibited in proliferation by R123 (42) or n-butyrate (50). Quiescent hypoxic cells from murine fibrosarcoma can be discriminated by density gradient centrifugation; those cells also have an RNA content similar to that of cycling cells (17).

Considering all the evidence on the RNA content of the quiescent cells listed above, the following generalizations can be drawn.

1. It is evident that, in the case of exponentially growing populations, the cells enter S-phase only when they

accumulate RNA up to a critical threshold level; the cells with a subthreshold RNA content remain in the growth or equalization subphase (G1A). Their residence times in G1A are widely distributed, with a characteristic exponential of quasiexponential component (*39,41*).

2. Normal cells may remain viable in quiescence for long periods of time, having markedly diminished RNA content (G1Q cells).

3. Tumor cells are predominantly arrested (e.g., by nutrient deprivation) in the growth (G1A) subphase, with RNA content only slightly lower than that of the cycling cells. Sometimes, tumor cells do arrest in S- or G2-phase.

4. Although extended quiescence of tumor cells leads to a further decrease in their RNA content to the level of the G1Q cells (e.g., in unfed cultures or multicellular spheroids), these cells lose viability with time.

The RNA content of isolated nuclei can also be measured by flow cytometry (*105*). G1Q cells (nonstimulated lymphocytes, noncycling hepatocytes) have distinctly lower nuclear RNA content in comparison with cells in the cycle (*68,105*).

5. MITOCHONDRIAL PROBES; ELECTRONEGATIVITY OF THE MITOCHONDRIAL MEMBRANE

Certain permeant cationic fluorochromes such as cyanine dyes (*25,71,120*) or R123 (*26,42,49,70–72*) are taken up specifically by mitochondria of living cells. Uptake of these fluorochromes is believed to reflect the transmembrane potential (*11,81,146*). Although the fluorescence intensity of individual mitochondria within a given cell appears to be uniform, large intercellular variations are observed in many cells systems. Cell-to-cell differences may be a result of either different numbers of mitochondria per cell, a difference in mitochondrial potential, or both. In either case, cellular staining with these fluorochromes appears to be related to the cells' metabolic state, i.e., energy requirements of the cell.

A severalfold increase in the cells' ability to accumulate R123 was observed during lymphocyte stimulation (*42*). Maximal dye uptake, seen on the third day of cell stimulation by phytohemag-glutinin, coincided in time with the peak of DNA synthesis and mitotic activity. A large intercellular variation among stimulated lymphocytes was observed, with some cells showing a fluorescence increase by as much as 15 times in comparison with G1Q lymphocytes (42). The increased uptake of R123 also correlated with the increased RNA content. Thus, G1Q cells bind distinctly less R123 than cells growing exponentially.

L1210 or FL cells arrested in G1A in stationary cultures exhibited only about a twofold decrease in R123 binding in comparison with cells growing exponentially, and there was a significant overlap between these populations (49). Erythroid differentiation of FL cells in cultures containing DMSO is paralleled by a decrease in R123 binding; fully differentiated and quiescent FL cells (G1D) had a twofold lower R123 fluorescence than their cycling counterparts (49). Likewise, cells of the human leukemic HL-60 cell line, differentiated and quiescent (G1D), exhibited 2–4-fold decrease in R123 fluorescence (26).

Binding of R123 by human fibroblasts in relation to cell senescence and proliferation was studied by Goldstein and Korczak (66). In exponentially growing cultures, young fibroblasts bound approximately four times more R123 than in confluent cultures (G1Q cells). Also, when exponentially growing cells were compared, young fibroblasts took up significantly more R123 than either senescent or progeric cells. In contrast, all fibroblasts in confluent cultures bound similar amounts of R123, regardless of age (66).

6. SENSITIVITY OF DNA *IN SITU* TO DENATURATION

The double helix of DNA in chromatin is locally stabilized by positively charged macromolecules and small ions that provide counterions for DNA phosphate (133). The strength and extent of interactions between DNA and these local counterions, as well as any DNA cross-linking that additionally stabilizes the double helix, may be evaluated by analysis of the profiles of DNA denaturation. Histone modifications that weaken interactions with DNA (e.g., phosphorylation, acetylation) are expected to decrease the stability of DNA in chromatin.

The assay of DNA stability *in situ* is based on the metachromatic properties of acridine orange. The dye differentially stains single- vs double-stranded DNA (*see* chapter 10 in this monograph). After partial DNA *in situ* denaturation induced by heat or acid treatments, the cells are stained with acridine orange and proportions of single- and double-stranded DNA are estimated from the extent of green and red fluorescence, respectively. In most cell systems studied, the stability of DNA *in situ* was found to be inversely correlated with the degree of chromatin condensation. Thus, cells in mitosis or quiescent cells with condensed chromatin can be easily distinguished by flow cytometry using this approach (*see* reviews in refs. 34,35,45). The molecular mechanisms responsible for the increased sensitivity of DNA *in situ* in condensed chromatin to denaturation are unknown, but are thought to be related to histone H1 modifications (34).

This technique was applied to a variety of cell systems to discriminate between cycling and noncycling cells (10,43,45,48,53, 148). It was consistently observed that the extent of DNA denaturation correlated with cell quiescence. Quiescent lymphocytes (G1Q cells) exhibited high sensitivity to denaturation, and their transition into the cycle triggered by mitogenic stimulation was accompanied by a substantial decrease in the sensitivity, already manifested prior to their entrance into S-phase. Thus, the G1Q cells could be easily discriminated from cycling cells in G1-phase (48). The discrimination of cells in transition (G1T) was also possible (45,48).

In certain cell systems, noncycling cells with condensed chromatin had S or G2 DNA values. Such cells were observed in chronic myeloid leukemia (33,43) and in FL cultures undergoing differentiation in the presence of DMSO (45,142). Dethlefsen et al. (53) and Bauer and Dethlefsen (10) measured the stability of DNA *in situ* to acid denaturation in HeLa-S3 cultures during their exponential and stationary phases of the growth, and were able to discriminate proliferative and quiescent populations based on differences in DNA sensitivity to denaturation. The identity of these populations was confirmed by thymidine incorporation studies and cell sorting. Among noncyling cells there were some dead or dying cells. Their data also suggest the presence of cells arrested in S-phase (SQ cells). Wallen et al. (148) detected quiescent cells in murine mammary tumor cultures by also using the DNA-acid

denaturation procedure. Brock et al. (17) analyzed DNA denaturation *in situ* in various fractions of mouse fibrosarcoma cells separated
by density-gradient centrifugation. The sensitivity to denaturation
of DNA from the fraction of quiescent hypoxic cells from these
tumors was only slightly increased in comparison with that of the
cycling cells. Many quiescent hypoxic cells had S or G2-DNA content (17).

Among cycling cells a threshold value of chromatin condensation in G1 is observed. This threshold discriminates G1A vs G1B
subpopulations. The technical aspects of the discrimination of G1A
and G1B compartments and their possible different functional roles
in the cell cycle are discussed in chapter 10 of this monograph.
Murine leukemic L1210 cells when treated with *n*-butyrate or the
mitochondrial antimetabolite R123 do arrest the phase that, according to the DNA-denaturation assay (and RNA content as well), is
identical with the G1A compartment (49,50).

7. ACCESSIBILITY OF DNA *IN SITU* TO THE INTERCALATING PROBES

It is well established that the binding of various fluorochromes
to DNA *in situ* is restricted by chromosomal proteins (reviewed in
refs. 34,44). The extent of DNA that is available for binding of the
ligand depends on the size of the probe and correlates in certain
cell systems with changes in chromatin structure that occur during
cell differentiation or quiescence.

In the case of small intercalating probes such as acridine orange
or ethidium bromide, a change in the extent of binding was
reproducibly demonstrated during cell differentiation in the following cell systems: (a) normal erythroid differentiation (78); (b) erythroid differentiation of FL cells (44); and (c) spermiogenesis (58,64).
In other cell systems, especially in studies of cycling vs quiescent
cells or cells progressing through the mitotic cycle (i.e., G2 vs M
cells), no significant change in binding of such probes under conditions of equilibrium could be observed (33,44). ''Supravital'' cell
staining with ethidium bromide, as shown by Böhmer (13) and confirmed in our laboratory (unpublished), results in artifacts unrelated
to cell position in the cycle, which can sometimes be erroneously
interpreted as indicating differences in stainability of quiescent vs
cycling cells.

In contrast to small intercalators, binding of larger probes, especially with bulky chains protruding into the grooves of the DNA helix (e.g., actinomycin D) was shown to be lowered in G1Q (*38*) and mitotic (*104*) cells. Also, the binding of quinacrine mustard was described recently to vary during the cell cycle and to correlate with chromatin condensation (*96*). These probes, however, have not been used yet in flow cytometry under equilibrium conditions for cell cycle analysis, although the fluorescent analog of actinomycin D (7-amino-actinomycin D) is now commercially available (*44*).

8. INCORPORATION OF BrdUrd

Several flow cytometric techniques have been developed as an alternative to thymidine-autoradiography to recognize cycling cells that incorporate the thymidine analog 5-bromodeoxyuridine (Brd-Urd) (*15,37,44,46,67,86–88,134*). Detection of cells incorporating this precursor is based either on immunochemical procedures utilizing BrdUrd antibodies (*55,67; see* also chapter 5 of this monograph) or on cytochemical techniques by measurement of the BrdUrd-induced quenching of Hoechst 33258 (*14,15,86–88*) or acridine orange (*37,46*) fluorescence following cell staining with these dyes. Enhancement of the fluorescence of the DNA-bound mithramycin by BrdUrd also provides a basis for flow cytometric recognition of cells replicating DNA (*134*).

The cytochemical techniques suffer in general from low sensitivity. Namely, a large portion of the genome has to be replicated in the presence of BrdUrd to observe a significant change in fluorescence intensity. In contrast, less than 1% of the replicated genome can be detected by the immunochemical approach utilizing monoclonal anti-BrdUrd antibodies (*55,67,* chapter 5 of this volume). The immunochemical technique, therefore, is ideally suited to measure the actual number of cells replicating DNA during a short exposure to the precursor. Furthermore, when cells are stained simultaneously with a DNA-specific fluorochrome and anti-BrdUrd antibodies, the cohort of pulse-BrdUrd-labeled cells can be followed when progressing through G2-, M-, and G1-phases (*55*). The technique, therefore, is applicable to the kinetic studies yielding data similar to the fraction of labeled mitoses (FLM) method. The combination of Hoechst 33258 and ethidium bromide staining (*15*) also permits measurement of total DNA content (i.e., cell position in the cell cycle) and detection of cells that incorporate BrdUrd, making it applicable for kinetic studies.

Progression of cells through the cell cycle can be correlated with their RNA content by using acridine orange to differentially stain cellular DNA vs RNA, in combination with BrdUrd (37). Two independent parameters (RNA content and ability to incorporate BrdUrd) are thus used to discriminate between cycling and noncycling cells. Cell staining with acridine orange following partial DNA denaturation allows detection of cells in mitosis (43,45). This type of staining in combination with the BrdUrd incorporation assay can be used to detect BrdUrd-labeled vs unlabeled mitotic cells (46) and, in principle, it can be applied in cultures or in vivo to discriminate between the cells that did not replicate DNA or replicated it once or twice during exposure to this precursor.

The scope of this article does not allow for a review of all the studies in which BrdUrd-labeling techniques were used in combination with flow cytometry. A very recent publication by Rabinovitch, however, should be mentioned (111), since it describes a novel and elegant approach to simultaneously detect and quantify noncycling cells and to measure the kinetics of cell exit from G1-phase in cultures of young and senescent fibroblasts.

9. IN SEARCH OF SPECIFIC MARKERS

There are many different causes of quiescence and correspondingly many different metabolic states characterizing cells that are withdrawn from the cell cycle. On one end of the spectrum there are normal cells in deep quiescence (G1Q), which have minimal RNA content, condensed chromatin, and low mitochondrial activity. These cells can be easily discerned from the cycling population. When stimulated, G1Q cells enter a long prereplicative phase. Likewise, quiescent and differentiated (G1D) cells exhibit a specific phenotype, which allows for their distinction. On the other end of the spectrum there are, for example, the hypoxic cells from solid tumors as described by Brock et al. (17), which by their RNA content or chromatin structure are undistinguishable from their cycling counterparts and which upon oxygenation can resume progression through the cycle momentarily. It is not surprising, therefore, that among the metabolic probes discussed above there is not a single one that would be universal and applicable to all cell systems.

Recognition of noncycling cells based on their inability to incorporate BrdUrd can be applied, of course, to a variety of cell systems, but as in the case of [^{3}H]-TdR incorporation, it requires long incubation times and thus has limited use for detection of quiescent cells in human tumors in vivo.

There are two types of quiescent tumor cells; namely, those that have irreversibly lost the ability to divide and those that are potentially clonogenic. From the practical point of view in chemotherapy, the most important to recognize is the latter. It is unclear at present to what extent the methods described can identify these cells. Since most techniques are based on staining the fixed cells, the clonogenicity of the selected and sorted population cannot be assayed. Some markers, however, (e.g., R123) can be used supravitally, which allows us to study survivability of the selected populations.

There is a group of cell constituents that is present in cells only at specific periods within the cell cycle or in close relationship with their proliferative potential. These constituents consist of three categories. The first includes those that are suspected of having regulatory function during the cell cycle. The second includes proteins or peptides directly involved in cell proliferation (e.g., enzymes of DNA replication). The third includes cell surface proteins or glycoproteins that exhibit cell cycle (phase) dependency. The majority of these products were characterized only very recently and their list is rapidly growing. The antibodies against these constituents or other means of their cytochemical detection can serve as markers of the cell proliferative potential and be of prognostic value in the clinic. The most notable products of this group are the following:

  a. *Transcripts of oncogenes.* Transcripts of oncogenes have either tyrosine-specific kinase activity (e.g., *scr* product) or growth factors receptor activity (e.g., *erb*-B) or mimic the activity of growth factors themselves (e.g., *sis*) or bind GTP in the cytoplasm (e.g., N-*ras*) or to DNA in the nucleus (e.g., the *myc*-transcript). The presence of the products of protooncogenes at abnormally high levels, for instance because of gene amplification or transposition, as well as their minor modification (e.g., *ras*-oncogene), correlate with uncontrolled cell proliferation. Transcription of the protooncogenes appears to be regulated by growth signals at specific points of the cell cycle. Thus, for instance, induction of the transcription of the c-*myc* gene occurs prior to DNA replication soon after stimulation of growth of lymphocytes or quiescent fibroblasts (77).

b. *The transformation-enhanced nuclear protein "cyclin."* The protein cyclin is present in very small amounts in normal noncycling cells. In transformed cells, as well as in normal cells in S-phase, the level of this protein is markedly increased (19,20). The "proliferating cell nuclear antigen" described by Tan and colleagues (136) and cyclin are most likely identical proteins (90).

c. *The species-specific family of proteins with molecular weights of approximately 53 kdalton (p53 protein).* p53 proteins are associated with virus-induced cell transformation and cell proliferation (28). Monoclonal antibodies against the p53 protein are available and are used to study regulation of cell proliferation (51,54,100). As shown recently by Mercer et al. (92,93), microinjection of anti-p53 antibodies inhibits transition of 3T3 cells from G0 to S, but has no effect on the cell progression through G1 and S. Induction of cell differentiation correlates with a decline of p53 (122). Using antibodies against p53, we have been able to discriminate cycling vs noncycling L1210 cells (36).

d. *Nonhistone proteins released from nuclear chromatin by limited digestion with DNase I.* Proteins with molecular weights of 37 and 100 kdalton were detected in proliferating cells, whereas the levels of proteins with molecular weights of 52 and 75 kdalton were higher in quiescent cells (16). Using the same strategy (DNase I digestion), Goldberger et al. (65) detected proteins associated with terminal cell division and myeloid differentiation of human leukemic cell lines.

e. *The nuclear antigens detected in proliferating cells by Gerdes et al. (62).* The monoclonal antibodies against this protein (Ki-67), induced by immunizing animals with nuclei of L-428 cells, were reactive toward Hodgkin lymphoma cells, as well as toward normal proliferating cells in interphase and mitosis, but not G0 cells. This antigen appears to be different from cyclin.

f. *Nuclear proteins associated with interchromatin granules or condensed or dispersed chromatin.* These proteins were described by Clevenger and Epstein (23,24) and were shown to be cell cycle specific. Here again monoclonal antibodies were induced by the immunization of mice with preparations of broken nuclei.

g. *Nuclear antigen with molecular weights of 55 kdalton (PSL).* This protein accumulates specifically in the nucleus during S-phase and appears to be distinct from p53 and cyclin (5).

h. *The highly phosphorylated nonhistone proteins B23 and C23.* These proteins are present in nucleoli or nuclear organizer regions of the metaphase chromatin. Location, quantity, and degree of phosphorylation (serine and threonine residues) of these proteins show clear cell cycle phase dependency (98).

i. *Adenosine (5')tetraphospho(5')adenosine (Ap₄A).* The quantity of Ap_4A varies over a thousandfold range depending on cell proliferation, growth, transformation, or differentiation (150). Ap_4A binds to DNA polymerase α (112), primes DNA (154), triggers DNA replication in quiescent cells, and may serve as the "second messenger" of the cell cycle (150).

j. *Enzymes associated with DNA replication.* DNA polymerase activity is higher in cycling than in noncycling cells (32). DNA polymerase α visualized by monoclonal antibodies was present in cycling cells, but undetectable in condensed chromatin of resting cells or during mitosis (91, 97). Recently, the biotinylated dUTP became available (85), and there is thus a possibility of measuring DNA polymerase activity using a fluorescence probe. The dihydrofolate reductase and the multi-functional protein associated with the "housekeeping" functions of the cell are both cell cycle regulated (83). The dansylated fluorescent derivatives of lysine or ornithine analogs of methotrexate were recently synthesized and found to bind with high affinity to dihydrofolate reductase (82). Tubulin synthesis is also cell cycle dependent and occurs predominantly in G2 (22). The rate of ribosomal protein synthesis is more accelerated during cell tran-

sition from G0 to G1 than the overall rate of protein synthesis (*143*).

k. *Cell surface glycoproteins.* Kehrl et al. (*76*) reported that
expression of two glycoproteins is cell cycle dependent.
These glycoproteins are nondetectable in quiescent B
lymphocytes. During cell stimulation one of them
becomes expressed early in G1 (G1A); this component
appears to be correlated with cell growth. The second
glycoprotein is detected later in G1 (G1B), and its
presence coincides with cell division.

l. *Probes of the hypoxic cells.* Hypoxic cells can be identified
by increased binding of the fluorescent nitroheterocycles
(*99*). This probe was recently used in flow cytometry
to detect the radioresistant and presumably noncyling
hypoxic cells from cellular spheroids (*99*). The
nitroheterocycles, therefore, could be used to identify
the noncycling hypoxic cells in solid tumors.

m. *Histone* $H1^0$. A striking feature of the chromatin of noncycling cells is the increase in quantity of a distinct protein, histone $H1^0$ (*21,31,34,101,102*). One would expect,
therefore, that antibodies against $H1^0$ could be specific
markers of noncycling cells.

With time, more and more specific cell constituents related to
cell progression through the cell cycle will be recognized and
isolated. Monoclonal or polyclonal antibodies against them promise to be more specific probes for detecting subpopulations of cells
with different cell cycle kinetics, potential for clonogenicity, or differentiation or sensitivity to certain drugs. Clinical application of
such probes will be of great help in designing tumor treatment
strategies.

ACKNOWLEDGMENTS

The author thanks Dr. Frank Traganos and Miss Robin Nager
for help in preparation of the manuscript. The work was supported
by Grants CA28704 and CA23296 from the National Cancer
Institute.

REFERENCES

1. Adam, G., Steiner, U., and Seuwen, K. Proliferative activity and ribosomal RNA content of 3T3 and SV40-3T3 cells. Cell Biol. Int. Rep., 7: 955–962, 1983.

2. Allison, D. C., Yuhas, J. M., Ridolpho, P. F., Anderson, S. L., and Johnson, T. S. Cytophotometric measurement of the cellular DNA content of ^{3}H-thymidine-labeled spheroids. Cell Tissue Kinet., 16: 237–246, 1983.

3. Anderson, E. C., Bell, G. I., Petersen, D. F., and Tobey, R. A. Cell growth and division. Determination of volume growth rate and division probability. Biophys. J., 9: 246–258, 1969.

4. Auer, G. and Zetterberg, A. The role of nuclear proteins in RNA synthesis. Exp. Cell Res., 75: 245–253, 1972.

5. Barque, J. P., Danon, F., Peraudeau, L., Yeni, P., and Larsen, C. J. Characterization by human autoantibody of a nuclear antigen related to the cell cycle. EMBO J., 2: 743–749, 1983.

6. Baserga, R. The Cell Cycle and Cancer. New York: Marcel Dekker, 1971.

7. Baserga, R. Multiplication and Division of Mammalian Cells. New York: Marcel Dekker, 1976.

8. Baserga, R. Resting cells and the G1 phase of the cell cycle. J. Cell Physiol., 95: 377–386, 1978.

9. Baserga, R. Growth in size and cell DNA replication. Exp. Cell Res., 151: 1–5, 1984.

10. Bashford, C. L. and Smith, J. C. The use of optical probes to monitor membrane potential. Meth. Enzymol., 15: 569–586, 1979.

11. Bauer, K. D. and Dethlefsen, L. A. Control of cellular proliferation in HeLa-S3 suspension cultures. Characterization of cultures utilizing acridine orange staining procedure. J. Cell Physiol., 108: 99–112, 1981.

12. Bauer, K. D., Keng, P., and Sutherland, R. M. Isolation of quiescent cells from multicellular tumor spheroids using centrifugal elutriation. Cancer Res., 42: 72–78, 1982.

13. Böhmer, R. M. Discrete changes of the fluorescence yield from cells stained vitally with ethidium bromide (EB) as determined by flow cytometry. Exp. Cell Res., 122: 407–410, 1979.

14. Böhmer, R. M. Flow cytometric cell-cycle analysis using the quenching of 33258 Hoescht fluorescence by bromodeoxyuridine incorporation. Cell Tissue Kinet., 12: 101–110, 1979.

15. Böhmer, R. M., and Ellwart, J. Combination of BUdR-quenched Hoechst fluorescence with DNA-specific ethidium bromide fluorescence for cell cycle analysis with a two-parameter flow cytometer. Cell Tissue Kinet., *14*: 653–658, 1981.

16. Briggs, R. C., Brewer, G., Goldberger, A., Wolff, S. N., and Hnilica, L. S. Antigens in chromatin associated with proliferating and non-proliferating cells. J. Cell Biochem., *21*: 249–262, 1983.

17. Brock, W. A. Swartzendruber, D. E., and Grdina, D. J. Kinetic heterogeneity in density-separated murine fibrosarcoma subpopulations. Cancer Res., *42*: 4499–5003, 1982.

18. Brown, R. A., McWalter, R., Slidders, W., Gibbs, J., and Swanson-Beck, J. Measurements by Quantimet 720 of the proportion of actively growing cells in tissue cultures of human lymphocytes. J. Microscopy, *115*: 51–63, 1978.

19. Celis, J. E., Fey, S. J., Larsen, P. M., and Celis, A. Expression of the transformation-sensitive protein "cyclin" in normal human epidermal basal cells and simian virus 40-transformed keratinocytes. Proc. Natl. Acad. Sci. USA, *81*: 3128–3132, 1984.

20. Celis, J. E., Bravo, R., Larsen, P. M., and Fey, S. J. Cyclin: A nuclear protein whose level correlates directly with the proliferative state of normal as well as transformed cells. Leukemia Res., *8*: 143–157, 1984.

21. Chabanas, A., Lawrence, J. J., Humbert, J., and Eisen, H. Cell cycle regulation of histone $H1^0$ in CHO cells: A flow cytofluorimetric study after double staining of the cells. EMBO J., *2*: 833–837, 1983.

22. Chang, M. T., Dove, W. F., and Laffler, T. B. The periodic synthesis of tubulin in the *Physarum* cell cycle. J. Biol. Chem., *25*: 1352–1356, 1984.

23. Clevenger, C. V. and Epstein, A. L. Use of immunogold electron microscopy and monoclonal antibodies in the identification of nuclear substances. J. Histochem. Cytochem., *32*: 757–765, 1984.

24. Clevenger, C. V. and Epstein, A. L. Identification of nuclear protein component of interchromatin granules using a monoclonal antibody and immunogold electron microscopy. Exp. Cell Res., *151*: 193–207, 1984.

25. Cohen, R. L., Muirhead, K. A., Gill, J. E., Waggoner, A. S., and Horan, P. K. A cyanine dye distinguishes between cycling and non-cycling fibroblasts. Nature, *290*: 593–595, 1981.

26. Collins, J. M. and Foster, K. A. Differentiation of promyelocytic (HL-60) cells into mature granulocytes: Mitochondrial-specific rhodamine 123 fluorescence. J. Cell Biol., *96*: 94–99, 1983.

27. Collste, L., Darzynkiewicz, Z., Traganos, F., Sharpless, T., Whitmore, W. F., and Melamed, M. R. Identification of polymorphonuclear leukocytes in cytologic samples for flow cytometry. J. Histochem. Cytochem., *27*: 390–393, 1978.

28. Crawford, L. The 53,000 dalton cellular protein and its role in transformation. Int. Rev. Exp. Pathol., *25*: 1–50, 1983.

29. Crissman, H. A. and Steinkamp, J. A. Rapid simultaneous measurement of DNA, protein and cell volume in single cells from large mammalian cell populations. J. Cell Biol., *59*: 766–769, 1973.

30. Crissman, H. A., van Egmond, J., Holdrinet, R. S., Pennings, A., and Haanen, C. Simplified method for DNA and protein staining of human hematopoietic cell samples. Cytometry, *2*: 59–62, 1981.

31. D'Anna, J. G., Gurley, L. R., and Becker, R. R. Histones H1⁰a and H1⁰b are the same as CHO histones H1 (iii) and H1 (iv): New features of H1⁰ phosphorylation during the cell cycle. Biochemistry, *20*: 4501–4505, 1981.

32. Darzynkiewicz, Z. Detection of DNA polymerase activity in fixed cells. Exp. Cell Res., *80*: 483–486, 1973.

33. Darzynkiewicz, Z. Drug effects on cell cycle. *In*: (H. Bush, S. T. Crooke, and Y. Daskal, eds.), Effects of Drugs on Cell Nucleus, New York: Academic, 1979.

34. Darzynkiewicz, Z. Molecular interactions and cellular changes during the cell cycle. Pharmacol. Ther., *21*: 143–188, 1983.

35. Darzynkiewicz, Z. Metabolic and kinetic compartments of the cell cycle distinguished by multiparameter flow cytometry. *In*: (P. Skehan and S. J. Friedman, eds.), Growth, Cancer, and the Cell Cycle. New Jersey: Humana, 1984.

36. Darzynkiewicz, Z., Staiano-Coico, L., DeLeo, A. B., and Old, L. J. p53 Content in relation to cell growth proliferation and RNA content in murine L1210 leukemia and normal thymocytes. Leukemia Res., (in press.)

37. Darzynkiewicz, Z., Andreeff, M., Traganos, T., Sharpless, T., and Melamed, M. R. Discrimination of cycling and noncycling lymphocytes by BUdR-suppressed acridine orange fluorescence in a flow cytometric system. Exp. Cell. Res., *115*: 31–35, 1978.

38. Darzynkiewicz, Z., Bolund, L, and Ringertz, N. R. Actinomycin binding of normal and phytohemagglutinin stimulated lymphocytes. Exp. Cell Res., *55*: 120–123, 1969.

39. Darzynkiewicz, Z., Crissman, H., Traganos, F., and Steinkamp, J. Cell heterogeneity during the cell cycle. J. Cell Physiol., *113*: 465–474, 1982.

40. Darzynkiewicz, Z., Dokov, V., and Pienkowski, M. Dry mass of lymphocytes during transformation after stimulation by phytohaemagglutinin. Nature, 214: 840–841, 1967.

41. Darzynkiewicz, Z., Sharpless, T., Staiano-Coico, L., and Melamed, M. R. Subcompartments of the G1 phase of cell cycle detected by flow cytometry. Proc. Natl. Acad. Sci. USA, 77: 6696–6699, 1980.

42. Darzynkiewicz, Z., Staiano-Coico, L., and Melamed, M. R. Increased mitochondrial uptake of rhodamine 123 during lymphocyte stimulation. Proc. Natl. Acad. Sci. USA, 78: 2383–2387, 1981.

43. Darzynkiewicz, Z., Traganos, F., Andreeff, M., Sharpless, T., and Melamed, M. R. Different sensitivity of chromatin to acid denaturation in quiescent and cycling cells as revealed by flow cytometry. J. Histochem. Cytochem., 27: 478–483, 1979.

44. Darzynkiewicz, Z., Traganos, F., Kapuscinski, J., Staiano-Coico, L., and Melamed, M. R. Accessibility of DNA in situ to various fluorochromes: Relationship to chromatin changes during erythroid differentiation of Friend leukemia cells. Cytometry, 5: 355–363, 1984.

45. Darzynkiewicz, Z., Traganos, F., and Melamed, M. R. New cell cycle compartments identified by flow cytometry. Cytometry, 1: 98–108, 1980.

46. Darzynkiewicz, Z., Traganos, F., and Melamed, M. R. Distinction between 5-bromodeoxyuridine labelled and unlabelled mitotic cells by flow cytometry. Cytometry, 3: 345–348, 1983.

47. Darzynkiewicz, Z., Traganos, F., Sharpless, T., and Melamed, M. R. Lymphocyte stimulation: A rapid multiparameter analysis. Proc. Natl. Acad. Sci. USA, 73: 2881–2886, 1976.

48. Darzynkiewicz, Z., Traganos, F., Sharpless, T., and Melamed, M. R. Cell cycle-related changes in nuclear chromatin of stimulated lymphocytes as measured by flow cytometry. Cancer Res., 37: 4635–4640, 1977.

49. Darzynkiewicz, Z., Traganos, F., Staiano-Coico, L., Kapuscinski, J., and Melamed, M. R. Interactions of rhodamine 123 with living cells studied by flow cytometry. Cancer Res., 42: 799–806, 1982.

50. Darzynkiewicz, Z., Traganos, F., Xue, S. B., and Melamed, M. R. Effect of n-butyrate on cell cycle progression and in situ chromatin structure of L1210 cells. Exp. Cell Res., 136: 279–293, 1981.

51. DeLeo, A., Jay, G., Apella, E., Dubois, G., Law, L. W., and Old, J. L. Detection of a transformation-related antigen in chemically induced sarcomas and other transformed cells of the mouse. Proc. Natl. Acad. Sci. USA, 76: 2420–2424, 1979.

52. Dethlefsen, L. A. In quest of quaint quiescent cells. In: (R. E. Meyn and H. R. Withers, eds.), Radiation Biology in Cancer Research. New York: Raven, 1980.

53. Dethlefsen, L. A., Bauer, K. D., and Riley, R. M. Analytical cytometric approaches to heterogeneous cell populations in solid tumors: A review. Cytometry, *1*: 89–97, 1980.

54. Dippold, W. G., Jay, G., DeLeo, A. B., Khoury, G., and Old, L. J. p53 Transformation-related protein. Detection by monoclonal antibody in mouse and human cells. Proc. Natl. Acad. Sci. USA, *78*: 1695–1699, 1981.

55. Dolbeare, F., Gratzner, H., Pallavicini, M. G., and Gray, J. W. Flow cytometric measurement of total DNA content and incorporated bromodeoxyuridine. Proc. Natl. Acad. Sci. USA, *80*: 5573–5577, 1983.

56. Doukas, J. G., Ruckdeschel, J. C., and Mardiney, M. R. Quantitative and qualitative analysis of human lymphocyte proliferation to specific antigen *in vitro* by use of helium mean laser. J. Immunol. Methods, *15*: 229–238, 1977.

57. Epifanova, O. I. Mechanisms underlying the differential sensitivity of proliferating and resting cells to external factors. Int. Rev. Cytol., (Suppl.) 5: 303–335, 1977.

58. Evenson, D. P., Darzynkiewicz, Z., and Melamed, M. R. Comparison of human and mouse sperm chromatin structures by flow cytometry. Chromosoma, *78*: 225–238, 1980.

59. Fujikawa-Yamamoto, K. RNA dependence in the cell cycle of V79 cells. J. Cell Physiol., *112*: 60–66, 1982.

60. Gelfant, S. A new concept of tissue and tumor cell proliferation. Cancer Res., *37*: 3845–3842, 1977.

61. Gelfant, S. Cycling-noncycling cell transitions in tissue aging, immunological surveillance, transformation and tumor growth. Int. Rev. Cytol., *70*: 1–25, 1981.

62. Gerdes, J., Schwab, U., Lemke, H., and Stein, H. Production of a mouse monoclonal antibody reactive with a human nuclear antigen associated with cell proliferation. Int. J. Cancer, *31*: 13–20, 1983.

63. Gibbs, J. H., Brown, R. A., Robertson, A. J., Potts, R. C., and Swanson, Beck, J. A new method of testing for mitogen-induced lymphocyte stimulation: Measurement of the percentage of growing cells and of some aspects of their kinetics with an electronic particle counter. J. Immunol. Methods, *25*: 147–158, 1979.

64. Gledhill, B. L., Gledhill, M. R., Rigler, R., and Ringertz, N. R. Changes in deoxyribonucleoprotein during spermatogenesis in the bull. Exp. Cell Res., *41*: 631–641, 1966.

65. Goldberger, A., Brewer, G., Hnilica, L. S., and Briggs, R. C. Nonhistone protein antigen profiles of five leukemic cell lines reflect the extent of myeloid differentiation. Blood, *63*: 701–710, 1984.

66. Goldstein, S. and Korczak, L. B. Status of mitochondria in living human fibroblasts during growth and senescence *in vitro* use of the laser dye rhodamine 123. J. Cell Biol., *91*: 393–398, 1981.

67. Gratzner, H. G. and Leiff, R. C. An immunofluorescence method for monitoring DNA synthesis by flow cytometry. Cytometry, *1*: 385–389, 1981.

68. Higgins, P. J., Piwnicka, M., Darzynkiewicz, Z., and Melamed, M. R. Multiparameter flow cytometric analysis of hepatic nuclear RNA and DNA of normal and hepatotoxin-treated mice. Am. J. Pathol., *115*: 31–35, 1984.

69. Hochhauser, S. J., Stein, J. L., and Stein, G. S. Gene expression and cell cycle regulation. Int. Rev. Cytol., *71*: 95–243, 1981.

70. James, T. W. and Bohman, R. Proliferation of mitochondria during the cell cycle of human cell line (HL-60). J. Cell Biol., *89*: 256–260, 1981.

71. Johnson, L. V., Walsch, M. L., Bockus, B. J., and Chen, L. B. Monitoring of relative mitochondrial membrane potential in living cells by fluorescence microscopy. J. Cell Biol., *88*: 526–535, 1981.

72. Johnson, L. V., Walsch, M. L., and Chen, L. B. Localization of mitochondria in living cells with rhodamine 123. Proc. Natl. Acad. Sci. USA, *77*: 990–994, 1980.

73. Johnson, T. S., Bain, E., Raju, M. R., and Martin, J. C. Correlation of the growth of Chinese hamster V79-171b multicellular spheroids with cytokinetic parameters. Cytometry, *1*: 65–70, 1980.

74. Johnson, T. S., Swartzendruber, D. E., and Martin, J. C. Nuclear size of G1/S transition cells measured by flow cytometry. Exp. Cell Res., *134*: 201–205, 1981.

75. Kachel, V. Electric resistance pulse sizing (Coulter sizing). *In*: (M. R. Melamed, P. F. Mullaney, and M. L. Mendelsohn, eds.), Flow Cytometry and Sorting, New York: John Wiley, 1979.

76. Kehrl, J. H. Differential expression of cell activation markers after stimulation of resting human B lymphocytes. J. Immunol., *132*: 2857–2861, 1984.

77. Kelly, K., Cochran, B. H., Stiles, C. D., and Leder, P. Cell-specific regulation of the c-myc gene by lymphocyte mitogens and platelet-derived growth factor. Cell, *35*: 603–610, 1983.

78. Kernell, A. M., Bolund, L., and Ringertz, N. R. Chromatin changes during erythropoiesis. Exp. Cell Res., *65*: 1–10, 1971.

79. Killander, D. and Zetterberg, A. Quantitative cytochemical studies on interphase growth. I. Determination of DNA, RNA and mass

content of age determined mouse fibroblasts *in vitro* and of intracellular variation in generation time. Exp. Cell Res., *38*: 272–284, 1965.

80. Killander, D. and Zetterberg, A. A quantitative cytochemical investigation of the relationship between cell mass and initiation of DNA synthesis in mouse fibroblasts *in vitro*. Exp. Cell Res., *40*: 12–20, 1965.

81. Kinnally, W. L., Tedeschi, H., and Maloff, B. L. Use of dyes to estimate the electrical potential of the mitochondrial membrane. Biochemistry *17*: 3419–3428, 1978.

82. Kumar, A. A., Kempton, R. J., Anstead, G. M., and Freishem, J. H. Fluorescent analogues of methotrexate: Characterization and interaction with dihydrofolate reductase. Biochemistry, *22*: 390–395, 1983.

83. LaBella, F., Brown, E. H., and Basilico, C. Changes in the levels of viral and cellular gene-transcripts in the cell cycle of SV-40 transformed mouse cells. J. Cell Physiol., *117*: 62–68, 1983.

84. Lajtha, L. G. On the concept of the cell cycle. J. Cell Comp. Physiol., *62* (Suppl. 1): 142–145, 1963.

85. Langer, P. R., Waldrop, A. A., and Ward, D. C. Enzymatic synthesis of biotin-labelled polynucleotides. Novel nucleic acid affinity probes. Proc. Natl. Acad. Sci. USA, *78*: 6633–6637, 1981.

86. Latt, S. A. Detection of DNA synthesis in interphase nuclei by fluorescence microscopy. J. Cell Biol., *62*: 546–550, 1974.

87. Latt, S. A. Fluorometric detection of deoxyribonucleic acid synthesis. Possibilities for interfacing bromodeoxyuridine dye techniques with flow cytometry. J. Histochem. Cytochem., *25*: 913–926, 1977.

88. Latt, S. A., George, Y. S., and Gray, J. W. Flow cytometric analysis of bromodeoxyuridine-substituted cells stained with 33358 Hoechst. J. Histochem. Cytochem., *24*: 24–33, 1976.

89. Ling, M. R. and Kay, J. E. Lymphocyte Stimulation. New York: North Holland-Elsevier, 1975.

90. Mathews, M. B., Bernstein, R. M., Franza, R. B., Jr., and Garrels, J. I. Identity of the proliferating cell nuclear antigen and cyclin. Nature, *309*: 374–376, 1984.

91. Matsukage, A., Yamamoto, S., Yamaguschi, M., Kusakabe, M., and Takahashi, T. Immunocytochemical localization of chick DNA polymerases α and β. J. Cell Physiol., *117*: 266–271, 1983.

92. Mercer, W. E., Avignolo, C., and Baserga, R. Role of the p53 protein in cell proliferation as studied by microinjection of monoclonal antibodies. Mol. Cell Biol., *4*: 276–281, 1984.

93. Mercer, W. E., Nelson, D., DeLeo, A. B., Old, L. J., and Baserga, R. Microinjection of monoclonal antibody to protein p53 inhibits serum-

induced DNA synthesis in 3T3 cells. Proc. Natl. Acad. Sci. USA, *79*: 6309–6312, 1982.

94. Mitchison, J. M. The Biology of the Cell Cycle, Cambridge, MA: The University Press, 1971.

95. Monroe, J. G. and Cambier, J. C. Level of mla expression on mitogen-stimulated murine B lymphocytes is dependent on position in cell cycle. J. Immunol., *130*: 626–631, 1983.

96. Moser, G. C., Fallon, R. J., and Meiss, H. K. Fluorimetric measurements and chromatin condensation patterns of nuclei from 3T3 cells throughout G1. J. Cell Physiol., *106*: 293–301, 1981.

97. Nakamura, H., Morita, T., Masaki, S., and Yoshida, S. Intracellular localization and metabolism of DNA polymerase α in human cells visualized by monoclonal antibody. Exp. Cell Res., *151*: 123–133, 1984.

98. Ochs, R., Lischwe, M., O'Leary, P., and Busch, H. Localization of nucleolar phosphoproteins B23 and C23 during mitosis. Exp. Cell Res., *146*: 139–149, 1983.

99. Olive, P. L. and Durand, R. E. Fluorescent nitroheterocycles for identifying hypoxic cells. Cancer Res., *43*: 3276–3280, 1983.

100. Oren, M., Reich, N. C., and Levine, A. J. Regulation of the cellular p53 tumor antigen in teratocarcinoma cells and their differentiated progeny. Mol. Cell Biol., *2*: 443–449, 1982.

101. Osborne, J. B. and Chabanas, A. Kinetics of histone H1⁰ accumulation and commitment to differentiation in murine erytholeukemia cells. Exp. Cell Res., *152*: 449–458, 1984.

102. Panyim, S. and Chalkley, R. A new histone found only in mammalian tissues with little cell division. Biochem. Biophys. Res. Commun., *37*: 1042–1043, 1969.

103. Pardee, A. B., Dubrow, R., Hamlin, J. L., and Kletzien, R. A. Animal cell cycle. Annu. Rev. Biochem., *47*: 715–750, 1978.

104. Pederson, T. Chromatin structure and the cell cycle. Proc. Natl. Acad. Sci. USA, *69*: 2224–2228, 1972.

105. Piwnicka, M., Darzynkiewicz, Z., and Melamed, M. R. RNA and DNA content of isolated cell nuclei measured by multiparameter flow cytometry. Cytometry, *3*: 269–275, 1983.

106. Pledger, W. J., Stiles, C. D., Antomiades, H. N., and Scher, C. D. An ordered sequence of events is required before BALB/c-3T3 cells become committed to DNA synthesis. Proc. Natl. Acad. Sci. USA, *75*: 2839–2843, 1978.

107. Pochron, S. F., Rossini, M., Darzynkiewicz, Z., and Baserga, R. Failure of accumulation of cellular RNA in hamster cells stimulated

to synthesize DNA by infection with adenovirus 2. J. Biol., Chem., 255: 4411–4413, 1980.

108. Pollack, A., Moulis, H., Block, N. L., and Irwin III, G. L. Quantitation of cell kinetic responses using simultaneous flow cytometry measurements of DNA and nuclear protein. Cytometry, 5: 473–481, 1984.

109. Preisler, H. and Darzynkiewicz, Z. Flow cytometric analysis of proliferating and quiescent human leukemia cells. J. Medicine, 12: 415–425, 1981.

110. Prescott, D. The cell cycle and the control of cellular reproduction. Adv. Gen., 18: 99–177, 1976.

111. Rabinovitch, P. S. Regulation of human fibroblast growth rate by both noncycling cell fraction and transition probability is shown by growth in 5-bromodeoxyuridine followed by Hoechst 33258. Proc. Natl. Acad. Sci. USA, 80: 2951–2955, 1983.

112. Rapaport, E., Zamecnik, P. C., and Baril, E. F. Association of diadenosine 5′,5′′′-P′P⁴-tetraphosphate binding protein with HeLa cell DNA polymerase. J. Biol. Chem., 256: 12148–12151, 1981.

113. Richman, P. D. Lymphocyte cell-cycle analysis by flow cytometry. Evidence for a specific postmitotic phase before return to G0. J. Cell Biol., 85: 459–465, 1980.

114. Rønning, O. W. and Lindmo, T. Progress through G1 and S in relation to net protein accumulation in human NHIK cells. Exp. Cell Res., 144: 171–179, 1973.

115. Rothstein, H. Regulation of the cell cycle by somatomedins. Int. Rev. Cytol., 78: 127–230, 1982.

116. Roti Roti, J. L. and Higashikubo, R. Simulation analysis of dual parameter DNA and protein distribution. In: (M. Rotenberg, ed.), Biomathematics and Cell Kinetics. New York: Elsevier/North Holland, 1981.

117. Salzman, G. C. Light scattering analysis of single cells. In: (N. Catsimpoolas, ed.), Cell Analysis, Vol. 1, New York: Plenum, 1982.

118. Seuwen, K., Steiner, U., and Adam, G. Cellular content of ribosomal RNA in relation to the progression and competence signals governing proliferation of 3T3 and SV40-3T3 cells. Exp. Cell Res., 154: 10–14, 1984.

119. Shapiro, H. M. Flow cytometric estimation of DNA and RNA content in intact cells stained with Hoechst 33342 and pyronin Y. Cytometry, 2: 143–150, 1982.

120. Shapiro, H. M., Natale, P. J., and Kamentsky, L. A. Estimation of membrane potentials of individual lymphocytes by flow cytometry. Proc. Natl. Acad. Sci. USA, 76: 5728–5730, 1979.

121. Sharpless, T. K. and Melamed, M. R. Estimation of cell size from pulse shape in flow cytofluorometry. J. Histochem. Cytochem., *24*: 257–264, 1976.

122. Shen, D. W., Real, F. C., DeLeo, A. B., Old, L. J., Mark, P. A., and Rifkind, R. A. Protein p53 and inducer-mediated erythroleukemia cell commitment to terminal cell division. Proc. Natl. Acad. Sci. USA, *30*: 5919–5922, 1983.

123. Smith, J. A. and Martin, L. Do cells cycle? Proc. Natl. Acad. Sci. USA, *70*: 1263–1267, 1973.

124. Stadler, B. M. and deWeck, A. L. Flow cytometric analysis of mouse peritoneal macrophages. Cell Immunol., *54*: 36–48, 1980.

125. Stanners, C. P., Adams, M. E., Harkins, J. L., and Pollard, J. W. Transformed cells have lost control of ribosome number through their growth rate. J. Cell Physiol., *100*: 127–138, 1979.

126. Steen, H. B. and Lindmo, T. The effect of colchicine and colcemid on the mitogen-induced blastogenesis of lymphocytes. Europ. J. Immunol., *8*: 667–671, 1978.

127. Steen, H. B. and Lindmo, T. Cellular and nuclear volume during the cell cycle of NHIK 3025 cells. Cell Tissue Kinet., *11*: 69–81, 1978.

128. Steen, H. B. and Nielsen, V. Lymphocyte blastogenesis studied by volume spectroscopy. Scand. J. Immun., *10*: 135–143, 1979.

129. Steffen, J. and Soren, L. Changes in dry mass of PHA-stimulated human lymphocytes during blast transformation. Exp. Cell Res., *53*: 652–659, 1968.

130. Steinkamp, J. A., Hansen, K. M., and Crissman, H. A. Flow microfluorometric and light-scatter measurement of nuclear and cytoplasmic size in mammalian cells. J. Histochem. Cytochem., *24*: 292–297, 1976.

131. Stohr, M., Vogt-Schaden, M., Knobloch, M., and Vogel, R. Evaluation of eight fluorochrome combinations for simultaneous DNA-protein analysis. Stain Technology, *53*: 205–215, 1978.

132. Sturani, E., Toschi, L., Zippel, R., Martegani, E., and Alberghina, L. G1 phase heterogeneity in exponentially growing Swiss 3T3 mouse fibroblasts. Exp. Cell Res., *153*: 135–144, 1984.

133. Subirana, J. A. Studies on the thermal denaturation of nucleo-histone. J. Molec. Biol., *74*: 363–385, 1973.

134. Swartzendruber, D. E. A bromodeoxyuridine (BUdR)-mithromycin technique for detecting cycling and noncyling cells by flow microfluorometry. Exp. Cell Res., *109*: 439–443, 1977.

135. Swartzendruber, D. E., Price, B. J., and Rall, L. B. Multiangle light-scattering analysis of murine teratocarcinoma cells. J. Histochem. Cytochem., *27*: 366–370, 1979.

136. Takaski, Y., Fishwild, D., and Tan, E. M. Characterization of proliferating cell nuclear antigen recognized by autoantibodies in lupus
sera. J. Exp. Med., *159*: 981–992, 1984.

137. Tanke, H. J., Niewehuis, I. A. B., Koper, G. J. M., Slats, J. C. M.,
and Ploem, J. S. Flow cytometry of human reticulocytes based on
RNA fluorescence. Cytometry, *1*: 313–320, 1980.

138. Taylor, I. W. and Tattersall, M. H. N. Methotrexate cytotoxicity in
cultured human leukemic cells studies by flow cytometry. Cancer
Res., *41*: 1549–1555, 1981.

139. Tobey, R. A. Production and characterization of mammalian cells
reversibly arrested in G1 by growth in isoleucine-deficient medium.
In: (D. M. Prescott, ed.), Methods in Cell Biology, Vol. 6, New York:
Academic, 1973.

140. Traganos, F., Darzynkiewicz, Z., and Melamed, M. R. The ratio of
RNA to total nucleic acid content as a quantitative measure of unbalanced growth. Cytometry, *2*: 212–218, 1982.

141. Traganos, F., Darzynkiewicz, Z., Sharpless, T., and Melamed, M. R.
Simultaneous staining of ribonucleic and deoxyribonucleic acids in
unfixed cells using acridine orange in a flow cytofluorometric system.
J. Histochem. Cytochem., *25*: 45–56, 1977.

142. Traganos, F., Darzynkiewicz, Z., Sharpless, T., and Melamed, M. R.
Erythroid differentiation of Friend leukemia cells as studied by
acridine orange staining and flow cytometry. J. Histochem. Cytochem., *27*: 382–389, 1979.

143. Tushinski, R. J., and Warner, J. R. Ribosomal proteins are synthesized preferentially in cells commencing growth. J. Cell Physiol., *112*:
128–135, 1982.

144. van den Engh, G., Visser, J., Bol, S., and Trask, B. Concentration of
hemopoietic stem cells using a light-activated cell sorter. Blood Cells.,
6: 1–12, 1980.

145. Visser, J. W. M., Cram, L. S., Martin, J. C., Salzman, G. C., and
Price, B. J. Sorting of a murine progenitor cell by use of laser light
scattering measurements. *In*: (D. Lutz, ed.), Pulse Cytophotometry,
vol. III. Ghent, Belgium: European Press, 1978.

146. Waggoner, A. S. Dye indicators of membrane potential. Annu. Rev.
Biophys. Bioeng., *8*: 47–68, 1979.

147. Wallen, C. A., Higashikubo, R., and Dethlefsen, L. A. Comparison
of two flow cytometric assays for cellular RNA-acridine orange and
propidium iodide. Cytometry, *3*: 155–160, 1982.

148. Wallen, A. C., Higashikubo, R., and Dethlefsen, L. A. Murine
mammary tumor cells *in vitro*. I. The development of a quiescent
state. Cell Tissue Kinet., *17*: 65–77, 1984.

149. Watson, J. V., and Chambers, S. H. Nucleic acid profile of the EMT6 cell cycle *in vitro*. Cell Tissue Kinet., *11*: 415–422, 1978.

150. Weinman-Dorsch, C., Hedl, A., Grummit, I., Albert, W., Ferdinand, F. J., Friis, R. R., Pierron, G., Moll, W., and Grummit, F. Drastic rise of intracellular adenosine (5′)tetraphospho(5′)adenosine correlates with onset of DNA synthesis in eukaryotic cells. Eur. J. Biochem., *138*: 179–185, 1984.

151. Wheeless, L. L., Hardy, J. A., and Balasubramanian, N. Slit-scan flow system for automated cytopathology. Acta Cytol., *19*: 45–52, 1975.

152. Yen, A., Fried, J., and Clarkson, B. Alternative modes of population growth inhibition in a human lymphoid cell line growing in suspension. Exp. Cell Res., *107*: 325–341, 1977.

153. Yen, A., and Pardee, A. B. Role of nuclear size in cell growth initiation. Science, *204*: 1315–1317, 1979.

154. Zamecnik, A. C., Rapaport, E., and Baril, E. F. Priming of DNA synthesis by diadenosine 5′-5′′′-P′,P^4-tetraphosphate with double-stranded octadecamer as a template and DNA polymerase. Proc. Natl. Acad. Sci. USA, *79*: 1791–1794, 1982.

155. Zetterberg, A., Engstrom, W., and Larsson, O. Growth activation of resting cells: Induction of balanced and unbalanced growth. Ann. NY Acad. Sci., *397*: 130–147, 1982.

Chapter 10

Assay of Cell Cycle Kinetics by Multivariate Flow Cytometry Using the Principle of Stathmokinesis

Zbigniew Darzynkiewicz, Frank Traganos,
and Marek Kimmel

1. INTRODUCTION

Many of the existing cytokinetic techniques are based on the use of radioisotope-labeled DNA precursors (*see* chapters 1–3 of this volume). Although numerous data obtained by these techniques contributed greatly to our present knowledge of the cell cycle, there are certain disadvantages related to the use of isotopes in general that limit their wider application, especially in the clinic. The drawbacks of autoradiography, which requires long exposure times and cumbersome grain-count analysis, are well known. There are also problems with the quantitative aspect of the autoradiography; especially of tritium detection (e.g., *see* ref 51). The alternative of autoradiography, liquid scintillation spectroscopy, precludes analysis of individual cells and thus restricts studies on cell heterogeneity. The variability in pools of endogenous nucleotides and changes in the accessibility of the precursor in vivo or in vitro create additional problems when any quantitative data on the rate of DNA replication (progression through S-phase) must be obtained based on radioactivity measurements. The most important limitation, however, especially when tritium-labeled thymidine is used, is the radiobiological effect of the precursor. Extensive evidence exists that tritium incorporated into DNA, even at low doses, may severely perturb the cell cycle, in particular, cell progression through G2 (e.g., 11,47,57). Thus the development of new techniques that may offer an alternative to the radioisotope-based method is of importance.

Kinetic methods based on cell measurements by flow cytometry provide such an alternative. So far, nearly all these techniques are based on measurements of a single parameter; namely DNA content. Recently, attempts have been made to measure DNA content and simultaneously detect incorporation of the DNA precursor 5-bromodeoxyuridine (BrdUrd), either by cytochemical (8,16,21,42, 58) or immunochemical (28,29,33) procedures (see chapters 5 and 7 of this volume). There is evidence, however, that BrdUrd itself may perturb the cell cycle (21).

In the present chapter we describe a technique that combines the principle of stathmokinesis as proposed by Puck and Steffen (50) with the multiparameter (multivariate) flow cytometric analysis of cell cycle distribution based on simultaneous measurement of DNA content and DNA sensitivity to denaturation (19,20,22,23,26) (for review, see ref. 15). The stathmokinetic or metaphase-arrest technique provides an estimate of the rate of entry of cells to mitosis or the cell birthrate (49,66). Because multivariate cell analysis makes it possible to detect numerous discrete compartments of the cell cycle, otherwise unrecognized by single parameter (DNA) measurements, the combination of stathmokinesis and multivariate analysis offers the possibility of studying in detail not only the rate of cell entry to mitosis, but also the rates of cell traverse through other portions of the cell cycle related to those compartments. The method may also be applied to investigate drug effects on the cell cycle (26).

In the present chapter we discuss, in general, the stathmokinetic techniques based either on measurements of mitotic indices or analysis of a single parameter, i.e., DNA content, by flow cytometry. We then describe the multiparameter approach. Substantial attention is given to the application of this technique in studies of drug effects on the cell cycle and on chromatin structure, as well as to mathematical techniques for analysis of stathmokinetic experiments.

2. STATHMOKINETIC APPROACH FOR ANALYSIS OF THE CELL CYCLE

2.1. Rate of Cell Entry Into Mitosis

The stathmokinetic metaphase-arrest techniques were developed to estimate the rate of entry of cells into mitosis ("mitotic rate") or the cell birthrate ("production rate") (2,50,59,65,66). The techniques are based on the use of agents that arrest cells in mitosis (at metaphase). These agents (such as colchicine, colcemid, or the periwinkle alkaloids vinblastine and vincristine) affect the mitotic spindle of the dividing cell and result in the inability of cells to complete mitosis. The slope of the plot representing the accumulation of cells arrested in mitosis vs time of stathmokinesis provides an estimate of the rate of cell entry into mitosis. If all cells are proliferating or if the growth fraction is known, the "cell cycle time" can be estimated from this slope (2,65). Mathematical analysis and interpretation of the data obtained from the stathmokinetic experiments are the subjects of several publications (30,40,41,43,53).

The stathmokinetic approach has been widely used during the past two decades to analyze cell kinetics, both in vitro and in vivo. The techniques based on this approach have also found application in clinical situations (37,48,66). An excellent review of stathmokinetic techniques in which their advantages and limitations are analyzed in detail, was recently written by Wright and Appleton (65). In the discussion that follows we outline their general limitations; some of which are pertinent to the method proposed in this chapter.

2.2. Restrictions of the Stathmokinetic Technique

Analyses of stathmokinetic experiments based on the "mitotic rate" curve are applicable to all populations that grow exponentially and asynchronously, and when the duration of individual phases do not change during the experiment. Although these techniques can be applied to studies of drugs that perturb asynchronous exponential growth, the results of drug effects can be meaningfully interpreted only by comparison with parallel, exponentially growing, asynchronous cultures. Of course the principle of stathmokinesis, in a broad sense, can be applied to specific experiments on synchronized cultures or in conjunction with cell labeling. In the present chapter, however, these situations will not be discussed.

In typical experiments mitotic cells are counted on smears or tissue sections. To obtain statistically accurate results, hundreds or thousands of cells per time-point of the experiment must be analyzed. This tedious analysis is one of the main restrictions of the technique.

Another restriction stems from the fact that stathmokinetic agents may exert undesirable effects manifesting as toxicity or perturbation of cell progression through other phases of the cell cycle besides mitosis. To ensure that these effects are minimal, and that cells do not escape mitotic arrest, a choice of the proper agent and its optimal dose (concentration) is of the utmost importance. Selection of such optimal conditions can generally be obtained in in vitro experiments. It is difficult, if not impossible, to obtain these conditions in in vivo studies. Opinions vary widely among investigators as to which of the stathmokinetic agents has minimal toxicity (10,56,59). Review of the literature on this subject indicates that different tissues or cell types show different sensitivities for a given agent. Hence, for each particular cell type the dose–response curve should be characterized and the optimal dose chosen, i.e., the lowest one that arrests all metaphases and prevents cell "leakage" through the block. In exponentially growing cell populations, the "leakage" can be detected as a lack of "emptying" of the G1 compartment. The presence of any·noncycling cells, however, restricts detection of the leakage. Effects of several stathmokinetic agents (each tested at the optimal dose) on the linearity of the metaphase collection curve, emptying of the G1 compartment, and mitotic cell disintegration (selective cell death during the arrest) should then be compared. After selection of the optimal agent and its dose, the stathmokinetic experiment aimed at analyzing cell kinetics of a given cell type can then be designed.

Even under optimal conditions, cells arrested in mitosis die after a certain period of time. This restricts the duration of the experiment. Depending on the cell type, the deficit in mitotic cells caused by their death (followed by disintegration) can be observed as early as 3.5 h (10,45) or as late as 12 h (32,56). This limitation creates problems when cells slowly progressing through the cycle are investigated, inasmuch as relatively few cells enter M-phase before the first arrested cells begin to disintegrate. Thus, rapidly growing cell populations may be studied with greater accuracy than cells growing at slower rates. Correction procedures estimating selective cell death in mitosis were proposed by Puck et al. (*49*) for Chinese hamster ovary (CHO) cells in cultures and by Aherne and Camplejohn (*1*) for mouse intestine cells in vivo.

The effect of the stathmokinetic agent on cell progression through G1-, S-, or G2-phases should be minimal. Generally, a delay is detected after addition of the agent before cells begin to accumulate in mitosis. The delay may indicate that there is either a lag between the addition of the mitotic blocker and its effect on the spindle (e.g., reflecting the time required by the agent to penetrate the cell membrane) or that the blocker induces a transient cell arrest in G2-phase. The evidence that colchicine induces a temporary arrest of lymphocytes in G2 was provided by Fitzgerald and Brehaut (*31*). Recently, we have observed that cells of one of the sublines of mouse-L cells undergo a transient G2-phase arrest lasting up to 4 h when treated with various concentrations of colcemid, vinblastine, or vincristine (*27*).

2.3. DNA Measurements by Flow Cytometry in Stathmokinetic Experiments

The cumbersome quantitation of mitotic figures by light microscopy contrasts dramatically with rapid, unbiased, and accurate measurements of large cell populations that can be performed by flow cytometry. The latter technique therefore offers an attractive alternative and has already been applied in stathmokinetic experiments. Barfod and Barfod (*4*) compared the rate of cell entry to mitosis as estimated by visual counts of mitotic figures, with cell entry to G2 + M measured by flow cytometry after staining cellular DNA. Their data demonstrate good correlation between the cell "production" rates evaluated by these two techniques in three different cell lines in vitro. The same authors (*5*) also applied flow cytometric measurements of DNA content for analysis of the stathmokinetic experiments performed on ascites tumors in vivo. Based on these measurements they calculated rates of cell entry into G2 + M in relation to tumor age and fraction of cells in S-phase (*5*).

Extended application of flow cytometric measurements in stathmokinetic experiments, as well as the in-depth mathematical analysis of the results, were presented by Dosik et al. (*30*). These authors, in addition to measuring cell accumulation in G2 + M during stathmokinesis, also analyzed cell number in G0/1 and S. This enabled them to estimate the rate of exit from G0/1 and to observe cell transit through S. Using this approach, Dosik et al. (30) established the cell cycle as well as the G1-, S-, and G2-transit times for CHO and human colon adenocarcinoma (LoVo) cell lines. These parameters were in accord with the data obtained by the classic technique based on the estimates of the fraction of labeled mitosis after pulse labeling with [³H]-thymidine. The approach proposed by Dosik et al. (30) offers a powerful method to rapidly analyze cell kinetics and can be applied to studies of drug effects on cell kinetics. Their method was mathematically analyzed by Macdonald (43), who developed a more general model for analysis of this type of data.

The method developed by us (26) and described in the following pages of this chapter has several elements similar to the techniques proposed by Barfod and Barfod (4,5) and Dosik et al. (30). The main distinction of our technique, however, is that it is based on recognition of additional compartments of the cell cycle. Recognition of these compartments can be accomplished only by a multivariate approach taking into account not only changes in DNA content during the cell cycle, but also changes in chromatin structure (most likely related to chromatin condensation), as reflected by DNA sensitivity to denaturation (19,22,23,26).

3. CELL CULTURES

The technique, in principle, could be applied in vivo provided that the presence of the stathmokinetic agent (colcemid, vinblastine, or vincristine) could be maintained in the investigated tissue during the experiment. So far, however, the method has only been applied in vitro and we will limit the discussion to this application.

Certain restrictions apply with respect to the choice of the cell type. To recognize cells in various phases of the cell cycle, the basic fluorochrome acridine orange (AO) is used under conditions in which the dye differentially stains single- vs double-stranded DNA in the cell. The specificity of AO toward nucleic acids, however, is not absolute, since the dye stains other polyanions as well. Therefore, cells that contain large quantities of glycosaminoglycans cannot be studied by the present technique because a significant component of the fluorescence measured may be nonspecific. It was observed, for instance, that certain lines of fibroblasts (known to produce large quantities of hyaluronic acid), some epithelial cells (synthesizing dermatan sulfate), chondrocytes (synthesizing chondroitin sulfate), or mast cells (containing heparin) have unacceptably high fluorescence unrelated to nucleic acids following staining with AO. Rapidly proliferating cell lines, however, especially of tumor origin, contain minimal amounts of these differentiation-related products. There is also good specificity in staining nucleic acids in cells of lymphocytic or monocytic origin, and in leukemias and lymphomas. The degree of nonspecific fluorescence can be estimated by control incubations of the cells with RNase (see further) and DNase I (1 mg/mL, 37 °C, 1 h) prior to staining with AO.

4. SCHEME OF THE STATHMOKINETIC EXPERIMENT

The stathmokinetic experiment should be performed on exponentially growing asynchronous cultures. At zero time, the stathmokinetic agent blocking cells in mitosis is added into the cultures and cell aliquots are withdrawn from these cultures periodically (every 30 or 60 min) thereafter. Because different cell types vary in sensitivity toward most common mitotic inhibitors (colcemid, vinblastine, or vincristine), pilot experiments should be run to establish the lowest concentration of inhibitor that is adequate to completely arrest cells in mitosis with minimal toxic effects. To investigate the effects of drugs on the cell cycle, culture(s) parallel to those treated with the mitotic inhibitor alone are treated with the inhibitor and with the drug to be investigated, either with both added at the same time or with the drug included 1 h following the mitotic inhibitor. Likewise, the effects of physical agents (e.g., X-rays, irradiation, hyperthemia) may be studied in similar experiments by subjecting cultures, to which the mitotic blocker was added at zero time, to the effects of these agents. One must be aware, however, of the possible synergistic effects between the studied drugs or physical agents and the mitotic blockers.

If the cells can grow in suspension it is advisable to perform the experiment using single cultures in large flasks (750–1000 mL) for each treatment. Thus, each individual flask is treated at zero time either with the mitotic blocker alone, or together with the drug under study. Cell aliquots (10–20 mL) are then withdrawn every 30 or 60 min, for the next 10–18 h. Care should be exercised to maintain constant pH and temperature during the experiment, which involves numerous samplings of the same flasks. Using this approach, however, all samples can be obtained from the same culture and thus the culture-to-culture variability is minimized, provided the cultures are stirred continuously or shaken each time prior to withdrawal of aliquots. Duplicate or multiple cultures serve to study variability between experiments and to establish reproducibility.

When cells grow attached to the tissue culture dishes, numerous dishes (plates) are seeded at the same time at low cell density; 12–24 h later, when exponential growth is established, a series of dishes are treated simultaneously with the mitotic inhibitor alone or in combination with the investigated drug. The cells from individual dishes are then trypsinized at 30–60-min intervals. Because cells arrested in mitosis have a tendency to detach and remain suspended in the medium, both the trypsinized cells as well as the cells floating in the medium should be collected from each dish and pooled prior to fixation.

The harvested cells are centrifuged and rinsed once with Hanks' salt solution. Each aliquot (containing 5×10^5–10^7 cells), representing an individual sample collected at a given time interval, is resuspended in 1 mL of Hanks' solution. The cells are then rapidly fixed by admixture of this suspension into 15-mL glass tubes containing 10 mL of 80% ethanol and absolute acetone mixed in a 1:1 proportion (v/v) at 4°C. Alternatively, cells can be fixed in 80% ethanol alone. Minimal cell clumping and minimal cell adherence to the tube walls is ensured if cells are uniformly suspended prior to fixation and if the suspension is rapidly injected (i.e., by Pasteur pipet) into the fixative, rather than layered on the surface of the fixative and then mixed. Reverse order, i.e., addition of fixative into tubes containing cell suspension in aqueous medium, results in extensive cell loss as a result of cell adherence to the surface of the tubes. The fixation time (at 4°C) may vary from a few hours to several months.

5. CELL STAINING

5.1. Reagents

(a) *Hanks' salt solution.* Should contain Mg^{2+}, but no phenol red. The solution may be substituted by phosphate buffered saline containing 1 mM $MgCl_2$.

(b) *Stock solution of AO.* 1 mg of AO per 1 mL of distilled water. The solution may be kept in the dark at 4°C for several months.

(c) *KCl/HCl Buffer.* Equal parts of 0.2M KCl and 0.2M HCl are mixed; the final pH is approximately 1.4. (For some cell types 0.1M HCl alone is preferred.)

(d) *Na$_2$HPO$_4$-citric acid buffer.* To prepare the buffer, 90 mL of 0.1M citric acid are mixed with 10 mL of 0.2M Na$_2$HPO$_4$. The final pH is approximately 2.6. The buffer may be kept at 4°C for several months.

(e) *AO stain solution.* Add 0.6 mL of AO stock solution (1 mg/mL) to 100 mL of the Na$_2$HPO$_4$-citric acid buffer, prepared as in step (d). The final AO concentration is 6 μg/mL (see, however, further text regarding AO concentration). The solution is stable for weeks when stored at 4°C in dark bottles.

(f) *RNase A solution.* DNase-free RNase should be used. Chromatographically pure RNase is offered by the Worthington Biochemical Corp., Freehold, NJ (brand name RASE). Less pure RNase preparations may be used after boiling to inactivate any DNase activity.

5.2. Staining Procedures

(a) Centrifuge fixed cells and remove the fixative. Suspend the pellet (10^6–10^7 cells) in 1 mL of Hanks' solution and add 2 × 10^3 units of RNase A (approximately 10 μL of RASE). Incubate at 37°C for 1 h.

(b) Following incubation with RNase rinse the cells and suspend them in Hanks' solution. Then withdraw 0.2 mL of cell suspension ($\sim 10^5$–5 × 10^5 cells) and mix with 0.5 mL of KCl-HCl buffer or with 0.1M HCl (solution c). After 30 s add 2 mL of AO solution (solution d). Measure cell fluorescence in suspension with the dye (do not centrifuge cells). Step (b) should be performed at room temperature, i.e., both KCl–HCl solution and AO solution should be adjusted to room temperature prior to use.

5.3. Fluorescence Measurements

Although any kind of flow cytometer can be used, there are certain differences in cell stainability with AO, depending on geometry and construction of the flow channel. A stable pattern of staining with AO requires equilibrium conditions; even small variations in AO concentration have deleterious effect on cell stainability. In flow cytometers in which cell measurement takes place outside the nozzle in air, as is the case with most Becton-Dickinson or Coulter cell sorters, a significant diffusion of the dye from the sample flow stream to the sheath stream takes place after the stream leaves the nozzle, prior to its intersection with the laser beam. This breaks the equilibrium and lowers the actual AO concentration in the sample at the time of measurement. Dye diffusion is also a problem in some "homemade" instruments that have flow channels characterized by an extremely narrow sample stream. This is generally the case when there are two sheath streams for the purpose of optimal hydrodynamic focusing of the sample stream. To counterbalance the diffusion effects, an increased concentration of AO (up to 16 μg/mL) in the staining solution (c) should be used. Also, addition of AO to the first (inner) sheath stream precludes diffusion of the dye from the sample stream and ensures staining under equilibrium conditions. Optimal stainability with AO, without any observable dye diffusion, was observed by many authors using instruments manufactured by Ortho (FC 4801; FC 200; ICP 22) or Becton-Dickinson (FACS Analyzer). The concentration of AO as listed above in this chapter provided optimal results when cell fluorescence was measured by these instruments (*14–27,62,63*). In the case of Becton-Dickinson and Coulter cell sorters, the optimal stainability was observed at 12–16 μg/mL AO (*6,7*).

Fluorescence of individual cells is measured at two wavelength bands, green (F_{530}, between 515 and 570 nm) and red ($F_{>600}$, above 620 nm). The optimal excitation wavelength is 480 nm. The blue line of the argon ion laser (488 nm) provides excellent excitation of AO. The green and red fluorescence values measured under these conditions allow discrimination between cells with different proportions of native and denatured DNA (*19,20,26*). The resolution of the measurements is higher when the integrated values of the pulses rather than the peak values are recorded. Pulsewidth measurements are helpful to eliminate cell doublets (*54*). The data are recorded either as a bivariate distribution of green and red fluorescence values or of total cell fluorescence (green + red fluorescence) and α_t value (red fluorescence/total fluorescence) (Fig. 1). Total cell fluorescence represents total cellular DNA, whereas α_t measures the extent of DNA denaturation in situ (*19,20,22,26*). The transformation of F_{530} and $F_{>600}$ to total fluorescence and α_t values can be done either in real time by an analog electronic circuit transforming green and red signals to their sums and ratios, or by a computer program designed to convert bivariate red/green fluorescence distributions to bivariate total fluorescence/α_t distributions.

6. DISCRIMINATION OF DIFFERENT COMPARTMENTS OF THE CELL CYCLE

Figure 1 shows fluorescence distributions measured for exponentially growing cells and illustrates the method of selection of various cell subpopulations. By locating the appropriate windows (thresholds) during the interactive computer analysis ("gating"), an estimate of the frequencies of all or some of the populations listed in the legend to Fig. 1 can be obtained from the single measurement. The percent of cells in these populations may thus be calculated for every time-point of the stathmokinetic experiment (*see* chapter 8 for additional details on subpopulation analysis.)

In comparison with analyses based on single parameter (DNA) measurements (*4,5,30*), the present model allows for discrimination of more subpopulations. Namely, a distinction can be made between G2 and M cells, as well as between G1A and G1B cells. The threshold dividing the G1 population into G1A and G1B is a continuation of the right borderline of the S-cell cluster. Thus, the G1B cells have α_t values (chromatin structure) similar to those of early S-phase cells, in contrast to G1A cells, which have higher α_t values. To have an objective criterion for discrimination between G1A and G1B cells, a mean value of α_t and standard deviation of that mean are calculated for early S cells, and a borderline discriminating G1A and G1B cells is placed two standard deviations above the mean value of the early S cells. As described before (*20,26*), G1A and G1B represent different compartments of the G1-phase. More detailed discussion on the presumptive function of G1A and G1B compartments as related to cell growth and the

DNA division cycle is given elsewhere (*14,15,17,18*) and also presented in chapter 9 of this volume.

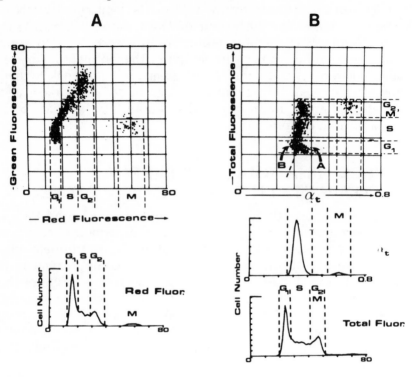

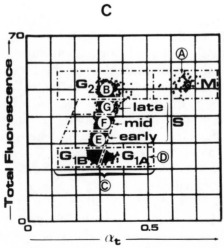

Fig. 1. (caption on facing page)

Fig. 1. Pattern of stainability with AO of cycling cell populations after partial DNA denaturation *in situ*. Exponentially growing L1210 cells were fixed, incubated with RNase, treated with acid (pH 1.4), and stained with AO at pH 2.6. Under these conditions, in absence of RNA, native (double-helical) DNA fluoresces green, whereas denatured (single-stranded) DNA fluoresces red. The relative intensities of red and green fluorescence for each cell correlated with the extent of DNA denaturation by acid, which in turn reflects the degree of chromatin condensation. Total cell fluorescence is proportional to the content of DNA per cell. Panel A. Distribution of cells with respect to green and red fluorescence. As is evident, mitotic cells (M) exhibit maximal denaturation of DNA *in situ* and can be easily distinguished from interphase cells based on their high red fluorescence. Panel B. Total fluorescence (red + green) and α_t values (α_t = red fluorescence/total fluorescence) of individual cells for the same cell population shown in panel A. Mitotic cells have the highest α_t index.

Single parameter frequency histograms of red fluorescence, α_t, and total fluorescence of these cells are below the A and B scattergrams. Panel C. The following subpopulations can be discriminated by the "gating" analysis based on either total fluorescence or α_t value. (A) Cells in mitosis (M) are distinguished as having the highest α_t values; (B) Cells in G2 + M form a typical "G2 + M" peak on the total fluorescence (DNA) histograms. After subtracting M cells, the number of cells in G2 can be estimated; (C) G1 cells form a typical "G1" peak on total fluorescence (DNA) histograms; (D) G1A cells are classified as cells having chromatin (α_t) significantly different relative to early-S cells. Thus, the gating window is at first located at the lowest one third (or quartile) of the S population and the mean α_t and SD value of these early S cells established. The threshold dividing G1A vs G1B is then located on the α_t coordinate at the α_t value two SD above the mean α_t of these early S-cells; (E, F, and G) Gating windows located along the S cluster (total fluorescence, DNA content) provide an estimate of the cell number in early, mid, and late S phase. The number of windows in S, and their size and location may vary depending on the accuracy required to measure perturbation of cell progression through S.

Although the distinction of cells in mitosis can be made based on analysis of either green vs red fluorescence or total fluorescence vs α_t histograms, optimal discrimination of G1A and G1B subcompartments requires the total fluorescence vs α_t plot.

Figure 2 illustrates changes in two-parameter (total vs α_t values) histograms of CHO cells during stathmokinesis. As evident, at 0 time, there are few cells in mitosis and the α_t of these cells, although distinctly higher than that of G2 cells, is lower than that of M cells in cultures treated with colcemid for 2 or 4 h. This sug-

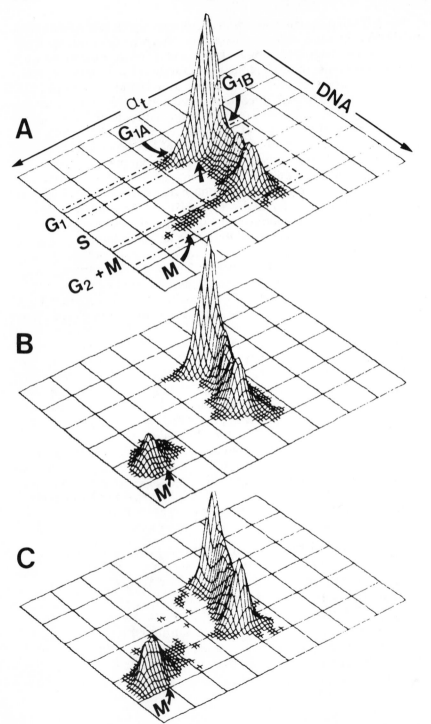

Fig. 2. Changes in bivariate (total fluorescence) DNA vs α_t histogram of CHO cells during stathmokinesis. (A) Exponentially growing CHO cells; (B) Exponentially growing cells treated with colcemid for 2 h; (C) Exponentially growing cells treated with colcemid for 4 h.

gests that with prolonged arrest of cells in mitosis, additional condensation of chromatin in chromosomes takes place. The G1A compartment is the first that shows a decrease in cell number during stathmokinesis.

It should be emphasized that in addition to the compartments listed in Fig. 1, this method of staining allows for the detection of dead cells with nuclei that undergo pyknosis (such cells have minimal green fluorescence and maximal α_t), as well as cells in deep quiescence *(19,20)*. Thus, cell death or their transition to quiescence (if associated with chromatin changes) can also be monitored during the experiment. Chapter 9 of this volume and refs. *14* and *15* describe in more detail the cell cycle compartments that can be detected by this method of cell staining.

In summary, the number of cells in the compartments discussed above may be estimated for each time-point during stathmokinesis and the data plotted with respect to time of cell arrest (Fig. 3). We will first discuss the type of raw data that can be obtained directly from such measurements. Mathematical analysis of those data will be presented later in this chapter.

7. RESULTS OF THE STATHMOKINETIC EXPERIMENT

7.1. Cell Entrance to M or G2 + M

In the majority of experiments performed on rapidly growing cells, the accumulation of cells in mitosis is represented by a straight exponential slope (Fig. 3A). The slope reflects the log-age distribution of the exponentially growing population *(50)*. A lag of approximately 1 h is often observed before the slope becomes exponential (dashed lines), which reflects either the time required for the mitotic inhibitor to penetrate the cell membrane and interact with the mitotic spindle or a transient arrest in G2 *(50)*. It is essential that

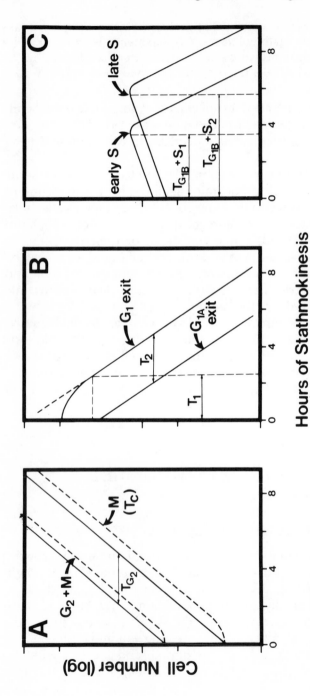

Hours of Stathmokinesis

Fig. 3. Scheme illustrating changes in cell number in the compartments of the cell cycle as distinguished in Fig. 1 during stathmokinesis. (A) Entrance to M and G2 + M; (B) Exit from G1A and G1; (C) Transit through S.

the slope of the M-entrance curve be exponential [straight line on the scale of log $(1 + f_M)$ where f_M = fraction of cells in mitosis] for most of the duration of the experiment, which should cover at least half of the cell cycle period. A deviation from that indicates that either the mitotic block is leaky, cells are not asynchronous or not in exponential growth during the experiment, or there is selective cell death in some part of the cycle. In these situations estimation of cell cycle kinetic parameters becomes much more difficult.

An example of actual data obtained for L-1210 cells treated with vinblastine, colcemid, or vincristine is displayed in Fig. 4. Note how the different stathmokinetic agents at the concentrations tested cause different lags in the mitosis accumulation curves and how vincristine effects the G2 + M accumulation curve as well (Fig. 4).

Cell entrance into G2 + M is also exponential and the lag period is usually shorter than the lag in the M-cell accumulation. This may indicate that the mitotic inhibitor induces a transient block in G2 prior to cell arrest in M. The slope of M or G2 + M curves provides an estimate of the duration of the cell cycle, as described by Puck and Steffen (50). The scale of log $(1 + f_M)$ covers the range between 0 and 0.301 for the full cell cycle. The time-distance between G2 + M vs M curves indicates the duration of G2.

7.2. Cell Exit From G1 and G1A

When cell re-entry into G1 is prevented by addition of the mitotic inhibitor, the rate of emptying of the G1 compartment can be measured. The curve representing cell exit from G1 during stathmokinesis has two distinct slopes (Fig. 3B). On a semilogarithmic scale, the first portion of the curve has a concave shape, whereas the second is a straight line. This biphasic character of the G1 exit curve was consistently observed in numerous experiments performed on CHO, L-1210, or L cell lines (25–27,62,63). Thus, two types of numerical data can be obtained from the G1 exit curve. The exponentially declining slope provides an estimate of the half-time of cell residence in G1 for a subpopulation of G1 cells ("exponential tail" of the G1 population). As will be discussed later in the chapter, in the "probablistic" model of the cell cycle (55) this parameter represents the probability of cell passage from the "indeterminate" to the "deterministic" compartment.

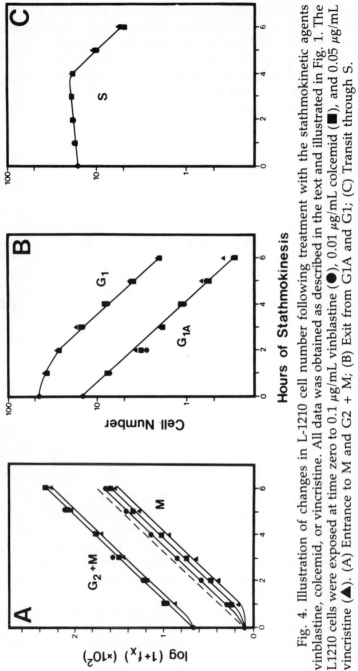

Fig. 4. Illustration of changes in L-1210 cell number following treatment with the stathmokinetic agents vinblastine, colcemid, or vincristine. All data was obtained as described in the text and illustrated in Fig. 1. The L1210 cells were exposed at time zero to 0.1 µg/mL vinblastine (●), 0.01 µg/mL colcemid (■), and 0.05 µg/mL vincristine (▲). (A) Entrance to M and G_2 + M; (B) Exit from G1A and G1; (C) Transit through S.

The second type of information that may be obtained from analysis of the G1 exit curve is the estimate of the duration of the cell transit times through G1 as represented by the concave portion of the curve, i.e., the parameter related to the linear or "deterministic" compartment of G1 (G1B). The mean duration of the G1B transit times can be estimated based on the detection of the first time point on the early portion of the G1 curve when the experimental data deflects significantly from the upward extension of the exponential portion of the curve (T_1). The time distance between the G1A and G1 slopes provides similar information (T_2).

The half-time of cell residence in the G1A compartment is established from the slope of the G1A-exit curve. This slope was found to be exponential for CHO (17), L (27), and L-1210 cells (25,62,63), and similar to the exponential portion of the G1-exit curve of these cells. Thus, it seems likely that the exponential portion of the G1 exit and G1A exit-curves represent the same kinetic event.

7.3. Cell Progression Through S-Phase

The number of cells progressing through S-phase may be counted in windows recognized on the basis of DNA content and localized, for example, at the early, mid, or late portion of S-phase. If the windows are of the same size, at the onset of the experiment fewer cells are present in the late-S window compared with the windows in early-S, which reflects the exponential age distribution of individual cells in the asynchronous, exponentially growing population. At first the number of cells measured in the windows increases slightly with time, which again reflects the log-age cell distribution. The second phase is characterized by an exponentially declining slope, which is similar to the G1A-exit curve. The deflection points appear later in windows localized later in S.

The length of the ascending portion of the curves provides a measure of the combined duration of the linear ("deterministic") portion of the G1-phase, plus that portion of S that precedes the lower threshold of the windows. Thus, when two narrow windows are located in very early S-phase and at the exit of S, respectively, the time distance of the deflection points measured between the respective curves may provide a rough estimate of the minimal duration of S-phase.

8. CELL CYCLE PERTURBATION
INDUCED BY VARIOUS DRUGS

8.1. Drug Effects on G1-, S-, and G2-Phases

Drug-induced changes in cell progression through the cycle can be easily recognized and quantified from the stathmokinetic experiment (Fig. 5). The analysis is based on comparison of the kinetic rates in control cultures vs cultures treated with the drug under investigation. Because kinetics of cell progression is measured at multiple points in the cell cycle simultaneously, even minor, secondary effects can be detected. To describe the character of the data and illustrate the potential of the method, some examples of the kinetics perturbation induced by various classes of drugs are presented in the discussion that follows. A detailed description of the experiments and effects of individual drugs are the subject of separate publications (*17,25–27,62,63*).

Figure 5 illustrates the type of changes in cell kinetics observed in experiments with various drugs. Changes in the rate of entrance to mitosis and G2 + M, resulting from perturbation of the late portion of the cell cycle, are shown in panel A. Cell arrest in G2 [observed for the intercalating drug dihydroxyanthraquinone (DHAQ) or the chelating agent ICRF], manifests as a rapid cessation of cell entrance into mitosis without significant change in the rate of cell entrance to G2 + M. Thus, these G2-blocking agents do not affect cell progression through S-phase, since the rate of their arrival into the G2 + M compartment remains unchanged. A partial block in G2 will cause a slowdown in the rate of cell entrance into mitosis, without an effect on the entrance to G2 + M (not shown).

Cell arrest in S-phase was observed in cultures treated with dihydro-5-azacytidine [5-AC(H)], for example, and the effects of such S-phase blocks are illustrated in Fig. 5. In this case, no change in the rate of cell entrance into mitosis is seen during the initial 2 h, i.e., time equivalent to the duration of G2. After 2 h, however, the entry into mitosis is halted. Cell entrance to G2 + M, on the other hand, is affected shortly after addition of the S-phase blockers, inasmuch as cells in late-S cannot now enter G2. A partial arrest in S, e.g., after addition of rhodamine 123 (R123) is reflected as

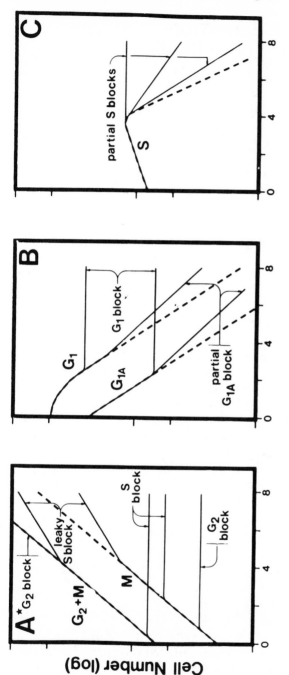

Hours of Stathkinesis

Fig. 5. Schematic illustration of the typical changes in cell kinetics induced by various drugs. (A) Cell entrance into G2 + M. Cell arrest in G2 is illustrated as it was observed in cultures treated with, for example, soluble ICRF 159 or DHAQ (mitoxantrone); cell arrest in S was induced by 5-AC(H) and a slowdown in the rate of cell progression through S and G2 was observed 4 h after addition of R123 (dashed lines, "control"). See text for details (refs. 25–27,62,63). (B) Cell exit from G1 and G1A. Suppression in the rate of cell exit from the portion of G1 characterized by exponentially distributed residence times and from G1A was observed in cultures treated with *n*-butyrate or R123. Cell exit from G1 was halted 2 h after addition of 5-AC(H). (C) Cell progression through S. The gating window is located in mid-S. 5-AC(H) arresting cells in S stops both cell entrance and exit from the mid-S window and thus the cell number remains constant. Decreased rate of progression was seen in cultures treated with *n*-butyrate or ICRF 159.

a slowdown rather than a total suppression of cell entrance to G2 + M (24).

The drug-induced changes in cell exit from the G1 or G1A compartments are shown in Fig. 5, panel B. In cultures treated with G1A-arresting agents, cells stop exiting G1A and G1 shortly after addition of the drug. If the number of cells remains constant in these compartments, one can conclude that cells are arrested in both G1A and G1B. Arrest of cells in G1B will be manifested as a suppression of cell exit from G1 without changes in the exit rate from G1A (not shown). Conversely, cell arrest in G1A will be reflected as a suppression of exit from G1A and only after a delay (approximately equal to the duration of G1B), as a suppression of exit from G1. A slowdown in the rate of cell exit from G1A is expressed as a change in the slope of the exponential portion of the exit curves, and was observed, for example, in cultures treated with n-butyrate or R123 (24,25).

Agents perturbing progression of cells through S or G1 will also produce immediate changes in the ascending and descending slopes, or in the position of the deflection, of the curves representing cells counted in the S-phase windows (Fig. 3C). Panel C in Fig. 5 illustrates changes observed in cell progression through S-phase, observed as a variation in the number of cells measured through a single window located in mid-S, such as is caused by agents that stop or slow cell transit through S (partial S block). In these cultures, the number of cells in the window decreases more slowly in time than in the control, though the ascending portion of the curve and the deflection point are similar. Thus a partial cell block occurs at about the fourth hour of stathmokinesis. Had the block occurred earlier, the ascending portion would be less steep and the deflection would appear later (not shown). Agents that completely block cell transit through S cause the number of cells in the window to remain constant throughout the experiment, reflecting the lack of cell progression. Agents that cause a block in G1, but have no effect on the rate of transit through S, cause the deflection on the curve to occur earlier than in the control.

Specific arrest of cells in early, mid, or late S, without any interruption of the transit through G1, will be reflected by steep, exponentially increasing slopes (similar to the control slopes of M or G2 + M curves in panel A), representing cell number counted in the early, mid, or late S windows, respectively (not shown). These slopes will reflect the influx of cells in the respective portions of S and their arrest therein.

It is a rather simple task to obtain numerical values related to drug effects on various parameters of the cell cycle from the respective slopes of the curves. Examples are shown elsewhere (24,25). More elaborate mathematical analysis of the experiment, such as given later in this chapter, may also be applied in studies of cell cycle perturbations in drug-treated cultures (40).

8.2. Terminal Point of Drug Action

A drug that is cell cycle-specific stops or slows progression of cells at a specific phase or time-point of the cycle. In asynchronous cultures, the cells that are beyond this point at the moment of drug addition progress to mitosis and divide. The "terminal point of drug action" represents the last time-point in the cell cycle prior to mitosis, in which cells are blocked or delayed by the drug. It is measured as the time interval between addition of the drug and a decrease in the mitotic accumulation rate (61).

The ability to estimate mitotic indices by flow cytometry makes it possible to measure with high statistical accuracy the terminal point of drug action (26). As shown in panel A (Fig. 5), the accumulation of cells in mitosis in control cultures is represented by a straight, exponential slope and the time-points at which the M-curves in drug-treated cultures deflect from the curves of control cultures is easy to estimate. The accuracy of that estimate relates to the frequency of cell sampling during the stathmokinetic experiment; withdrawal of samples every 5 or 10 min can make the estimate very accurate. Because a lag time of up to 1 h is sometimes observed before the M-curve becomes exponential, addition of the drug 1 h following the mitotic inhibitor (rather than at the same time) allows one to study drugs that affect G2-phase with more precision, especially in situations when the terminal point of drug action is close to mitosis.

When drugs affect cells in S-phase rather than in G2, a more accurate estimate of the "terminal" point of action can be obtained by the detection of the perturbation in the rate of cell entry into G2 + M, rather than to M. In this case, of course, the "terminal" point will be calculated as a time-point in relation to cell entrance to G2.

9. DETECTION OF CHROMATIN CHANGES

The sensitivity of DNA *in situ* to denaturation represented by the α_t index may be used as a parameter reflecting changes in nuclear structure that are related to chromatin condensation, histone modification, or binding of certain drugs to DNA (23,24). In the present method the α_t index is used to discriminate G1A from G1B and mitotic from G2 cells (Figs. 1 and 2). Exposure of cells to certain drugs during the stathmokinetic experiment alters their α_t values and this information, in addition to the cell kinetics data, may provide clues as to the mechanism of drug action. The examples of such changes are illustrated in Fig. 6. As evident from this figure, *n*-butyrate decreases the α_t values of interphase and mitotic cells. On the other hand, in cultures exposed to the intercalating drug, dihydroxyanthraquinone (DHAQ, mitoxanthrone), the interphase cells exhibit decreased α_t whereas the α_t of the mitotic cells is increased. These changes in α_t are statistically significant and reproducible.

In cultures treated with DHAQ, cells are arrested in G2-phase and thus no new cells accumulate in M during the stathmokinetic experiment. As a consequence, an "aged" mitotic cell population is present in the M cluster, in contrast to the variable age of mitotic cells in control cultures at an equivalent time after the vinblastine-induced block. It is likely that cells maintained for extended periods of time in mitosis undergo additional condensation of chromatin,

Fig. 6. Effects of *n*-butyrate and DHAQ on chromatin stability to denaturation measured during the stathmokinetic experiment. The scattergrams represent the distribution of control and drug-treated L-1210 cells with respect to their total fluorescence (ordinate) and α_t (abscissa) values, plotted as in Fig. 1C. The first, second, and third rows show cells from cultures treated with vinblastine for 1, 4, and 8 h, respectively. A slowdown in cell progression is observed in cultures grown with *n*-butyrate (middle column) and a complete block is evident in the presence of DHAQ (right column) with few cells in mitosis even at 8 h. A shift in the stainability of all cells toward low α_t values is seen in *n*-butyrate-treated cultures and in the interphase cells in the DHAQ-treated cultures. Mitotic cells, however, in contrast to interphase cells, increase their α_t values in DHAQ cultures during the duration of the experiment. The changes in α_t although numerically small, are statistically significant and reproducible.

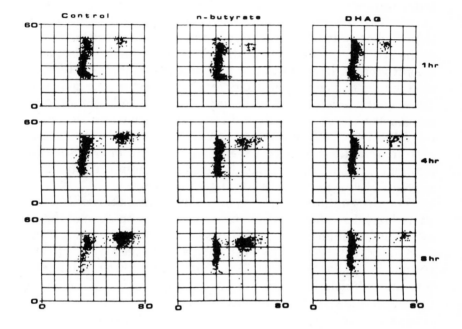

which accounts for the progressive increase in their α_t values. The changes in α_t of mitotic cells in DHAQ cultures, therefore, may be reflecting both drug effects, as well as chromatin "supercondensation" during extended mitosis (10). In contrast, the α_t changes of the interphase cells can only be explained as being caused by stabilization of DNA by the intercalating agent (26).

The mechanism that may be responsible for the observed effects of *n*-butyrate on DNA stability *in situ* is unknown. This agent induces hyperacetylation of the core particle histones, phosphorylation of H2A histone, and appearance of histone $H1^0$, which characterizes quiescent cells (13–15,25). These histone changes coincide in time with cell arrest in G1A and reversion of the morphology of tumor cells toward a normal appearance (25). However, regardless of the specific mechanisms of drug action on nuclear chromatin, the present method allows one to detect chromatin alterations in the course of screening of drugs for their effect on the cell cycle. These characteristics may be of importance in understanding the molecular basis of drug action on cells and may stimulate further research of drug effects on chromatin. Furthermore, if the severity of the chromatin changes correlates with the efficiency of the drug treatment, this parameter may be used as a rapid prognostic test reflecting cell sensitivity to the particular drug.

The drug-induced changes in nuclear chromatin reflected by altered α_t values may, in principle, affect the distinction of the cell cycle compartments recognized based on differences in α_t (e.g., G1A and G1B). So far, however, all drugs studied by us affected chromatin of all interphase cells in a rather uniform fashion, shifting the population toward higher or lower α_t, but not affecting distinction between the compartments. This probability, however, should be considered with drugs that have a high affinity toward chromatin of cells at specific portions of the cell cycle.

Nonviable cells, especially with pyknotic nuclei, can be recognized by this staining technique as having minimal green and high-red fluorescence and α_t values markedly higher than those of mitotic cells (27). The method thus permits estimation of the cytotoxicity of the drugs studied, in addition to the cytostatic- and chromatin-modifying effects. Such analysis has been made, for example, when the cell cycle-specific effects of tumor necrosis factor were studied (27).

10. MATHEMATICAL ANALYSIS OF THE STATHMOKINETIC EXPERIMENT

10.1. General Assumptions

With the reservation that the "probabilistic" event(s) assumed to exist in the cell cycle may have a defined metabolic cause(s) (see further), in the discussion that follows the parameters of cell kinetics are analyzed based on the probabilistic model of Smith and Martin (55) [also, cf. (38)]. Namely, it is assumed that after the division of the mother cell, each of the daughter cells enters an indeterminate phase (AG1), the duration of which is a random variable τ, with exponential distribution:

$$\text{Prob}\{\tau \leqslant t\} = 1 - e^{-\alpha\tau} \quad [1]$$

Parameter α is equal to the reciprocal of the mean time spent in AG1. After leaving AG1, the cell progresses through phases BG1, S, G2, and M (Fig. 7). The residence times in these phases are deterministic and denoted by T_2, T_3, T_4, and T_5, respectively. The dividing cell in phase 5 (mitosis) will always be counted as a single cell.

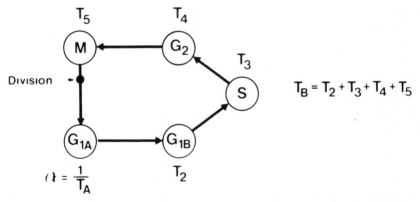

Fig. 7. The basic model of the cell cycle. The cell cycle phases are numbered from 1 to 5. The residence times in phases 2, 3, 4, and 5 (i.e., BG1, S, G2, and M, respectively) are considered to be deterministic and denoted by T_2, T_3, T_4, and T_5. The residence time in phase 1 (AG1) is treated as an exponentially distributed random variable with the mean value $T_A = \frac{1}{\alpha}$.

In fact, the probabilistic model that is used need not be necessarily true from the biological viewpoint. The hypothesis of Smith and Martin (55) is far from being proven (46). The source of the randomness of cell cycle events may be the result of a phenomenum preceding cell entry into the indeterminate AG1 phase. There is strong evidence, for instance, that the unequal division of RNA or other metabolic constituents during cytokinesis causes the heterogeneity of cell progression through the cell cycle (17). Nevertheless, the approach of Smith and Martin accurately describes many facets of the cell cycle and remains a good and simple mathematical tool for practical computations connected with cell cycle kinetics.

It should be emphasized that the "indeterminate" and "deterministic" phases of G1 are denoted here as AG1 and BG1, respectively. These phases, although related to the metabolic compartments of G1A and G1B described earlier in this chapter (and discussed elsewhere; *see* refs. 15–18,20), are not equivalent to these compartments, but represent mathematically estimated kinetic states of G1. The G1A compartment is recognized here based on differences in chromatin structure, as reflected by α_t (17). The cell residence times in G1A exhibit a characteristic exponential "tail," and during stathmokinesis, cell exit from G1A is represented by an exponentially declining slope. The G1A compartment may be considered, therefore, to be the locus of the indeterminate state (AG1). Although the "indeterminate" character of cell residence time in G1A is most likely determined by metabolic differences between the cells (3,9,17,18), in the discussion that follows we refer to the kinetics of cell transit through G1A as reflecting the kinetics of cells in the indeterminate state, i.e., AG1.

The present model describes in simple terms the transition of cells between indeterminate and deterministic phases. $N_i(t)$ denotes the number of cells in phase i and $n_i(t)$ denotes the efflux of cells from this phase, at time t (Fig. 8). The efflux $n_i(t)$ from a phase with deterministic residence time T_i is equal to the efflux from the preceding phase i–1, delayed by T_i:

$$n_i(t) = n_{i-1}(t - T_i), \quad i = 2, 3, 4, 5 \qquad [2]$$

To obtain the number of cells in one of the deterministic phases of the cycle, it is enough to note that $N_i(t)$ is equal to the efflux $n_i(t)$ accumulated over the interval $(t, t + T_i)$,

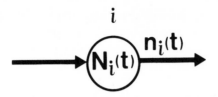

Fig. 8. Number of cells $[N_i(t)]$ and efflux $[n_i(t)]$ of the arbitrary i-th phase of the cell cycle.

$$N_i(t) = \int_t^{t+T_i} n_i(\tau)d\tau, \; i = 2, 3, 4, 5 \qquad [3]$$

For indeterminate phase 1 (i.e., AG1), it can be proven (41) that $N_1(t)$ satisfies the differential equation:

$$\dot{N}_1(t) = -\alpha N_1(t) + n^+(t)$$

where $n^+(t)$ is the influx to phase 1, equal to $2n_5(t)$: the efflux from mitosis multiplied by two. Therefore,

$$\dot{N}_1(t) = -\alpha N_1(t) + 2n_5(t) \qquad [4]$$

while the efflux from phase AG1 is equal to

$$n_1(t) = \alpha N_1(t) \qquad [5]$$

Equations [1] through [5] provide us with the complete kinetic description of the cell cycle model (41).

10.2. Exponential and Asynchronous Growth

The assumption is made that at the onset (at $t = 0$) of the stathmokinetic experiment, the cell population was growing exponentially and that the proportions of cells in all phases were constant. Thus, for $t < 0$:

$$N_i(t) = N_i(0) \, e^{\lambda t} \qquad [6]$$

$$n_i(t) = n_i(0) \, e^{\lambda t} \qquad [7]$$

for $i = 1,...,5$. The constant λ is called the Malthusian growth parameter (39). Similarly:

$$N(t) = N(0) \, e^{\lambda t} \qquad [8]$$

where $N(t) = N_1(t) + N_2(t) + N_3(t) + N_4(t) + N_5(t)$ is the total number of cycling cells. The Malthusian parameter λ is related to the time T_d of population doubling: $N(T_d) = 2N(0) = N(0)e^{\lambda T_d}$, which implies

$$\lambda = \frac{\ln 2}{T_d} \qquad [9]$$

Substituting exponential functions [7] and [8] into the model equations [1] through [5], performing the necessary manipulations and dividing the equations by $e^{\lambda t}$, the constants $N_i(0)$, $n_i(0)$ can be found:

$$\begin{cases} N_1(0) = \dfrac{2\lambda}{\lambda + \alpha}\, N(0) \\[3em] N_i(0) = \dfrac{2\alpha}{\lambda + \alpha}\, N(0)e^{-\lambda(T_2 + \ldots + T_i)}(e^{\lambda T_i} - 1),\ i = 2, 3, 4, 5 \end{cases} \qquad [10]$$

$$\begin{cases} n_1(0) = \dfrac{2\alpha\lambda}{\alpha + \lambda}\, N(0), \\[3em] n_i(0) = \dfrac{2\alpha\lambda}{\alpha + \lambda}\, N(0)e^{-\lambda(T_2 + \ldots + T_i)},\ i = 2, 3, 4, 5 \end{cases} \qquad [11]$$

The Malthusian parameters λ must satisfy the relation, called "the characteristic equation":

$$e^{\lambda(T_2 + T_3 + T_4 + T_5)} = \frac{2\alpha}{\alpha + \lambda} \qquad [12]$$

This equation leads to an interesting conclusion that can be stated using formula [9]. Namely:

$$e^{-\frac{T_B}{T_d}\ln 2} = \frac{1}{2}\left(\frac{T_A}{T_d}\ln 2 + 1 \right) \qquad [13]$$

where $T_A = \dfrac{1}{\alpha}$ is the average duration of AG1, and $T_B = T_2 +$ $T_3 + T_4 + T_5$ is the duration of the deterministic part of the cycle. If our cell cycle model is correct, then any values of T_A, T_B, and T_d estimated using any method must satisfy Eq. [13]. The doubling time is, in general, shorter than the sum $T_A + T_B$. In fact, it is possible to check using Eq. [13] that the value of $\dfrac{T_A + T_B}{T_d}$ depends on the ratio $\dfrac{T_A}{T_A + T_B}$; it is equal to 1 for $T_A = 0$ and then increases up to $\dfrac{1}{\ln 2}$ for $\dfrac{T_A}{T_A + T_B} = 1$ (i.e., for $T_B = 0$). This relation is depicted in Fig. 3. In the pioneer work by Puck and Steffen (50), all the cycle phases were assumed to be deterministic ($T_A = 0$), so the doubling time was equal to the sum of the residence times in the consecutive phases of the cycle.

10.3 Stathmokinetic Experiment

In the idealized stathmokinetic experiment, the cell population that was previously in the state of exponential and asynchronous growth loses the possibility of producing new cells at $t = 0$. The cells that enter mitosis are "trapped" there, so no new cells re-enter AG1. Some of our model equations must be changed to cover the new situation. The arrest in mitosis is equivalent with "cutting" to the efflux from this phase:

$$n_5(t) = 0, \, t > 0 \qquad [14]$$

Also, the cell number in mitosis is equal to their number at $t = 0$, plus the efflux from G2, accumulated from 0 to t:

$$N_5(t) = N_5(0) + \sum_0^t n_4(\tau)d\tau \qquad [15]$$

Since no cells are re-entering AG1 after $t = 0$, Eq. [4] must be replaced by the following:

$$\dot{N}_1(t) = -\alpha N_1(t) \qquad [16]$$

Using standard mathematical methods, Eqs. [1]–[5] with modifications [14]–[16] and with initial conditions [10]–[11] can be solved. The details are not given here since they are based on rather elementary mathematical computations. For clarity of exposition we will define:

$$f_i(t) = N_i(t)/N(0) \qquad [17]$$

to be equal to the fraction of the initial cell population that remains in the *i*-phase of time *t*. Using this notation, final results can be stated in the following form:

$$f_1(t) = (2 - e^{\lambda T_B})e^{\alpha t}, \; t > 0 \tag{18}$$

for phase AG1, then:

$$f_i(t) = \begin{cases} e^{\lambda(T_B - T_2 - \dots - T_i)}(e^{\lambda T_i} - 1)e^{\lambda t}; \; 0 < t \leqslant T_2 + \dots + T_{i-1} \\[2mm] 2 - e^{\lambda(T_B - T_2 - \dots - T_i)}e^{\lambda t} - (2 - e^{\lambda T_B})e^{\alpha(T_2 + \dots + T_{i-1})}e^{-\alpha t}; \\[1mm] T_2 + \dots + T_{i-1} < t \leqslant T_2 + \dots + T_i, \\[2mm] (2 - e^{\lambda T_B})e^{\alpha(T_1 + \dots + T_{i-1})}(e^{\alpha T_{i-1}}e^{-\alpha t}; \; t > T_2 + \dots + T_i \end{cases} \tag{19}$$

for *i* = 2, 3, 4 (i.e., for phases BG1, S, and G2) and finally:

$$f_5(t) = \begin{cases} e^{\lambda(t + T_5)} - 1; \; 0 < t \leqslant T_2 + T_3 + T_4 \\[2mm] 1 - (2 - e^{\lambda T_B})e^{\alpha(T_B - T_5)}e^{-\alpha t}; \; t > T_2 + T_3 + T_4 \end{cases} \tag{20}$$

for mitosis. For *i* = 2, i.e., phase BG1, the symbol $T_2 + \dots + T_{i-1}$ should be interpreted as 0; therefore, the first equation of [19] is not applicable in this case.

These equations may be combined to obtain theoretical formulae for the empirical curves of the stathmokinetic experiment. We will present them in the order in which they appear in Fig. 9.

We begin with the case of mitotic and G2 cells. The most classic. way of estimating the doubling time was introduced by Puck and Steffen (50), (*see* also ref. 2). We draw the plot of ln [1 + $f_5(t)$], i.e., of the natural logarithm of the fraction of cells in mitosis plus 1, vs time. Formulae [19] and [20] imply that:

$$\ln[1 + f_5(t)] = \lambda(T_5 + t), \; t < T_2 + T_3 + T_4 \tag{21}$$

Similarly, for the fraction $f_4(t) + f_5(t)$ in G2 + M, it holds:

$$\ln[1 + f_4(t) + f_5(t)] = \lambda(T_4 + T_5 + t), \; t < T_2 + T_3 \tag{22}$$

Both curves are (theoretically) straight lines in the initial time interval, after which they exhibit curvature. The slopes of both straight line intervals are equal to λ. Thus, estimating the slope of the experimental graph (using linear regression) (2), we can find λ, and as a consequence, the doubling time T_d by application of formula

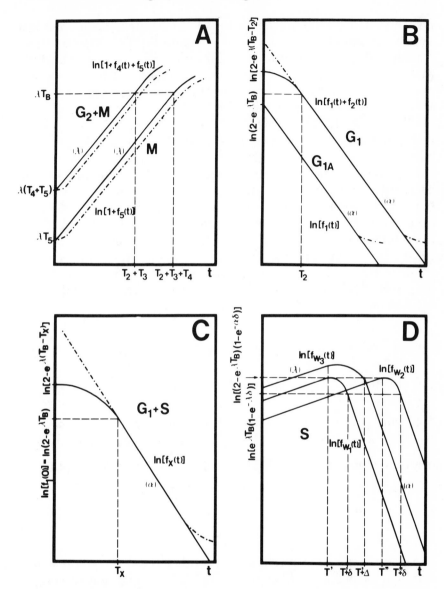

Fig. 9. Dependence of the doubling time T_d on the average duration T_A of the AG1 phase and the duration T_B of the "deterministic" portion of the cell cycle. Dimensionless variables $\dfrac{T_A + T_B}{T_d}$ and $\dfrac{T_A}{T_A + T_B}$ are introduced in order to make the graph independent of the absolute values of T_A, T_B, and T_d.

[9]. The situation is depicted in Fig. 9A, in which some geometrical details are added. The curve representing Eq. [22] may be more convenient, since it is only the total number of cells in M and G2 that can be found using the usual one-dimensional DNA histogram analysis. The examples of experimental curves describing Eqs. [21] and [22] can be found in refs. *25,26,62,63*.

Theoretically, the values of $T_2 + T_3$ and $T_2 + T_3 + T_4$ could be estimated from the same graphs, as the times at which curvature begins (*see* Fig. 10A). This estimate, however, is biased with considerable errors because of uncertainty of determination of the precise point of deflection of these curves. Furthermore, the deflection point sometimes cannot be estimated at all because of the need for extended duration of stathmokinetic (e.g., in the case of slowly cycling cells), which may result in deterioration of metaphase cells. The average residence time T_A in the AG1 state can be estimated by analyzing the logarithm of the fraction of cells in G1A (i.e., $\ln[f_1(t)]$) or in the whole G1 = G1A + G1B (i.e., $\ln[f_1(t) + f_2(t)]$). From [18] and [19] one can find that:

$$\ln[f_1(t)] = \ln(2 - e^{\lambda T_B}) - \alpha t; \ t > 0 \qquad [23]$$

$$\ln[f_1(t) + f_2(t)] = \ln(2 - e^{\lambda T_B}) - \alpha(t - T_2); \ t > T_2 \qquad [24]$$

Both curves are straight lines with slopes equal to α; the second one after initial curvature (Fig. 9B). Again T_A can be estimated from the slope; $T_A = \dfrac{1}{\alpha}$. The time $T_{1/2}$ after which 50% of the initial number of cells remains in G1A can also be found:

$$T_A = \frac{\ln 2}{\alpha} \qquad [25]$$

The distinction between cells in G1A and G1B is not possible when one-dimensional DNA histograms are analyzed and only the graph of $\ln[f_1(t) + f_2(t)]$ can be drawn. It seems worth noting, however, that even in such a case the initial cell fraction in G1A can be found by localizing the "deflection point" (i.e., the point in which the curvature ends) of the curve $\ln[f_1(t) + f_2(t)]$, as evident in Figs. 4B and C.

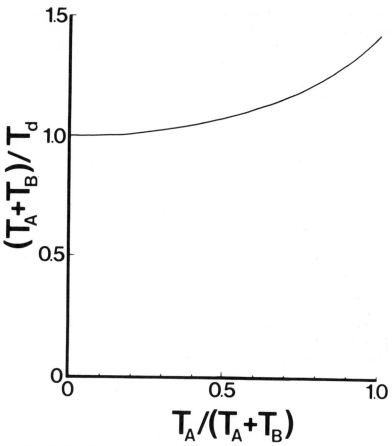

Fig. 10. Analysis of the curves obtained in the stathmokinetic experiment. The graphs are scaled arbitrarily. The symbols in brackets denote the slopes of straight line intervals. More details are found in the text. (A) The accumulation of cells in M and G2 + M according to Eqs. [21] and [22]. The theoretical curves are represented by solid lines. The dot-dash lines characterize the usual shape of experimental curves caused by a delay in the action of the stathmokinetic agent. (B) The exit curves from G1A and G1 according to Eqs. [23] and [24]. The dot-dash lines in the lower parts of both graphs (*see* also panel C) symbolize the pattern occasionally observed in experiments reflecting the presence of the non-cycling cells. (C) The exit curve from G1 + early S according to Eqs. [26] and [27] (also *see* Fig. 11). (D) Cell number in the W_1, W_2, and W_3 windows located in S, according to Eq. [28]. The positions of the windows are depicted in Fig. 11.

The exact boundary between G1 and S is not always well defined. In such cases, it should be recognized that the time coordinate of the "deflection point" of the exit curve is not localized at $t = T_2$ but at $t = T_X$, where,

$$T_{\frac{1}{2}} < T_X < T_2 + T_3 \qquad [26]$$

If the fraction of cells present in G1 plus the fraction present in "early" S is denoted by $f_X(t)$, then:

$$f_X(t) = \begin{cases} 2 - e^{\lambda(T_B - T_X)}e^{\lambda t}; & t < T_X \\ (2 - e^{\lambda T_B})e^{\alpha T_X}e^{-\alpha t}; & t \geqslant T_X \end{cases} \qquad [27]$$

The logarithm of this exit curve is depicted in Fig. 9C. Equation [27] is derived by subdividing the deterministic phase B into two "phases" of length T_X and $T_B - T_X$, respectively (Fig. 11), and by using formulae analogous to Eqs. [18]–[20]. Sharpless and Schles-

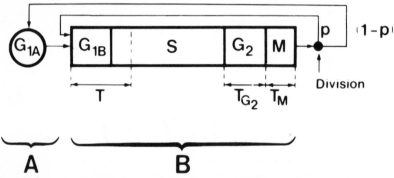

Fig. 11. The subdivision of the "deterministic" part of the cycle used to derive Eq. [27]. See the text for details, as well as Fig. 10C.

inger (53) developed a very accurate computer method of fitting curves of form of Eq. [27] to the experimental data.

Another experimental technique is to observe the cell number in the "window" in S-phase, i.e., by counting only part of the cells present in this phase. Using windows of different widths and locations, various experimental curves can be produced, all of them being special cases of the following pattern: Suppose that the window W, begins T' units of time after the cells enter G1B and its width is equal to δ. This situation is depicted in Fig. 12. Then the following formulae for the fraction $f_{W1}(t)$ of cells counted in the window can be obtained:

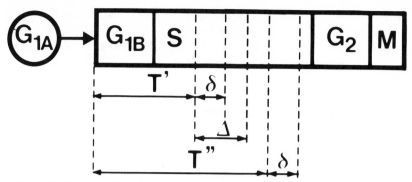

Fig. 12. The subdivision of the "deterministic" part of the cycle, used to derive Eq. [28] and to draw graphs in Fig. 10. Details are found in the text.

$$f_{W1}(t) = \begin{cases} e^{\lambda(T_B - T' - \delta)}(e^{\lambda\delta} - 1); \; t < T' \\ 2 - e^{\lambda(T_B - T' - \delta)}e^{\lambda t} - (2 - e^{\lambda T_B})e^{-\alpha t}; \; T' \leqslant t < T' + \delta \\ (2 - e^{\lambda T_B})e^{\alpha T'}(e^{\alpha\delta} - 1)e^{-\alpha t}; \; t > T' + \delta \end{cases}$$

As evident, the graph of $\ln (f_{W1}(t))$ is composed of two intervals of straight lines with slopes λ and α, respectively, connected by a curvilinear interval of length δ (Fig. 9D). In Figs. 9D and 12, other windows are also shown: W_2 of the same width as W_1, but beginning at $T'' > T'$ and W_3 beginning at T' but wider than W_1. In this way the respective locations of graphs resulting from cell counting in different windows are shown.

Graphs of type "window in S" are a good tool for confirming the information about parameters λ and α, and in analysis of the cell cycle perturbation (e.g., induced by the drugs). They can also help in estimating the S-phase duration. If the window covers S-phase, Eq. [28] coincides with Eq. [19] for $i = 3$, $T_3 = \delta$, $T_2 = T'$.

10.4. Extensions of the Model

For simplicity, the model of the cell cycle discussed above was limited to situations in which both daughter cells after division entered the AG1 phase. This conforms with the classical form of the A to B-transition hypothesis (55).

However, as was pointed out by Nedelman and Rubinow (46), such simplification cannot always be justified. In fact, no experimental evidence exists that both daughter cells indeed enter AG1. A more general model of the cell cycle, therefore, should be considered in which each daughter cell enters AG1 with some fixed probability $(1 - p)$ and bypasses G1A with another probability (p), as depicted in Fig. 13.

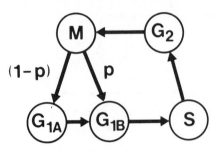

Fig. 13. The model of the cell cycle in which any daughter cell may bypass AG1 subphase with probability p.

To describe this new model, a slightly different equation can be used. First, only a fraction of the cells enters phase AG1, hence:

$$\dot{N}_1(t) = -\alpha N_1(t) + 2(1 - p)n_5(t)$$

Second, part of the outflux from mitosis contributes directly to the influx into BG1 and thus to the outflux from BG1:

$$n_2(t) = n_1(t - T_2) + 2pn_5(t - T_2)$$

By inserting these relations into our mathematical model, the new formulae can be found. The main difference is in the form of the characteristic Eq. [12]. Now it holds:

$$e^{-\lambda T_B} = \frac{\lambda + \alpha}{2(p\lambda + \alpha)} \qquad [29]$$

and this reduces to Eq. [12] if $p = 0$.

The final equation for fractions $f_1(t), \dots f_5(t)$ are formally identical with Eqs. [18]–[28]. Also, all the details of Fig. 9 remain valid in the new situation.

11. CONCLUSIONS: NOVEL ELEMENTS OF THE TECHNIQUE

The technique described in this chapter has many elements similar to those of other stathmokinetic methods. It also offers some new solutions. In the discussion that follows, the novel aspects are emphasized by comparison of the type of cell kinetic information that can be obtained by various techniques.

The novelty of the proposed method stems from the fact that the flow cytometric measurements allow us to discern G1, S, G2, and M compartments. Hence, during stathmokinesis, cell number in all these compartments can be estimated. In addition, a distinction is made between G1A and G1B cells. The threshold discriminating these two compartments is believed to indicate a metabolic landmark in G1 and is based on chromatin conformation (*15–18*). As discussed elsewhere (*15,16*), differences in phosphorylation of histone H1 and changes in the ratio of nonhistone/histone proteins may be responsible for the observed differences in DNA sensitivity to denaturation between G1A and G1B cells. Cells in G1A are characterized by widely distributed residence times, with a typical exponential, or quasi-exponential, tail of the distribution. Accepting Mitchison's concept of the cell cycle (*44*), it is most likely that G1A represents an interruption in the "DNA division cycle"; during which the "growth cycle" continues. The DNA division cycle is resumed when cells enter the G1B subphase. Cell progression through G1A thus reflects cell growth rates, up to the threshold value ("cell equalization times"), as suggested before (*17,18*). Specifically, when G1A cells are recognized based on chromatin structure (α_t value), the G1A residence times represent the times required by individual postmitotic cells to decondense their chromatin to the level of that of the early S cells (minimal α_t value). Further discussion about the distinctive metabolic features of G1A vs G1B can be found elsewhere (*15–18*).

During stathmokinesis, therefore, when the cell number in all subphases is counted, the following rates of transition and other parameters can be estimated either in control or drug-treated cultures:

(a) the rate of cell entry into M ("M curve"). This parameter is identical to the classic mitotic or cell production-rates, as proposed by Puck and Steffen (*50*). Similarly, to estimate the kinetic parameters, Scheiderman et al. (*52*) proposed collection of mitotic cells in drug-treated cultures, based on their selective detachment. There are several mathematical models for analysis of such synchronized populations (e.g., *36,60,64*) or for analysis of perturbed populations (e.g., *34*).

(b) the time of appearance of the perturbation in the "M curve" in the drug-treated cultures. This allows one to estimate the "terminal point of drug action," as originally proposed by Tobey (61). One novelty of our approach is that the measurements and discrimination of mitotic cells are done rapidly, by flow cytometry.

(c) the rate of cell entry into the G2 + M compartment. This parameter is identical to the one proposed by Barfod and Barfod (4,5) and Dosik et al. (30) and indicates a kinetic event similar to (b), i.e., the cell "production" rate. However, when compared with the simultaneously measured "M-rate," the time-distance between the "M" vs "G2 + M"-curves and their respective shapes enable one to: estimate the duration of G2 (which is a novel approach); and discriminate between cell arrest in M vs G2 in the drug-treated cultures, as well as to measure a delay in cell transit through G2 (also a novel approach).

(d) the rate of cell exit from G1. Flow cytometric measurements of cell exit from G1 were introduced by Coffino and Gray (12) and adapted to stathmokinetic experiments by Dosik et al. (30) and the authors (26). The novelty of the present approach is that the discrimination is made between the linear vs exponential slope of the G1-exit curve and thus two distinct kinetic parameters of G1 are estimated. We want to emphasize that the parameter related to the "exponential tail" of the G1 population (which may be expressed as the half-time of cell residence in G1) may be of importance in the clinic inasmuch as it provides a more accurate kinetic characterization of cells that are the most likely to escape effects of the cell cycle-specific drugs.

(e) The rate of cell exit from G1A, expressed as the half-time of residence in G1A, is also a novel parameter. This parameter represents the cell growth rate (or cell "equalization time" in G1) rather than progression through the "DNA division-cycle." The ability to recognize and enumerate G1A cells makes it possible to discriminate between effects of drugs that arrest cells in growth phase (e.g., n-butyrate, R123, retinoic acid, actinomycin D), compared with agents that arrest cells in the DNA-division cycle, without suppressing their growth (e.g., hydroxyurea or aphidocolin). The rate at which cells in G1 pass through the metabolic threshold introduces a new yardstick for cell cycle analysis.

(f) progression of cells through S. Analysis of cells progressing through narrow windows in S was introduced by Gray et al. (35). The novelty of the present technique is that it measures the time of appearance of the perturbation in cell number (within the windows), induced by the stathmokinetic agent, or the drug under investigation. No cell prelabeling and sorting are therefore required. Cell number in the window represents the actual equilibrium between the inflow and outflow. Thus, the method is very sensitive to drug effects that result in cell arrest in S, i.e., when the outflow is stopped. In addition, it allows for discrimination between drugs that affect either early, mid, or late S-phase. If the windows are located in very early and very late S, the difference in time when the perturbation in rate of cell transit induced by the stathmokinetic agent is detected in these windows, respectively, provides a measure of the duration of S-phase.

(g) chromatin structure. Because the cell staining pattern is based both on DNA content and chromatin structure (sensitivity of DNA to denaturation) the method allows for the detection of drug-induced chromatin alterations. Such changes, related to drug intercalation to DNA, chemical modification of histones, and cell differentiation or death can be easily recognized and estimated with respect to the cell position in the cell cycle based on its DNA content.

ACKNOWLEDGMENTS

We thank Miss Robin Nager for help in preparation of the manuscript. The work was supported by Grants CA28704 and CA23296 from the National Cancer Institute.

332 Darzynkiewicz, Traganos, and Kimmel

REFERENCES

1. Aherne, W. A. and Camplejohn, R. S. On correcting the error due to metaphase degeneration in stathmokinetic studies. Exp. Cell Res., 74: 496–501, 1972.

2. Aherne, W. A., Camplejohn, R. S., and Wright, N. A. An Introduction to Cell Population Kinetics. London: Edward Arnold, 1977.

3. Alberghina, L., Merran, L., and Mortegani, E. Cell cycle variability: Modeling and Simulation. In: (M. Rottenberg, ed.), Biomathematics and Cell Kinetics. Amsterdam: Elsevier, 1981.

4. Barfod, I. J. and Barfod, N. M. Cell-production rates estimated by the use of vincristine sulphate and flow cytometry. I. An in vitro study using murine tumour cell lines. Cell Tissue Kinet., 13: 1–8, 1980.

5. Barfod, I. H. and Barfod, N. M. Cell-production rates estimated by the use of vincristine sulphate and flow cytometry. II. Correlation between the cell-production rates of aging ascites tumours and the number of S phase tumour cells. Cell Tissue Kinet., 13: 9–19, 1980.

6. Bauer, K. D. and Dethlefsen, L. A. Total cellular RNA content: Correlation between flow cytometry and ultraviolet spectroscopy. J. Histochem. Cytochem., 28: 493–498, 1980.

7. Bauer, K. D., Keng, P. C., and Sutherland, R. M. Isolation of quiescent cells from multicellular tumor spheroids using centrifugal elutriation. Cancer Res., 42: 75–78, 1982.

8. Bohmer, R. M. Flow cytometric cell-cycle analysis using the quenching of 33258 Hoechst fluorescence by bromodeoxyuridine incorporation. Cell Tissue Kinet., 12: 101–112, 1979.

9. Castor, L. N. A G1-rate model accounts for cell cycle kinetics attributed to "transition probability." Nature, 287: 76–79, 1980.

10. Clarke, R. M. A comparison of metaphase arresting agents and tritiated thymidine in measurements of the rate of entry into mitosis in the crypts of Lieberkuhn of the rat. Cell Tissue Kinet., 4: 263–272, 1971.

11. Cleaver, J. E. Thymidine Metabolism and Cell Kinetics. Amsterdam: North Holland, 1967.

12. Coffino, P. and Gray, J. W. Regulation of S49 lymphoma cell growth by cyclic adenosine 3':5'-monophosphate. Cancer Res., 38: 4285–4288, 1978.

13. D'Anna, J. A., Tobey, P. A., and Gurley, L. R. Concentration-dependent effects of sodium butyrate in Chinese hamster cells: Cell cycle progression, inner-histone acetylation, histone H1 dephosphorylation and induction of an H1-like protein. Biochemistry, 19: 2656–2671, 1980.

14. Darzynkiewicz, Z. Molecular interactions and cellular changes during the cell cycle. Parmac. Therap., *21*: 143–188, 1983.

15. Darzynkiewicz, Z. Metabolic and kinetic compartments of the cell cycle distinguished by multiparameter flow cytometry. *In*: (P. Skehan and S. Friedman, eds.), Growth, Cancer and the Cell Cycle. New Jersey: Humana Press, 1984.

16. Darzynkiewicz, Z., Andreeff, M., Traganos, F., Sharpless, T., and Melamed, M. R. Discrimination of cycling and non-cycling lymphocytes by BUdR-suppressed acridine orange fluorescence in a flow cytometric system. Exp. Cell Res., *115*: 31–35, 1978.

17. Darzynkiewicz, Z., Crissman, H., Traganos, F., and Steinkamp, J. Cell heterogeneity during the cell cycle. J. Cell Physiol., *113*: 465–474, 1982.

18. Darzynkiewicz, Z. and Traganos, F. RNA content and chromatin structure in cycling and noncycling cell populations studied by flow cytometry. *In*: (G. M. Padilla and K. S. McCarty, eds.), Genetic Expression in the Cell Cycle. New York: Academic Press, 1982.

19. Darzynkiewicz, Z., Traganos, F., Andreeff, M., Sharpless, T., and Melamed, M. R. Different sensitivity of chromatin to acid denaturation in quiescent and cycling cells as revealed by flow cytometry. J. Histochem. Cytochem., *27*: 478–485, 1979.

20. Darzynkiewicz, Z., Traganos, F., and Melamed, M. R. New cell cycle compartments identified by multiparameter flow cytometry. Cytometry, *1*: 98–108, 1981.

21. Darzynkiewicz, Z., Traganos, F., and Melamed, M. R. Distinction between 5-bromodeoxyuridine labelled and unlabelled mitotic cells by flow cytometry. Cytometry, *3*: 345–348, 1983.

22. Darzynkiewicz, Z., Traganos, F., Sharpless, T., and Melamed, M. R. Cell cycle-related changes in nuclear chromatin of stimulated lymphocytes as measured by flow cytometry. Cancer Res., *37*: 4635–4640, 1977.

23. Darzynkiewicz, Z., Traganos, F., Sharpless, T., and Melamed, M. R. Recognition of cells in mitosis by flow cytofluorometry. J. Histochem. Cytochem., *25*: 875–880, 1977.

24. Darzynkiewicz, Z., Traganos, F., Staiano-Coico, L., Kapuscinski, J., and Melamed, M. R. Interactions of rhodamine 123 with living cells studied by flow cytometry. Cancer Res., *42*: 799–806, 1982.

25. Darzynkiewicz, Z., Traganos, F., Xue, S. B., and Melamed, M. R. Effect of *n*-butyrate on cell cycle progression and *in situ* chromatin structure of L1210 cells. Exp. Cell Res., *136*: 279–293, 1981.

26. Darzynkiewicz, Z., Traganos, F., Xue, S. B., Staiano-Coico, L., and Melamed, M. R. Rapid analysis of drug effects on the cell cycle. Cytometry, *1*: 279–286, 1981.

27. Darzynkiewicz, Z., Williamson, B., Carswell, E. A., and Old, L. J. The cell cycle specific effects of tumor necrosis factor analysed by multiparamter flow cytometry. Cancer Res., *44*: 83–90, 1984.

28. Dean, P. N., Dolbeare, F., Gratzner, H., Rice, G. C., and Gray, J. W. Cell-cycle analysis using a monoclonal antibody to BrdUrd. Cell Tissue Kinet., *17*: 427–436, 1983.

29. Dolbeare, F., Gratzner, H., Pallavicini, M. G., and Gray, J. W. Flow cytometric measurements of total DNA content and incorporated bromodeoxyuridine. Proc. Natl. Acad. Sci. USA, *80*: 5573–5577, 1983.

30. Dosik, G. M., Barlogie, B., White, A. R., Gohde, W., and Drewinko, B. A rapid automated stathmokinetic method for determination of *in vitro* cell cycle transit times. Cell Tissue Kinet., *14*: 121–134, 1981.

31. Fitzgerald, P. H. and Brehaut, L. A. Depression of DNA synthesis and mitotic index by colchicine in cultured human lymphocytes. Exp. Cell Res., *59*: 27–35, 1970.

32. Frei, E., Whang, J., Scoggins, R. B., Van Scott, E. J., Rall, D. P., and Ben, M. The stathmokinetic effect of vincristine. Cancer Res., *24*: 1918–1928, 1964.

33. Gratzner, H. G. and Leif, R. C. An immunofluorescence method for monitoring DNA synthesis by flow cytometry. Cytometry, *1*: 385–389, 1981.

34. Gray, J. W. Cell-cycle analysis of perturbed cell population. Computer simulation of sequential DNA distribution. Cell Tissue Kinet., *9*: 499–510, 1976.

35. Gray, J. W., Carver, J. H., George, Y. S., and Mendelsohn, M. L. Rapid cell cycle analysis by measurement of the radioactivity per cell in a narrow window in S phase (RCS). Cell Tissue Kinet., *10*: 97–107, 1977.

36. Hahn, G. M. State vector description of the proliferation of mammalian cells in tissue culture. I. Exponential growth. Biophys. J., *6*: 275–286, 1966.

37. Iversen, O. H., Iversen, U., Ziegler, J. L., and Bluming, A. Z. Cell kinetics in Burkitt lymphoma. Eur. J. Cancer, *10*: 155–163, 1974.

38. Jagers, P. Branching Processes With Biological Applications. New York: Wiley and Sons, 1975.

39. Kimmel, M. General theory of cell cycle dynamics based on the branching processes in varying environment. *In*: (M. Rotenberg, ed.), Biomathematics and Cell Kinetics. Amsterdam: Elsevier, 1981.

40. Kimmel, M. and Traganos, F. Kinetics of drug induced G2 block *in vitro*. Mathematical analysis of stathmokinesis and continuous exposure. Cell Tissue Kinet., *18*: 91–110, 1985.

41. Kimmel, M., Traganos, F., and Darzynkiewicz, Z. Do all daughter cells enter the indeterminate ("A") state of the cell cycle? Analysis of stathmokinetic experiments on L1210 cells. Cytometry, *4*: 191–201, 1983.

42. Latt, S. A., George, Y. S., and Gray, J. W. Flow cytometric analysis of bromodeoxyuridine-substituted cells stained with 33258 Hoechst. J. Histochem. Cytochem., *25*: 927–934, 1977.

43. Macdonald, P. D. M. Towards an exact analysis of stathmokinetic and continuous-labelling experiments. *In:* (M. Rottenberg, ed.), Biomathematics and Cell Kinetics. Amsterdam: Elsevier, 1981.

44. Mitchison, J. M. The Biology of the Cell Cycle. Cambridge: The University Press, 1981.

45. Morris, W. T. *In vivo* studies on the optimum time for the action of colchicine on mouse lymphoid tissues. Exp. Cell Res., *48*: 209–217, 1967.

46. Nedelman, J. and Rubinow, S. I. Investigation into the experimental kinetics support the two-state model of the cell cycle. Cell Biophys., 2: 207–231, 1980.

47. Pollack, A., Bagwell, C. B., and Irvin III, G. L. Radiation from tritiated thymidine perturbs the cell cycle progression of PHA stimulated lymphocytes. Science, *203*: 1025–1026, 1979.

48. Pritchett, C. J., Senior, P. V., Sunter, J. P., Watson, A. J., Appleton, D. R., and Wilson, R. G. Human colorectal tumours in short-term organ culture. A stathmokinetic study. Cell Tissue Kinet., *15*: 555–564, 1982.

49. Puck, T. T., Sanders, P., and Petersen, D. Life cycle analysis of mammalian cells. II. Cells from the Chinese hamster ovary grown in suspension culture. Biophys. J., *4*: 441–455, 1964.

50. Puck, T. T. and Steffen, J. Life cycle analysis of mammalian cells. I. A method for localizing metabolic events within the life cycle and its application to the action of Colcemid and sublethal doses of X-irradiation. Biophys. J., *3*: 379–397, 1963.

51. Rogers, A. W. Techniques of Autoradiography. Amsterdam: Elsevier, 1973.

52. Scheiderman, M. H. Scheiderman, G. S., Rusk, C. H. A cell kinetic method for the mitotic selection of drug treated G2 cells. Cell Tissue Kinet., *16*: 41–49, 1983.

53. Sharpless, T. K. and Schlesinger, F. H. Flow cytometric analysis of G1 exit kinetics in asynchronous L1210 cell cultures with the constant transition probability model. Cytometry, 3: 196–200, 1982.

54. Sharpless, T., Traganos, F., Darzynkiewicz, Z., and Melamed, M. R. Flow cytofluorimetry: Discrimination between single cells and cell aggregates by direct size measurements. Acta Cytol., 19: 577–581, 1975.

55. Smith, J. A. and Martin, L. Do cells cycle? Proc. Natl. Acad. Sci. USA, 70: 1263–1267, 1973.

56. Smith, R. S., Thomas, D. B., and Riches, A. C. Cell production in tumour isografts measured using vincristine and colcemid. Cell Tissue Kinet., 7: 529–535, 1974.

57. Staiano-Coico, L., Darzynkiewicz, Z., Hefton, J. M., Dutkowski, R., Darlington, G. J., and Weksler, M. E. Increased sensitivity of lymphocytes from people over 65 to cell cycle arrest and chromosomal damage. Science, 219: 1335–1337, 1983.

58. Swartzendruber, D. G. A bromodeoxyuridine (BUdR)-mithromycin technique for detecting cycling and noncycling cells by flow microfluorometry. Exp. Cell Res., 109: 439–443, 1977.

59. Tannock, I. F. A comparison of the relative efficiencies of various metaphase arrest agents. Exp. Cell Res., 47: 345–356, 1967.

60. Thames, H. D. and White, R. A. State-vector models of the cell cycle. I. Parametrization and fits to labelled mitoses data. J. Theoret. Biol., 67: 733–743, 1977.

61. Tobey, R. A. Different drugs arrest cells at a number of distinct stages in G2. Nature, 254: 245–247, 1975.

62. Traganos, F., Staiano-Coico, L., Darzynkiewicz, Z., and Melamed, M. R. Effects of dihydro-5-azacytidine on cell survival and cell cycle progression of cultured mammalian cells. Cancer Res., 41: 780–789, 1981.

63. Traganos, F., Staiano-Coico, L., Darzynkiewicz, Z., and Melamed, M. R. Effects of aclacinomycin on cell survival and cell cycle progression of cultured mammalian cells. Cancer Res., 41: 2728–2737, 1981.

64. White, R. A., Grdina, D. J., Meistrich, M. L., Meyn, R. A., and Johnson, T. S. Cell synchrony techniques. II. Analysis of cell progression data. Cell Tissue Kinet., 17: 237–245, 1984.

65. Wright, N. A. and Appleton, D. R. The metaphase arrest technique. A critical review. Cell Tissue Kinet., 13: 643–663, 1980.

66. Wright, N. A., Britton, D. C., Bone, G., and Appleton, D. R. A stathmokinetic study of cell proliferation in human gastric carcinoma and gastric mucosa. Cell Tissue Kinet., 10: 429–437, 1977.

Chapter 11

Flow Cytometric Studies on Intracellular Drug Fluorescence

Awtar Krishan

1. INTRODUCTION

Flow cytometry has rapidly evolved from a technique for detecting and sorting cells on the basis of their DNA content or immunological markers into a useful tool for detection and quantitation of intracellular drug fluorescence (2,25,30,37). Recent studies have shown that intracellular content of fluorescent drugs can be rapidly quantitated on a cell-to-cell basis by this sophisticated analytical method. Thus, one can analyze intracellular drug transport (influx, efflux), retention, and/or binding, and correlate these parameters with effects on cellular metabolism and proliferation (21,22).

Several well-known drugs are intrinsically fluorescent (e.g., tetracyclines, anthracyclines), whereas others can either be directly labeled with a fluorescent marker or have a semisynthetic fluorescent analog. The use of flow cytometry for measurement of intracellular drug fluorescence offers certain unique advantages not available in analytical procedures based on quantitation of drug content in body fluids, tissue extracts, or biopsies. Unlike the biochemical procedures, which may involve either spectrophotometric or fluorometric assays for quantitation of free drug or drug-target complexes, flow cytometry offers the advantage that individual cells in a population can be detected on the basis of their intracellular drug fluorescence. Subsequently, these cells can be sorted and used for correlative studies that may involve other biochemical or cytokinetic parameters. The availability of multiparameter analysis in which a number of different cellular markers can be studied simultaneously in a flow system, makes this approach tempting for studies involving heterogeneous cell populations and individual response of subpopulations to drugs (37).

2. INSTRUMENTATION AND SAMPLE CONSIDERATIONS

Several commercial flow cytometers have recently become available. Some of these systems use excitation from a high-pressure mercury bulb, whereas others utilize excitation from a single- or dual-laser system. Most of the commercially available units use an argon or krypton laser, and recently, tunable dye lasers have been incorporated in dual-laser excitation systems. In a typical flow cytometer, cells naturally occurring as single cell suspensions (e.g., leukemic blasts) or dissassociated in a monodisperse suspension after either enzymatic digestion or by mechanical means, are exposed to a focused narrow laser beam. Hydrodynamic focusing through the use of precision sheath flow is used to achieve passage of cells single-file in front of the focused laser beam either in a quartz cuvet or in a midair jet. Sample flow rate is controlled to allow for excitation of a single cell at a time. Light scatter detectors, (forward and/or right angle), dichroic mirrors, and beam splitters are used to collect light scatter and emitted fluorescence. Signals collected from photomultipliers are transmitted to analog–digital converters (ADC) for further processing (amplification, gating) and generation of frequency distribution single- or multiparameter histograms. These signals can further be used to activate the cell-sorting function, which by electrostatic deflection can collect droplets containing the cells of interest for further studies. Figure 1 shows a schematic for a commercially available cell sorter. The reader is referred to the textbook on flow cytometry edited by Melamed et al. (30), for excellent articles on flow instrumentation, principles, and applications.

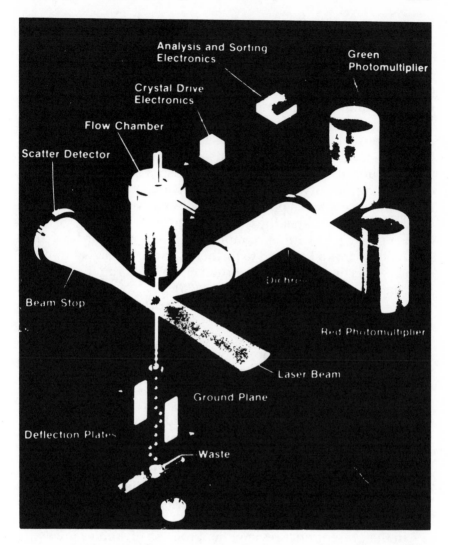

Fig. 1. Coulter Epics V cell sorter showing excitation of a sample in a sheath flow with a laser beam separation of emitted fluorescence by dichroic mirror and by green and red photomultipliers. Laser scatter is measured by a forward angle scatter detector. A piezoelectric transducer is used to break the sheath into cells containing droplets that may be further sorted by electrostatic deflection to collect the cells of interest (photograph courtesy of Mr. W. Gutierrez, Epics Division of Coulter Corporation, Hialeah, Florida).

Since intracellular drug fluorescence in cells exposed to therapeutic drug concentrations in most cases is very low, laser excitation with large 4–5 W argon lasers has been favored over excitation by high-pressure mercury or xenon sources. Several earlier studies attempted detection of intracellular adriamycin fluorescence with excitation from the small laser units such as the Bio/Physics cytofluorograph or the Coulter Electronics TPS-1 instruments. However, in most of these studies, unusually large drug concentrations had to be used to achieve any satisfactory measurement of intracellular fluorescence. The superior optics and light-collecting efficiency of the mercury source units (32) have been offset by the need for maximum excitation power of a laser beam. Most of our reported studies have used either the Coulter Electronics TPS-1 (large laser) or the Epics V unit with Spectraphysics 4- or 5-W argon laser. The Los Alamos group has tried to further enhance the signal-to-noise ratio by collecting a larger proportion of emitted drug fluorescence in a special parabolic flow chamber (38).

In a typical flow cytometric analysis for drug content, cells may be incubated in vitro with the fluorescent drug, and aliquots analyzed for intracellular fluorescence. For in vivo studies, it is essential to tansport the material (blood, bone marrow, or tumor samples) to the flow cytometry laboratory under ideal conditions for maintenance of cellular viability and drug-retention characteristics with proper regard for temperature, diluents, and anticoagulants. In some of our cellular pharmacokinetic studies, we have mixed cell suspensions directly with the drug-containing medium in the sampling cuvet of the cell sorter (Coulter, Epics V) maintained at 37°C with constant stirring. In this manner, we were able to generate multiparameter histograms of cellular drug fluorescence in relation to length of incubation, without the need for transfer of samples from a test tube to the cell sorter.

In the present brief review, the progress of studies involving flow cytometry for quantitation of intracellular drug fluorescence will be discussed. Since most of the agents studied in our and other laboratories have been cancer chemotherapeutic drugs (anthracyclines and methotrexate), we will focus primarily on studies related to these important drugs. Other fluorescent drugs (e.g., rhodamine 123) and chemicals (e.g., carcinogens) and noncancer drugs may also be analyzed by this procedure, but are not covered in this review.

3. ANTHRACYCLINES

3.1. Anthracycline Fluorescence and Flow Cytometry

Anthracyclines are important fungal antibiotics with significant antitumor activity in leukemias and solid tumors (6). Daunomycin and adriamycin are currently used in treatment of cancer. The dose-limiting cardiotoxicity of these drugs has stimulated the search for newer, naturally occurring anthracyclines and the development of semisynthetic analogs. Several new anthracyclines have been recently isolated and new semisynthetic analogs prepared in the hope of increasing the therapeutic efficacy of these important antibiotics (6).

The proposed mechanisms of action for anthracyclines include intercalation with nucleic acids (6), binding to cell surface (42), and free radical production (1). Although several earlier published studies on anthracyclines have used flow cytometry for cell cycle analysis (20), this methodology has been particularly useful in studies related to drug transport, (influx, efflux), retention, and binding. Anthracyclines are generally highly fluorescent and can be excited with the 488 nm argon laser line. As shown in Table 1, their excitation characteristics are within the usual optical meas-

Table 1
Fluorescence Characteristics of Anthracyclines

	Excitation, max	Emission, max	Fluorescent,[a] quenching
Aclacinomycin	452	534	1.2X
Adriamycin	470	566	8X
Adriamycin-14-Octanoate	476	560	
AD 32	476	560	0X
AD 41	466	560	
AD 143	466	560	2X
Carminomycin	470	554	
Cinerubin	470	553	21X
Daunomycin	470	560	
4'-Demethoxydaunomycin	470	545	12X
4'-Epiadriamycin	466	558	8X
5'-Iminodaunorubicin	462	553	0.6X

[a]On binding to calf thymus DNA (*see* ref. 21 for details).

urement parameters of flow cytometers and cell sorters. Thus one can use flow cytometry for quantitation of anthracycline content on a single-cell basis in a heterogeneous population, in relation to cell cycle phase and other morphological and kinetic characteristics. This method can be used for analysis of drug transport and binding characteristics in different cell types and subpopulations (22). Recently, flow cytometry has been used to compare newly synthesized analogs of various anthracyclines for differences in their drug transport characteristics (22), as well as to study effects of agents that alter anthracycline transport characteristics (19,26).

We, along with other investigators, have screened several clinically important anthracyclines for their excitation and emission spectra and changes in these profiles on binding to DNA (6,21,22). In cells incubated with anthracyclines, most of the drug fluorescence is initially localized in cytoplasm followed by binding to chromosomes and nuclei (11). The condensed portion of chromatin especially fluoresces brightly. In contrast to daunomycin and adriamycin, which preferentially bind to DNA and stain nuclei, some of the semisynthetic analogs (for example, AD 32) do not bind to nuclei (24).

Intracellular detection of anthracyclines by flow cytometry can be achieved either by direct excitation of intracellular drug fluoresence (direct method) or by measuring reduction in fluorescence of a second fluorochrome after exposure of cells to an anthracycline (indirect method). The indirect method is based on our earlier observations that in cells exposed in vitro or in vivo to a DNA intercalating anthracycline, the subsequent binding of a DNA-binding fluorochrome such as propidium iodide is reduced (23). Anthracycline fluorescence on binding to isolated DNA or nuclei is quenched, whereas that of propidium iodide is enhanced approximately 20-fold. Exposure of cells to a DNA-binding anthracycline before staining with propidium iodide decreases the amount of nuclear fluorescence. This reduction in propidium iodide/nuclear fluorescence is dependent on and influenced by drug concentration, influx, efflux, retention, and binding. Direct correlation between extracellular anthracycline concentration, length of exposure, and the reduction in fluorescence of nuclei stained with propidium iodide can be demonstrated. This indirect method for anthracycline detection can be used for detecting the intracellular content of a variety of DNA-binding anthracyclines (e.g., rubidazone, nogalamycin, daunomycin, carminomycin). In contrast to DNA-binding anthracyclines, no reduction of propidium iodide/nuclear fluorescence is seen in cells incubated with non-DNA binding anthracyclines such as AD 32. The propidium iodide/nuclear fluorescence reduction by adriamycin can be demonstrated by laser flow cytometry both in vitro and in vivo. For example, in P388 tumor-bearing mice injected with 4 mg/kg of adriamycin (Fig. 2), a prominent subpopulation of tumor cells with reduced PI/nuclear fluorescence can be seen in ascites 24 h after drug administration. A similar quenching of Hoechst 33342 fluorescence by adriamycin in human leukemic cells has been described by Preisler (33).

The direct excitation method has been more extensively used than the indirect method. Figure 3 illustrates the potential of this (direct excitation) method for monitoring cellular drug fluorescence in adriamycin-sensitive and -resistant murine leukemic lymphoblasts of the P388 cell line. The adriamycin-resistant cell line used in these studies was isolated from surviving clones of the P388 cell line growing in methyl cellulose in the presence of increasing drug concentrations (up to 2 μg/mL). In soft agar assays, this cell line shows > 84-fold resistance to adriamycin and daunomycin compared to that of the parent cell line (26). Cells from this or the P388/S cell line maintained in suspension cultures were mixed with an equal amount of drug-containing medium in the glass specimen chamber of the Coulter Epics V cell sorter. The argon laser line (488 nm, 500 mW) was used to excite the cellular drug fluorescence and cells were analyzed at the flow rate of 1000/s. Multivariate histograms A and B in Fig. 3 are 64 × 64 channel dot plots of P388/S and P388/R cells incubated with adriamycin (2 μg/mL). The abcissa records cellular drug fluorescence; the ordinate records forward angle light scatter (an approximate measure for cell size). The total number of cells in a particular channel can be interpreted either from the density of the pattern (as in Fig. 3A or B) or from the height of the vertical bars (as in Fig. 3C). As seen in these histograms, one can correlate cell size (based on light scatter) with cellular drug fluorescence and thereby identify and sort subpopulations of interest from a heterogeneous population. Aside from using forward angle light scatter or any other of the several available variables (such as right angle light scatter, Coulter volume measurement, or the presence of a second fluorochrome), in some of the available commercial units one can use time as a variable. In this analytical

ADR / PI IN VIVO

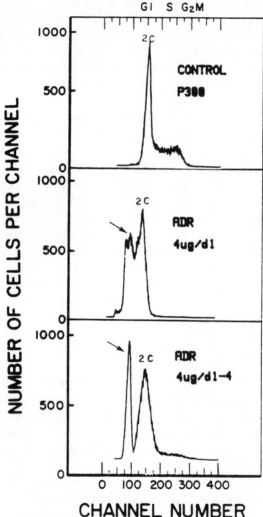

CHANNEL NUMBER

Relative Amount of DNA

Fig. 2. DNA distributions of propidium iodide-stained P388 leukemic ascites from animals injected with saline (control) or 4 μg adriamycin on d 1 (histogram in center) or 4 μg on d 1–4. Note the emergence of cells (arrow) with reduced propidium iodide fluorescence in animals injected with adriamycin.

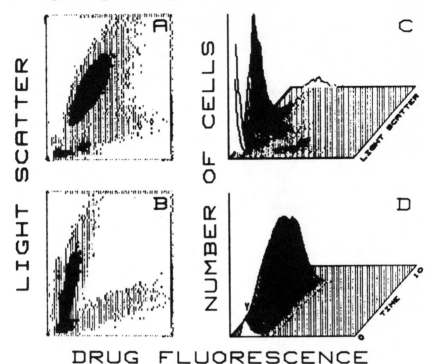

DRUG FLUORESCENCE

Fig. 3. Shows dot plots (A,B) of P388-drug sensitive (A) or P388-resistant cells (B). Histogram (C) is of resistant cells, but shows number of cells with bar height. Abscissa records cellular drug fluorescence; ordinate records light scatter. In histogram (D) ordinate records time (length of incubation) vs cellular drug fluorescence or abscissa.

mode, the multivariate histogram (Fig. 3D), records number of fluorescent cells (indicated by height of bars), cellular drug fluorescence (abcissa recorded in 64 channels of increasing linear value), and length of drug exposure (on ordinate in 64 channels covering a total timespan of 10 min). Note the increase with time in both number of fluorescent cells recorded and the amount of per cell fluorescence. In most currently available commercial flow cytometers, electronic gating can be used to isolate or focus on a particular subpopulation of interest for further analysis or sorting.

3.2. Drug Transport Studies

Direct laser excitation for quantitation of intracellular anthracyclines has been used by several workers for monitoring drug transport (influx, efflux, retention), differences in distribution and transport in relation to cell cycle or proliferation status of a population. In most of the studies reported, the standard 4–5 W argon lasers have been used. Thus one can study not only differences between subpopulations, but also compare transport characteristics of closely related anthracyclines. In an earlier study, we used this method to elucidate differences in transport characteristics of adriamycin and its rapidly transported analog AD 32 and metabolite AD 41 (21). Histograms in Fig. 4 compare the cellular drug fluorescence of P388 cells exposed to four related anthracyclines, adriamycin (Fig. 4A, E), daunomycin (Fig. 4B, F), AD 32 (Fig. 4C, G), and THP-adriamycin (Fig. 4D, H). The abcissa in these histograms records cellular drug fluorescence, whereas on the ordinate, time is recorded for 0–10 min (Fig. 4A–D) or 10–20 min (Fig 4E–H) of incubation. Note the differences not only in the amount of intracellular fluorescence content, but also the rate at which cells with drug fluorescence appear in cultures exposed to the different anthracyclines. In cells exposed to adriamycin (Fig. 4A, E), fluorescence gradually increases both in number of fluorescent cells (height of bars), as well as in per cell fluorescence content (abscissa). In contrast daunomycin fluorescence appears rapidly and per cell fluorescence reaches a peak within the first 15 min (Fig. 4B, F). AD 32 fluorescence profiles, as shown in Fig. 4C and G, are similar to those of daunomycin, whereas cells exposed to THP-ADR do not seem to increase their cellular fluorescence content with time (Fig. 4D, H).

In an earlier study (22), we have used univariate flow cytometry for monitoring intracellular drug transport (influx, retention, and efflux) of adriamycin and several related anthracyclines in relation to length of exposure and drug concentration (22). Figure 5 from this study compares the effect of extracellular drug concentration vs length of incubation on the cellular drug fluorescence of cells incubated with three anthracyclines—carminomycin, rubidazone, and nogalamycin. Observations from this study demonstrated the value of flow cytometry for elucidating the relation between drug concentration and length of exposure in anthracyclines known to be slowly transported (e.g., adriamycin) in contrast to anthracyclines that are rapidly transported or enter by passive diffusion (e.g., AD 32). As shown in this figure, carminomycin- and rubidazone-related intracellular drug fluorescence was directly influenced by extracellular drug concentration and length of exposure. In contrast,

NUMBER OF CELLS

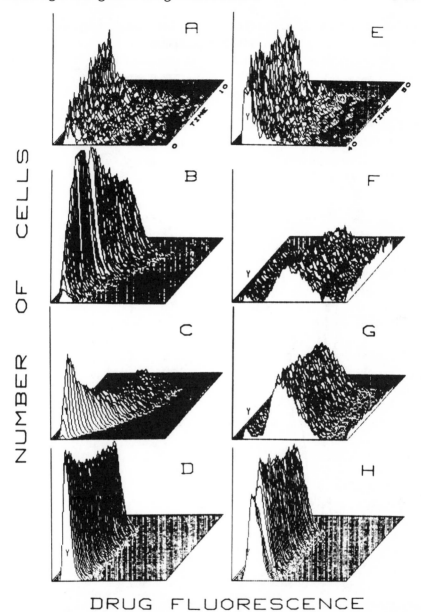

DRUG FLUORESCENCE

Fig. 4. Abscissa records cellular drug fluorescence of P388 cells incubated with 2 µg/mL of adriamycin (A), daunomycin (B), AD 32 (C), or THP-adriamycin (D). On ordinate, the length of incubation is recorded; (A)–(D) are 0–10 min and (E)–(H) are 40–50 min of incubation in drug-containing medium.

LASER EXCITATION OF INTRACELLULAR
ANTHRACYCLINES

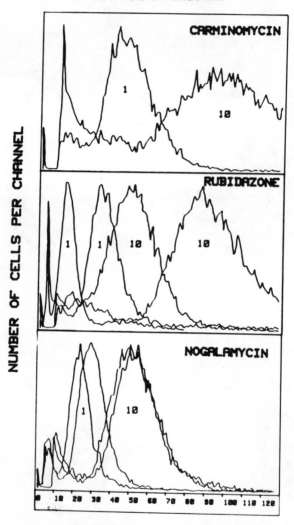

RELATIVE FLUORESCENCE
(CHANNEL NUMBER)

Fig. 5. Univariate histograms of CCRF-CEM cells incubated with 1 or 10 μg/mL of carminomycin (A), rubidazone (B), or nogalamycin (C). In (B) and (C), successive curves are after 15 min and 2 h of drug incubation.

for anthracyclines such as nogalamycin, the differences between intracellular fluorescence after incubation with 1 or 10 μg/mL for 1 or 3 h was not significant.

Flow cytometry is a useful tool for studying differential drug transport characteristics of a cell population, intrinsic differences between cell lines and among subpopulations of a heterogeneous population, and the effect of different drugs on anthracycline cellular transport and retention. Some of the selected examples presented below will serve to further illustrate the capability of flow cytometry for analyzing these characteristics.

3.3. Drug Resistance

Earlier studies have shown that cellular adriamycin transport is slow, but a major cause for drug resistance is the presence of a rapid, energy-dependent drug efflux mechanism in resistant cells (7,17,39). Appearance of cell surface-specific glycoproteins has been reported in a variety of adriamycin-resistant cells (3,15,18). However, no interrelation between these glycoproteins and cellular drug transport has been established. In an earlier study, we have used flow cytometry, fluorometric quantitation, and soft agar assays to compare adriamycin and daunomycin transport in drug-sensitive and -resistant cells (13). Histograms in Fig. 6 will serve to illustrate major differences between the appearance of cellular daunomycin fluorescence in P388/S and P388/R cells as a function of length of incubation in drug-containing medium. On the abscissa of these histograms, cellular drug fluorescence (linear, 64 channel) is recorded; on the ordinate, length of incubation for a total span of 10 min (per histogram) is measured.

Histograms on the left (Fig. 6A, B, C) are of the P388 drug-sensitive cells, and those on the right (Fig. 6D, E, F) are of the drug-resistant cells. In P388/S cultures, cells with cellular drug fluorescence appear as soon as the sample reaches the laser intersection point (within less than a minute of sample mixing), and the number of fluorescent cells as well as their fluorescence gradually increases (as shown by the increase on the abcissa and the height of the bars in Fig. 6A), reaching a peak between 10 to 20 min of incubation (Fig. 6B). The histogram in Fig. 6C (40–50 min) also shows that at least two distinct populations (arrows) on the basis of their high and low cellular drug fluorescence can be identified in P388/S cells.

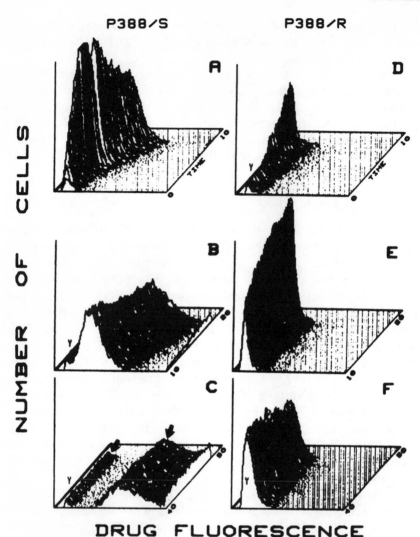

P388/S   P388/R

NUMBER OF CELLS

DRUG FLUORESCENCE

Fig. 6. Histograms compare uptake of daunomycin in P388-sensitive and -resistant cells. Ordinate records the length of incubation, and abscissa records drug fluorescence.

Histograms in Figs. 6D, E, and F show drug-resistant P388/R cells after incubation with daunomycin for 1–10, 10–20, and 40–50 min. In contrast to the P388/S cells, the number of fluorescent cells gradually increases (compare height and width of bars in Fig. 6A and D) and the amount of per cell drug fluorescence does not increase significantly with time. After 40 min of incubation (compare abscissa of Fig. 6C with F), the cellular drug fluorescence in P388/S cells is much higher than that of the comparable P388/R cells.

The use of flow cytometry for rapid analysis of intracellular drug binding in cells isolated from various tissues of animals was also demonstrated in an earlier study (21). For example, in nuclei isolated from spleen, bone marrow, and heart of mice and incubated with adriamycin (1μg/mL) for 15 min, maximum fluorescence was seen in heart and bone marrow nuclei representing organs that show the most cytotoxic response to adriamycin in vivo and are mostly responsible for the dose-limiting toxicity. In some of our ongoing clinical studies, we use flow cytometry for monitoring drug transport in leukemic cells, ascites, and pleural effusion of patients treated with various anthracyclines (Krishan et al., manuscript in preparation).

Several other workers have used flow cytometry for monitoring anthracycline fluorescence and resistance in a variety of cell lines (monolayers, suspension cultures, or spheroids). Some of these studies are summarized below.

Tapiero et al. (40) monitored differential affinity of drug-sensitive and -resistant Friend leukemia cells to adriamycin and daunomycin. They reported effect of temperature and metabolic poisons on intracellular drug fluorescence. The bimodal fluorescence of isolated nuclei was related to size differences as determined by laser scatter. In resistant cells, a heterogeneous fluorescence distribution (multimodal) was seen, whereas in the drug-sensitive cells and cells exposed to lower daunomycin concentrations, fluorescence distribution was unimodal.

Durand (9) and Durand and Olive (10) have used flow cytometry to quantitate the intracellular fluorescence of mouse L-929 and V-79 Chinese hamster fibroblasts and correlated drug fluorescence with DNA damage and effects on clonogenicity. The effect of various growth parameters on drug uptake and correlation of extracellular drug concentration with intracellular drug fluorescence were reported. Heterogeneity of drug uptake was noted and cell sorting experiments confirmed a positive correlation between drug fluorescence and cytotoxicity. Thus, lowest survival was observed for highly fluorescent small cells, whereas the larger, lightly fluorescent cells had greater survival in soft agar assays. These authors particularly noted that cells in suspension had greater intracellular drug fluorescence than cells from monolayer cultures.

Muirhead et al. (*31*) used flow cytometry to measure intracellular fluorescence in dissassociated cells from EMT/RO tumor spheroids. Significant differences in mean fluorescence for intact cells and isolated nuclei from inner and outer regions of the spheroids were described. These differences were not related to cell size or caused by lack of drug diffusion in the spheriods. These authors concluded that factors other than lack of drug access to inner cells are significant determinants of drug resistance in spheroids.

The Los Alamos group (Drs. Raju, Johnson, and Tokita) have used flow cytometry to correlate intracellular adriamycin fluorescence with radiation response of V-79 (drug-sensitive and -resistant) and Chinese hamster ovary (CHO) cells. Their data indicate that detectable intracellular fluorescence was linearly dependent on drug concentration. Drug fluorescence was higher in drug-sensitive than in drug-resistant cells and correlated with cytotoxicity in soft agar assays and cell cycle traverse perturbations (*34*).

3.4. Transport and Cell Cycle

Several earlier studies have attempted correlation of adriamycin cytotoxicity and cell cycle phase position and proliferation status on a cell population. We have analyzed the differential influx and efflux of adriamycin and daunomycin in cell cycle phase-enriched subpopulations from a log phase population. In cell cycle phase-enriched subpopulations of L-1210 cells obtained by centrifugal elutriation and incubated with adriamycin or daunomycin, multivariate analysis shows differential drug uptake in the various subpopulations. Multivariate histograms in Fig. 7 are from such an experiment in which daunomycin fluorescence in elutriated cell cycle phase-enriched populations was quantitated by flow cytometry.

Univariate DNA distributions on the left show the cell cycle distribution of the elutriated population used in this experiment. The histograms in Fig. 7A¢D are, respectively of total log phase population (Fig. 7A), and enriched populations of G1 (Fig. 7B), mid- and late S (Fig. 7C), and late S, G2M (Fig. 7D). Bivariate dot plots (in the middle row) and bar histograms (on the right) show the increased amount of daunomycin fluorescence in S and G2M cells as compared to the G1 cells. Further, in the G1 cells (Fig. 7B), two distinct subpopulations (on the basis of their daunomycin fluorescence and light scatter) can be identified, as indicated by arrows.

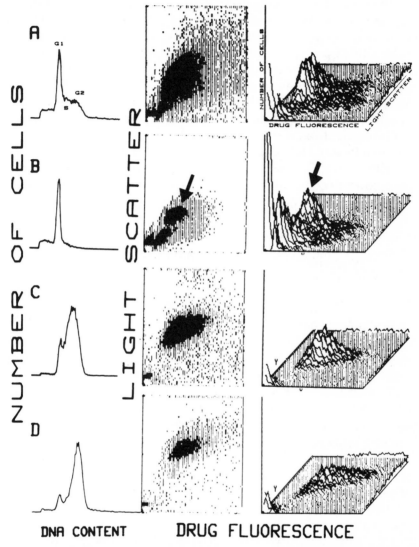

Fig. 7. Intracellular daunomycin fluorescence histograms of cell cycle-enriched subpopulations of elutriated L-1210 cells incubated with daunomycin (1 µg/mL) for 1 h. Histograms (A–D) on the left are of DNA distributions showing successively log-phase, G1-, S-, and G2M-enriched populations. Scattergrams (center) and the bar histograms (right) show that the intracellular drug fluorescence is higher in S-phase than in G1 cells. In G1 cells, two distinct peaks of fluorescent cells can be recognized (in addition to the dead cells), whereas the S and G2M cells have a unimodal fluorescent distribution.

In other unpublished observations on cell cycle phase-enriched populations, we have observed that drug influx and efflux in S and G2M, part of the cell cycle is more pronounced than that of cells in the G0/G1-phase. In comparing drug uptake in relation to proliferation status, greater amounts of intracellular drug fluorescence were seen in log than in plateau phase cells. These observations may partially account for the enhanced sensitivity of log phase and S and G2M cells to adriamycin (20). In a recent study, Tokita et al. (41) have studied the cell cycle phase-dependent transport of adriamycin in CHO cells.

The relation between cell cycle position and the effect of phenothiazines on adriamycin transport as monitored by flow cytometry is discussed in the following section.

3.5. Drug Interactions

Anthracyclines differ in their cellular transport characteristics. For example, anthracyclines such as adriamycin are slowly transported, in contrast to analogs such as AD 32 or THP-AdR, which enter the cell rapidly. In drug-resistant cells, a rapid energy-dependent mechanism is believed to efflux the drug and thereby lead to reduced cytotoxicity. Several agents (e.g., anticalmodulin drugs and calcium channel blockers) can interfere with this drug efflux mechanism and thereby enhance anthracycline cytotoxicity in drug-resistant cells (12,43). Flow cytometry is an ideal tool for studying the effects of these drugs on adriamycin transport in selected subpopulations from a heterogeneous population. Bivariate histograms in Fig. 8 are of P388 drug-sensitive (Fig. 8A, B) and drug-resistant (Fig. 8C, D) cells incubated with adriamycin alone (5 μg/mL, Fig. 8A, C) or a combination of adriamycin (5 μg/mL) and phenothiazine, (chlorpromazine, 50 μM) for 2 h (Fig. 8B, D). In these bivariate histograms, abscissa records cellular drug fluorescence, and forward angle light scatter (approximate cell size) is recorded on the ordinate. In P388/S cells incubated with adriamycin (Fig. 8A), cellular drug fluorescence is greater than that of the P388/R cells (Fig. 8B). The reduced cellular drug fluorescence in the P388/R cells (Fig. 8C) is presumably a result of rapid drug efflux, which can be blocked by coincubation of cells in the presence of chlorpromazine. Note that in both the P388/S and R cells incubated with chlorpromazine, cellular drug fluorescence is approximately similar (Fig. 8B, D). The enhanced adriamycin sensitivity of the P388/R cells in-

P388/S   P388/R

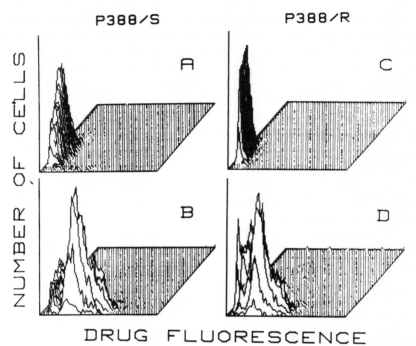

DRUG FLUORESCENCE

Fig. 8. Effect of coincubation with chlorpromazine (10^{-5} *M* on the intracellular adriamycin fluorescence of P-388/S and P-388/R cells. (A) P-388/S plus adriamycin; (B) same as (A), plus chlorpromazine; (C) P-388/R plus adriamycin; (D) same as (C), plus chlorpromazine. Comparison of histograms (A) and (C) (control) with (B) and (D) show that in both P-388/S and P-388/R cells, a major peak of cells with enhanced intracellular drug fluorescence appears on coincubation with chlorpromazine. Presumably this is a result of decreased drug efflux from the treated cells.

cubated with chlorpromazine is confirmed by cytotoxicity assays in soft agar (*26*). In a recent study, we have used flow cytometry to monitor effect of Ca^{2+} channel blockers (e.g., verapamil) and anticalmodulin drugs (e.g., phenothiazines), on anthracycline transport in a variety of cell lines (*26*).

3.6. Conclusions

It is clear from the examples shown above and those in literature that flow cytometry can be a powerful tool for detection and quantitation of intracellular anthracycline content. Several workers have used it for measurement of parameters that affect drug transport (influx, efflux, and binding), characterization of subsets in a hetero-geneous population, drug transport in relation to cell cycle and cellular proliferation, and in cells exposed to drug combinations. This valuable technique has been used for monitoring differences between various anthracyclines and their semisynthetic analogs. However, laser excitation of intracellular anthracycline content has some inherent drawbacks and one must bear in mind that for a majority of anthracyclines, quantitation of intracellular fluorescence may not accurately reflect intracellular drug content, since several anthracyclines quench their fluorescence on binding to DNA. However, there are exceptions (e.g., AD 32), and these nonquench-ing anthracyclines may be particularly suited for laser excitation studies. In crucial studies, it may be important to correlate data from laser excitation with standard high-pressure liquid chromato-graphic or fluorometric assays for anthracyclines. It is conceivable that a cell with more binding sites that quench drug fluorescence may emit lower amounts of total fluorescence than a cell with a lower number of fluorescence-quenching binding sites. This may be especially true of heterogeneous populations in which different subpopulations may have different potentials for drug transport, binding, and the resulting fluorescence quenching.

4. METHOTREXATE

Methotrexate (MTX), a folic acid antagonist, is one of the earliest known drugs with significant antitumor activity in several human tumors (4). The mechanism of MTX cytotoxicity involves binding to dihydrofolate reductase (DHFR), which in turn affects thymidy-late and purine biosynthesis, resulting in arrest of cells in early S-phase, unbalanced growth, and ultimate cell death. In an earlier study, Darzynkiewicz et al. (8) used autoradiography with tritium-labeled methotrexate as a cytochemical method for detection of folate reductase in animal tissues. They reported firm binding and prolonged retention with preferential localization of the labeled MTX in kidney proximal tubules, intestinal epithelium, and liver parenchymal cells. Several studies have shown that various phar-macokinetic conditions related to drug transport, retention, and efflux of bound vs free drug can modulate response of a tumor cell population to MTX. The possibility that higher MTX concentrations can be used if citrovorum factor or folinic acid was given for rescue/ protection of normal tissues has further led to development of high-dose methotrexate/rescue protocols (5). In drug-resistant cells, reduced intracellular drug transport, as well as enhanced cellular levels of DHFR, have been shown to confer resistance. With the availability of labeled MTX analogs, it is possible to study the presence of either of these mechanisms alone or in combination in drug-resistant cells. Because of its clinical importance, the relative ease with which MTX levels can be monitored and the wealth of published data on MTX pharmacokinetics and mechanism of resistance, several workers have used MTX analogs with fluorescent label and flow cytometry to study transport and resistance in vitro.

4.1. Fluorescence Labeling

MTX by itself is not fluorescent, and several workers have used a MTX–FITC conjugate for monitoring cellular drug transport and retention. In an earlier study, Kaufman et al. (27) conjugated MTX and FITC using the procedures originally described by Gapski et al. (14). The peak fluorescence of the conjugated MTX product was 17% of that of uncoupled FITC. On binding to DHFR, however, fluorescence intensity increased 4.5-fold. The fluorescent compound inhibited DHFR activity stoichiometrically, although it was less effective than nonlabeled MTX. On the basis of high affinity of the fluorescent MTX to DHFR and the assumption that this compound was efficiently transported, Kaufman et al. reported that it was reasonable to assume that by measuring intracellular fluorescence one could possibly quantitate DHFR levels in individual cells. The Gapski procedure used for MTX–FITC conjugation has a low yield of the fluorescent MTX product, and DHFR nonrelated fluorescence may be released by amidase cleavage of a potential amide bond in this conjugate. Doctor Rosowsky at Dana Farber Cancer Center has taken a different approach to the synthesis of a fluorescent MTX analog (35). He has first replaced the glutamate sidechain of MTX with lysine, then attached FITC directly to the amino groups of lysine. This fluorescent MTX analog (PT 430) has one tenth the binding capacity of MTX to DHFR and in in vitro assays is 100-fold less toxic. Fluorescence of PT 430 is pH dependent and enhanced 3–5-fold on binding to DHFR.

4.2. DHFR Levels and MTX Resistance

Kaufman et al. (*28*) used the MTX–FITC conjugate to measure intracellular levels of enzyme dihydrofolate reductase in drug-sensitive and drug-resistant cells. In drug-resistant cells, bright fluorescence indicated the presence of MTX–DHFR conjugates, whereas no fluorescence was seen in parental cells that presumably had lower levels of intracellular DHFR. On the basis of experiments devised to measure DHFR levels in drug-sensitive and -resistant sublines by flow cytometry, a linear correlation between level of the fluorochrome and the emitted fluorescence was established. On the basis of results from cell sorting experiments, these authors concluded that fluorescence intensity per cell was directly related to intracellular DHFR content and can be accounted for by variations in the rate of enzyme synthesis. Large variations in drug fluorescence seen in resistant populations were not related to proliferation or cell cycle status of cells grown in the absence of MTX. Reversion to a median drug-sensitive status with intermediate levels of DHFR and not to the drug-sensitive state was seen.

In a subsequent study, Kaufman and Schimke (*29*) used flow cytometry to measure intracellular DHFR levels in CHO cells grown in successively increasing concentrations of methotrexate. By utilizing double-staining procedures for flow cytometry, Mariani et al. (*29*) reported that in CHO cells, DHFR biosynthesis took place during a S-phase-specific period. Within 2 h of the start of S-phase and continuing through S-phase, there was a 90% increase in DHFR specific activity, whereas the total protein content increased by only 50%. The maximum peak of DHFR synthesis coincided with the maximum rate of DNA synthesis, both of which took place within the last stages of the 6.5-h S-phase.

In a subsequent paper, Haber and Schimke (*16*) followed with flow cytometry the progressive emergence of MTX resistance in 3T6 mouse fibroblasts. In this cell line initial resistance resulted from gene amplification, resulting in increased amounts of DHFR production. Growth of these drug-resistant cells at higher MTX concentrations results in the evolution of cells with high DHFR levels accompanied by reduced affinity to MTX.

Rosowsky and colleagues (*35*) have used their lysine-substituted MTX–FITC analog (PT 430) for flow cytometric analysis of DHFR over-production in a 5000-fold MTX-resistant subline of 3T3 fibroblasts. The drug-resistant cells under UV microscopy were brightly fluorescent, thus confirming that overproduction of DHFR is the basis for MTX resistance in this cell line. For flow cytometry, cells grown in the presence of 60–90 μM of PT430 were analyzed. Intracellular fluorescence of PT430–DHFR complex was time-dependent and significantly greater in the resistant cells. After 6 h, the fluorescence per cell was 2.5 times higher in the resistant than in the drug-sensitive cells. These authors point out that the magnitude of increased fluorescence in the resistant cells did not reflect the 5000-fold higher drug resistance in vitro.

In a recent study (36), these authors have used the MTX–FITC conjugate to analyze the DHFR enzyme levels in MTX-sensitive and -resistant human lymphoid cells. Figure 9 shows two-parameter histograms of methotrexate-sensitive and -resistant CCRF-CEM human lymphoblasts incubated with 30 μM of MTX–FITC conjugate (PT 430). These histograms show that the resistant cells have a smaller amount of light scatter, as well as intracellular PT 430 fluorescence, than the methotrexate-sensitive cells.

MTX is extensively used in chemotherapy of squamous cell carcinomas of head and neck. In recent unpublished studies, Rosowsky and colleagues (36) used PT430 to study DHFR levels in a human squamous cell carcinoma of head and neck cell line and its 15- and 90-fold drug-resistant sublines. As shown in Fig. 10 from his study, in the 15-fold resistant (R-15) cell line, drug uptake is initially slower, but after 2 h reaches the same level as that of the parental drug-sensitive cell line. This would indicate impaired drug transport, but normal level of DHFR in the R-15 cell line. In the R-90 (90-fold resistant cell line), the initial drug uptake is also slower, but the final steady-state level is higher than that of R-15 or the drug-sensitive cell line. These observations indicate the presence of a dual mechanism of drug resistance involving both impaired drug transport and increased intracellular DHFR content in the resistant cells.

Studies summarized above serve to illustrate the usefulness of flow cytometry in studying drug transport and mechanism of drug resistance related to enhanced enzyme levels and reduced transport in MTX-sensitive and -resistant populations. The fluorescent MTX conjugates prepared either by the Gapski or the Rosowsky method can identify either or both of these mechanisms. It is hoped that these studies will be extended to clinical specimens, in which one may be able to correlate chemosensitivity with known mechanism of MTX resistance and clinical response.

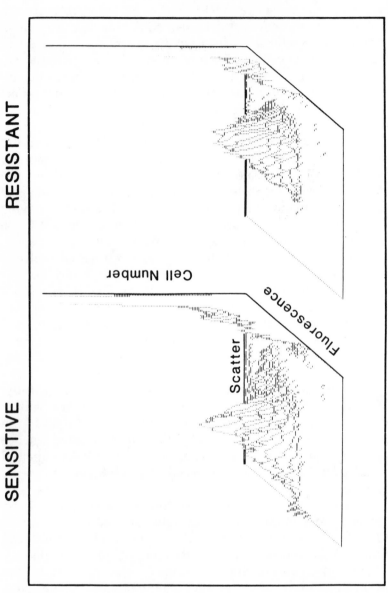

Fig. 9. Histograms of methotrexate-sensitive and -resistant CEM cells after incubation with 30 μM MTX–FITC complex (PT430) for 20 min (from Rosowsky et al., personal communication).

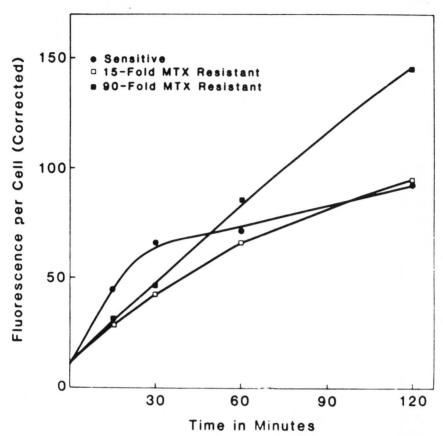

Fig. 10. Uptake of PT430 in methotrexate-sensitive and -resistant human squamous cell carcinoma of the head and neck cells. Cells were incubated with 15 μM PT430 at 37°C before flow cytometry. ●, parental and sensitive line; □, SCC15/R14 cells with decreased MTX transport; ■, SCC15/R90 cells with decreased MTX transport, increased DHFR content (from J. E. Wright, C. Cucchi, A. Rosowsky, and E. Frei, unpublished data, courtesy of Dr. A. Rosowsky, Dana Farber Cancer Center).

5. CONCLUSIONS

Flow cytometry offers the unique advantage of speed and analysis of a statistically significant sample in the shortest possible time. However, the procedure does need single cell isolates and in most solid tumors, this may be achieved only after vigorous enzymatic or mechanical disaggregation, which, in turn, may interfere with drug transport. However, this should not pose a problem in leukemias, lymphomas, and other tumors in which preparation of single cell suspensions does not require vigorous or traumatic handling of cells. Information obtained by flow cytometry may be of prognostic value and possibly allow for proper selection of chemotherapeutic protocols. In fact, several clinical investigators, especially Dr. H. D. Priesler at Roswell Park Memorial Institute, Buffalo, Dr. B. Barlogie at M.D. Anderson Cancer Institute, Houston, and our clinical associates at the University of Miami Comprehensive Cancer Center (Drs. K. Sridhar and E. Davila) are actively looking for a correlation between flow cytometric analysis of anthracycline cellular content (by either the direct or the indirect method) and clinical outcome of selected patients treated with anthracyclines or m-Amsa. However, it is clear that flow cytometry data cannot stand by itself, but must be confirmed and calibrated against other standard or accepted analytical procedures. The inherent problems related to quenching or enhancement of fluorescence, shift in excitation or emission wavelengths on binding to target molecules, energy transfer between dyes, and the low signal-to-noise ratios that may be caused by use of high laser power and PMT noise must be kept in mind in any interpretation of flow cytometric data.

From the technical point, the flow systems currently available collect approximately 10% of the total emitted fluorescence. The parabolic light collection chambers designed by the Los Alamos group (34,38 and Coulter Electronics (US patent 4,188,543, 1980, Brunsting et al.) are not available in commercial instruments. It is hoped that future units with better light-collecting efficiency will minimize the need for high laser beam power and use of higher than physiological or therapeutic drug concentrations for flow cytometric quantitation of intracellular drug fluorescence.

The studies summarized in this review on anthracyclines and methotrexate can be extended to a variety of other cancer chemotherapeutic agents that are either inherently fluorescent, may interfere with binding of other fluorochromes, or may have fluorescent analogs available. Current work in our laboratory is focused on fluorescent labeling of some important antimetabolites, alkaloids, and antibiotics used in cancer chemotherapy. We hope to use these for flow cytometric quantitation of intracellular drug transport and its correlation with drug cytotoxicity.

ACKNOWLEDGMENTS

Technical assistance of Kristie Gordon, Antonieta Sauerteig, and Larry L. Wellham is gratefully acknowledged. I am grateful to Dr. Robert T. Schimke (Stanford University), Dr. Ralph E. Durand (British Columbia Cancer Research Center, Canada), Dr. Andre Rosowsky (Dana Farber Cancer Center), and Dr. Tod Johnson (M.D. Anderson Hospital and Tumor Institute) for providing me with copies of their published and unpublished data. These studies were partially supported by US PHS-NIH grant CA-29360.

REFERENCES

1. Bachur, N., Gordon, S., and Gree, M. V. A general mechanism for microsomal activation of quinone anticancer agents to free radicals. Cancer Res., *38*: 1745–1750, 1978.
2. Barlogie, B., Raber, M. N., Schumann, J., Johnson, T. S., Drewinko, B., Swartzendruber, D. E., Gohde, W., Andreeff, M., and Freireich, E. J. Flow cytometry in clinical cancer research. Cancer Res., *43*: 3982–3997, 1983.
3. Beck, W. T., Mueller, T. J., and Tanzer, L. R. Altered surface membrane glycoproteins in vinca alkaloid-resistant human leukemic lymphoblasts (CCRF-CEM). Cancer Res., *39*: 2070–2076, 1976.
4. Bertino, J. R. The mechanism of action of the folate antagonists in man. Cancer Res., *23*: 1286–1306, 1963.
5. Bertino, J. R. "Rescue" techniques in cancer chemotherapy: Use of leucovorin and other agents after methotrexate treatment. Semin. Oncol., *4*: 203–216, 1977.
6. Crooke, S. T. and Reich, S. D., eds. Anthracyclines: Current Status and New Developments. New York: Academic, 1980.
7. Dano, K. Active outward transport of daunomycin resistant Ehrlich ascites tumor cells. Biochem. Biophys. Acta, *323*: 466–483, 1973.
8. Darzynkiewicz, Z., Rogers, A. W., Barnard, E. A., Wang, D., and Werkheiser, W. C. Autoradiography with tritiated methotrexate and the cellular distribution of folate reductase. Science, *151*: 1528–1530, 1966.
9. Durand, R. E. Flow cytometry studies of intracellular adriamycin in multicell spheroids in vitro. Cancer Res., *41*: 3495–3498, 1981.

10. Durand, R. E. and Olive, P. L. Flow cytometry studies of intracellular adriamycin in single cells in vitro. Cancer Res., *41*: 3489–3494, 1981.

11. Egorin, M., Hildebrand, R. C., Cimino, E. F., and Bachur, N. Cytofluorescence localization of adriamycin and daunomycin. Cancer Res., *34*: 2243–2245, 1974.

12. Ganapathi, R. and Grabowski, D. Enhancement of sensitivity to adriamycin in resistant P388 leukemia by the calmodulin inhibitor trifluoperazine. Cancer Res., *43*: 3696–3699, 1983.

13. Ganapathi, R., Reiter, W., and Krishan, A. Comparative studies on intracellular adriamycin levels and cytotoxicity in sensitive and resistant P388 leukemia cells. J. Natl. Cancer Inst., *68*: 1027–1031, 1982.

14. Gapski, G. R., Whitely, J. M., Rader, J. J., Gramer, P. L., Henderson, G. B., Neef, V., and Huennekins, F. M. Synthesis of a fluorescent derivative of amethopterin. J. Med. Chem., *18*: 526–528, 1975.

15. Garman, D. and Center, M. S. Alterations in cell surface membranes in Chinese hamster lung cell resistant to adriamycin. Biochem. Biophys. Res. Commun., *105*: 157–163, 1982.

16. Haber, D. A. and Schimke, R. T. Unstable amplification of an altered dihydrofolate reductase gene associated with double-minute chromosomes. Cell, *26*: 355–362, 1981.

17. Inaba, M. and Johnson, R. K. Uptake and retention of adriamycin and daunorubicin by sensitive and anthracycline resistant sublines of P388 leukemia. Biochem. Parmacol., *27*: 2123–2130, 1978.

18. Juliano, R. L. and Ling, V. A surface glycoprotein modulating drug permeability in Chinese hamster ovary cell mutants. Biochem. Biophys. Acta, *455*: 152–162, 1976.

19. Krishan, A. and Bourguignon, L. Y. W. Cell cycle phenothiazine effects on adriamycin transport. Cell Biol. Int. Rep., *8*: 449–456, 1984.

20. Krishan, A. and Frei, E. III. Effect of adriamycin on the cell cycle traverse and kinetics of cultured human lymphoblasts. Cancer Res., *36*: 143–150, 1976.

21. Krishan, A. and Ganapathi, R. Laser flow cytometry and cancer chemotherapy: Detection of intracellular anthracyclines for flow cytometry. J. Histochem. Cytochem., *27*: 1655–1656, 1979.

22. Krishan, A. and Ganapathi, R. Laser flow cytometric studies on intercellular fluorescence of anthracyclines. Cancer Res., *40*: 3895–3900, 1980.

23. Krishan, A., Ganapathi, R., and Israel, M. Effect of adriamycin and analogs on nuclear fluorescence of propidium iodide stained cells. Cancer Res., *38*: 3656–3662, 1978.

24. Krishan, A., Israel, M., Modest, E. J., and Frei III, E. Differences in cellular uptake and cytofluorescence of adriamycin and *N*-trifluoro-acetyladriamycin-14-valerate. Cancer Res., *36*: 2114–2116, 1976.

25. Krishan, A., Pitman, S. W., Tattersall, M. H. N., Paika, K. D., Smith, D. C., and Frei III, E. Flow microfluorometric patterns of human bone marrow and tumor cells in response to cancer chemotherapy. Cancer Res., *36*: 3813–3820, 1976.

26. Krishan, A., Sauerteig, A., and Wellham, L. L. Flow cytometric studies on modulation of cellular adriamycin retention by phenothiazines. Cancer Res., *45*: 1046–1051, 1985.

27. Kaufman, R. J., Bertino, J. R., and Schimke, R. T. Quantitation of dihydrofolate reductase in individual parental and methotrexate-resistant murine cells. J. Biol. Chem., *253*: 5852–5860, 1978.

28. Kaufman, R. J. and Schimke, R. T. Amplification and loss of dihydrofolate reductase genes in a Chinese hamster ovary cell line. Mol. Cell Biol., *1*: 1069–1076, 1981.

29. Mariani, B. D., Slate, D. L., and Schimke, R. T. S phase-specific synthesis of dihydrofolate reductase in Chinese hamster ovary cells. Proc. Natl. Acad. Sci. USA, *78*: 4985–4989, 1981.

30. Melamed, M. R., Mullaney, P. F., and Mendelsohn, M. L., eds. Flow Cytometry and Sorting. New York: John Wiley, 1979.

31. Muirhead, K. A., Freyer, J. P., and Sutherland, R. M. Distribution of adriamycin within tumor spheroids. Cytometry, *2*: 115, 1981.

32. Peters, D. C. A comparison of mercury arc lamp and laser illumination for flow cytometers. J. Histochem. Cytochem., *27*: 241–245, 1979.

33. Preisler, H. D. Alteration of binding of the supravital Dye Hoechst 33342 to human leukemic cells by adriamycin. Cancer Treatment Rep., *62*: 1393, 1978.

34. Raju, M. R., Johnson, T. S., Tokita, N., and Gillette, E. L. Flow cytometric applications to tumour biology: Prospects and pitfalls. Brit. J. Cancer, *41*: 171–176, 1980.

35. Rosowsky, A., Wright, J. E., Shapiro, H., Beardsley, P., and Lazarus, H. A new fluorescent dihydrofolate reductase probe for studies of methotrexate resistance. J. Biol. Chem., *257*: 14162–14167, 1982.

36. Rosowsky, A., personal communication.

37. Shapiro, H. M. Multistation multiparameter flow cytometry: A critical review and rationale. Cytometry, *3*: 227–243, 1983.

38. Skogen-Hagenson, M. J., Salzman, G. C., Mullaney, P. F., and Brockman, W. H. A high efficiency flow cytometer. J. Histochem. Cytochem., *25*: 784–789, 1977.

39. Skovsgaard, T. and Nissen, N. Adriamycin, an antitumour antibiotic: A review with special reference to daunomycin. Dan. Med. Bul., *22*: 62–73, 1975.

40. Tapiero, H., Fourcade, A., Vaigot, P., and Farhi, J. J. Comparative uptake of adriamycin and daunorubicin in sensitive and resistant Friend leukemia cells measured by flow cytometry. Cytometry, *2*: 298–302, 1982.

41. Tokita, N. and Raju, M. R. Cell-cycle dependency of adriamycin uptake in Chinese hamster cells. Eur. J. Cancer Clin. Oncol., *19*: 547, 1983.

42. Tritton, T. R., Murphree, S. A., and Sartorelli, A. C. Adriamycin: A proposal on the specificity of drug action. Biochem. Biophys. Res. Commun., *84*: 802–808, 1978.

43. Tsuruo, T., Lida, H., Yamashiro, M., Tsukagoshi, S., and Sakurai, Y. Enhancement of vincristine- and adriamycin-induced cytotoxicity by verapamil in P388 leukemia and its sublines resistant to vincristine and adriamycin. Biochem. Pharmacol., *31*: 3138–3140, 1982.

Chapter 12

Cell Synchrony Techniques

A Comparison of Methods

David J. Grdina, Marvin L. Meistrich,
Raymond E. Meyn, Tod S. Johnson,
and R. Allen White

1. INTRODUCTION

With the development of the technique of autoradiography by Howard and Pelc (*16*), the mammalian cell cycle has become a topic of considerable interest and study. To facilitate this effort, investigators have directed their attention to the development of techniques for use in obtaining cell populations enriched in selected phases of the cell cycle. The technique of mitotic selection (*35*) described by Terasima and Tolmach in 1961 is one such approach that has enjoyed widespread use and success. By taking advantage of the property of cells from certain cell lines to detach from the surface of culture vessels during mitosis, cell populations enriched in the mitotic phase could be collected through a gentle shaking process. These cells then progress through the cell cycle in a synchronous manner. The major limitation of this procedure, however, is that it is applicable only to selected cell lines that grow in vitro.

Subsequent to the development of this procedure, synchrony methods were developed that involved the blockage of cells in a specific cell cycle phase by either cytostatic and/or cytotoxic agents or the deprivation of essential nutrients from the culture medium (6,28–34,37). These methods suffered the limitation of possibly inducing perturbations both in the biochemical balance within cells and in the subsequent progression of cells through the cell cycle. An additional complicating factor in the use of some of these techniques in biochemical or cell function studies has been the presence of dead or damaged cells.

Because of these limitations and the need for greater applicability, speed, and efficiency in isolating cell populations unique with respect to their position within the division cycle, new technologies have been developed. For the most part they have been designed to exploit differences with respect to biophysical parameters between cells as a function of their position in the cell cycle (2,19,27, 29). Two such technologies currently evoking the most interest are centrifugal elutriation (10,11,14,23,24,26) and cell sorting with flow systems (3,9). The former is a continuous-flow centrifugation system that exploits the fact that cell volume increases steadily in an exponential manner with age during the cell cycle and that the sedimentation rate of a cell is proportional to the two thirds power of its volume, as well as to its shape and density (37). The latter is a cell sorter that can separate cells according to differences in their DNA contents, and hence, as a function of position in the cell cycle.

The use of these technologies and their relative merits have been extensively described in the literature. However, no attempt has been made to directly compare these approaches using the same cell line. In addition it is difficult to draw conclusions from the published data because of the different culture conditions used and the different analytical techniques employed to assess cell synchrony. For these reasons we initiated a study to compare selected methods of cell synchrony using the same cell culture and analytical techniques in order to identify and compare the relative advantages and disadvantages of each (14,38). Cognizant of the existence of a plethora of biophysical techniques such as velocity sedimentation, isokinetic gradient sedimentation, isopycnic centrifugation, electrophoresis, and isoelectric focusing, that can be applied to synchronizing cells, we have limited our studies to comparing only the methods of mitotic selection, hydroxyurea (HU) blocking, centrifugal elutriation, and flow sorting. These procedures are either currently in widespread use and/or of timely interest.

2. MATERIALS AND METHODS

2.1. Cell System and Methods of Flow Cytometry Analysis

Chinese hamster ovary (CHO) cells obtained from the American Type Culture Collection (#CCL 61) were grown in monolayer cultures at 37°C in a humidified atmosphere of 95% air and 5% CO_2. Culture conditions included the use of McCoy's 5A medium supplemented with 15% fetal bovine serum and antibiotics.

To assess the degree of synchrony obtained with each method, as well as to monitor the progression of the cells through the first cell cycle following synchronization, a Phywe ICP-II flow cytometer was used. Prior to analysis by flow cytometry (FCM), cells were treated with pepsin, then stained with a mixture of ethidium bromide and mithramycin (4).

FCM profiles describing the DNA distributions of cells were analyzed by a method described in detail elsewhere (17). DNA distribution analysis is described in more detail in chapter 5 (13). Briefly, cells in G1- and G2 + M-phases were assumed to form Gaussian distributions, and the S-phase fraction was represented by a sum of equispaced Gaussian distributions. The coefficient of variation (CV) was assumed to be constant (i.e., independent of cell age) at approximately 2%. This assumption is based on the use of relatively pure G1- and G2 + M-phase populations obtained following cell synchronization, both of which exhibited a CV of 2%. Examination of the relatively pure G2 + M-phase populations also indicated that the G2 + M-phase peak in the DNA profile was located at a distance from the origin of twice that observed for the G1-phase peak. These data made it possible to define the positions of the G1- and the G2 + M-phase peaks in those profiles, which described S-phase-enriched populations.

2.2. Mitotic Selection and HU Synchronization

Mitotic selection and HU synchronization methods are described in detail elsewhere (25). Briefly, asynchronous CHO cells were grown and maintained as described above. In each culture bottle 10^7 cells were seeded and allowed to grow for 18 h. To induce a relatively large mitotic index, the standard culture medium was removed and medium containing 7.5 mM thymidine (TdR) was added. Cells were incubated for an additional 9 h. These conditions produced a parasynchronous culture of S-phase cells. The excess TdR medium was then removed. The mitotic index of the culture approached 15% 6 h after addition of the standard culture medium. Mitotic cells (MI = 95%) were selectively detached from the monolayer by gentle agitation and were removed from the culture bottle by decanting the medium. Fresh medium was added and the mitotic selection repeated at 15-min intervals during the peak mitotic period for a total of four collections. Mitotic cells were held on ice until after the last collection, at which time they were plated for the cell progression studies.

HU synchronization of cells in early S-phase was accomplished by incubating mitotic cells as described above in the presence of medium containing 2 mM HU for 9 h. The HU medium was removed from the attached cells and fresh medium without HU was added in order to allow cell progression.

2.3. Centrifugal Elutriation

Asynchronous CHO cells were separated by centrifugal elutriation using a Beckman JE-6 elutriator rotor and a standard Beckman chamber (23,25). The system was sterilized by pumping 70% ethanol through it the night before the run, keeping the system closed, and then pumping ethanol through it again just prior to the run. The entire system was kept at room temperature. The elutriator medium used was McCoy's 5A medium supplemented with 5% fetal bovine serum.

The rotor speed was kept constant at 1790 rpm (23). Cells (2 × 10^8 in 50 mL) were introduced into the chamber at a flow rate of 9.4 mL/min, after which 50 mL of the medium was pumped through the chamber at the same flow rate. Flow rates were varied from 12.6 to 37.8 mL/min in 1.8 mL/min increments except for the last fraction (i.e., fraction 12), which was collected by raising the flow rate from 30.6 to 37.8 mL/min. A volume of 50 mL was collected in each fraction. The volume distribution of cells was determined by Coulter volume spectrometry using a ZBI Coulter counter and a multichannel analyzer (Channelyzer II, Coulter Electronics, Hialeah, FL). Based on past experience, the cell counts and modal volumes, fractions 3, 8, and 11, were chosen to represent G1-, S-, and G2-phase-enriched populations. Cells were maintained at room temperature for about 1.5 h: 30 min for the preparation and washing of the cell suspension, 30 min for elutriation, and 30 min for replating.

2.4. Cell Sorting

Exponentially growing asynchronous CHO cells were harvested and stained (1.5×10^8 cells in 100 mL) for 100 min in the dark using a spinner flask with 1 μg/mL Hoechst 33342 dye. Immediately after staining, the spinner flask was placed on ice and 10–20-mL aliquots of cells were removed at various times over a 3-h period. Cells were resuspended in cold phosphate buffered saline (PBS) and filtered using a 65-μm diameter nylon mesh. Cells were then analyzed by forward light scatter and Hoechst blue fluorescence with an EPICS V cell sorter (Coulter Electronics) using the 351.1 and 363.8 nm lines (100 mW power output at these wavelengths) from a Spectra Physics 164 Argon-ion laser. The fluorescence emission was measured at wavelengths above 418 nm using a barrier filter. A 3–14° light scatter signal was processed simultaneously to use for size gating to eliminate debris. To confirm that the droplet brakeoff and droplet charging deflection were correct, we determined the sort fidelity [number of sorted particles of cells/number of EPICS V electronically registered sorted counts $\times$ 100] by sorting 200–300 10-μm diameter plastic microspheres (Coulter Electronics) and Hoechst 33342 vitally stained G1-CHO cells onto microscope slides and then counting. The sort fidelity was determined to be at least 98% in each experiment. The sample flow rate was 2.0–2.5 $\times$ 10^3 cells/s. The DNA distributions were measured and a 10-channel-wide window centered on the modal G1 peak channel was used for electrostatic sorting. Approximately 1–1.5 $\times$ 10^3 Hoechst-stained G1 cells were sorted per s into sterile medium on ice. Twenty minutes were required per sample to sort 2 $\times$ 10^6 G1 CHO cells.

2.5. Cell Progression Analysis

To study cell progression, selected aliquots of cells obtained by mitotic selection, HU synchrony, centrifugal elutriation (fractions 3, 8, and 11), and cell sorting were plated (5×10^5 cells) into fresh complete medium in 60-mm plastic culture dishes. Cells were allowed to progress for up to 14 h. At selected times samples were harvested by trypsinization and fixed for FCM analysis.

Details of the mathematical analysis of these data are presented in the appendix of this chapter, and are described elsewhere (38). Mathematical cell cycle analysis techniques are also described in chapter 9 (8). Briefly, DNA distributions were analyzed to determine the fractions of cells in G1, S, and G2 + M at each time point. A mathematical model was used to simulate the mean transit times through G1, S, and G2 + M. S-phase was not subdivided, since that would require estimating more parameters (i.e., transit times in each subdivision of S), which would approach the number of data points obtained. Thus we used the percentage of cells in S-phase and did not try to reproduce all the details of the observed distribution of cells within S-phase. A least squares analysis was employed to minimize the mean square error between observed and fitted data to provide the best estimate for these transit times. The best fit procedure depends strongly on the mean transit times of the cells and only weakly on the variance in transit time. Preliminary analysis of progression of synchronized CHO cells showed that the following variances, $\sigma^2(\text{G1}) = 3 \text{ h}^2$, $\sigma^2(\text{S}) = 1 \text{ h}^2$, $\sigma^2(\text{G2M}) = 1 \text{ h}^2$, were consistent with all data used. Therefore the variances were fixed at these values throughout these analyses and the optimal values of the means were computed.

In one experiment cell progression was analyzed by following the division of cells after plating. Replicate samples of about 3000 cells were plated in 60-mm culture dishes and incubated. Dishes were removed at various times, fixed, stained, and analyzed for the average number of cells per clone in 200 clones.

3. RESULTS

3.1. Empirical Determinations of Cell Progression Parameters

FCM profiles describing the DNA distributions of asynchronously and exponentially growing CHO cells are presented in Fig. 1. These cells were found to have a doubling time of from 12 to 14 h under the growth conditions used. Cell progression through the cell cycle was monitored for populations initially enriched in G1-phase cells by either mitotic selection or automated

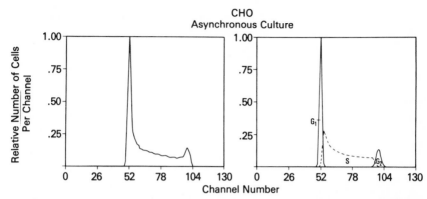

Fig. 1. A representative histogram describing the DNA content of an asynchronous population of CHO cells is presented in the panel. The regions of the histogram representing G1-, S-, and G2-phase cells are indicated in the right panel. G1 = 36%; S = 55%; and G2 = 9%; coefficient of variation (CV) of the G1 peak = 2%.

cell sorting. FCM profiles describing the DNA distributions of CHO cells as a function of time following synchronization by these methods are presented in Figs. 2 and 3, respectively. Although mitotic cells, as determined by FCM analysis, appeared to complete their progression through the cell cycle by 12 h, a significant proportion of the sorted cells remained in G2 even after 14 h had elapsed.

Following centrifugal elutriation of asynchronously growing CHO cells, 12 fractions were routinely collected. FCM profiles of selected but representative elutriator fractions are presented in Fig. 4. Maximum enrichment of cells in the various phases of the cell cycle were 92% for G1-phase in fraction 2, 84% for S-phase in fraction 8, and 50% for G2-phase in fraction 12. These values compare favorably with data presented in Figs. 2 and 3. Based on cell progression data, the G1 cells in fraction 3 appeared to be mainly in early G1 (*see* Fig. 5). Cells from fraction 11 were chosen for progression studies on the basis of their volume distribution and were thus believed to be maximally enriched in G2. Unfortunately this fraction contained only 34% G2-phase cells with 63% of the remaining cells in late S-phase.

Two approaches were tested for obtaining relatively pure populations of cells in S-phase. Using a HU block of mitotically selected cells, a relatively homogeneous population of S-phase cells

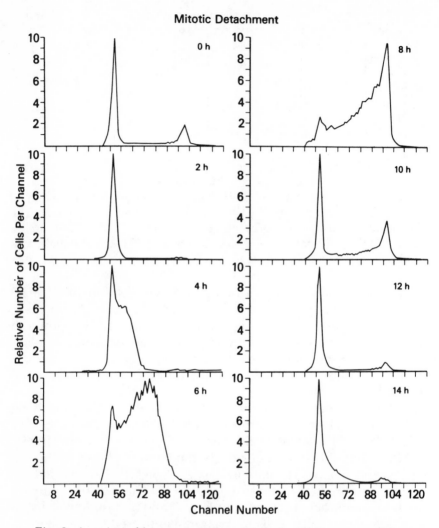

Fig. 2. A series of histograms describing the DNA content of CHO populations as a function of time following their synchronization by mitotic detachment. The relative percents of G1-, S-, and G2M-phase cells are presented in Fig. 8. CV of G1 peaks ranged from 1.8 to 2%.

was obtained. There was little dispersion in this population during its traverse through S-phase following its release from the HU block (*see* Fig. 6). In contrast, the starting population of S-phase enriched cells obtained by centrifugal elutriation was much less homogene-

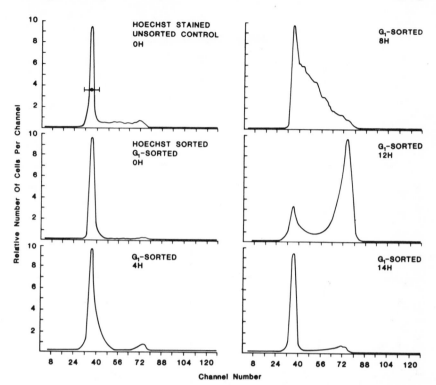

Fig. 3. A series of histograms describing the DNA content of CHO populations as a function of time following synchronization using an EPICS V cell sorter. The upper left panel describes the sort window in sorting G1 Hoechst-stained cells. The relative percents of G1-, S-, and G2M-phase cells are presented in Fig. 9. CV of G1 peaks ranged from 3 to 4%. All cells were stained with Hoechst 33258 and subsequently sorted, ethanol fixed, and reanalyzed following staining with ethidium bromide and mithramycin using an ICP-II instrument.

ous and contained only 84% cells in S-phase compared to 100% with the former method. This dispersion was also reflected in the diffuse progression of this population as a function of time through S-phase (see Fig. 7). Diagrammatic representations describing the initial cell cycle distributions of cells prepared by these synchronization procedures are presented for comparison in Fig. 8.

The fractions of cells in G1-, S-, or G2M-phases as a function of the synchrony method used and the time following synchrony are presented in Figs. 5, 9, and 10. Of special interest are the progression data on G1 populations obtained by the three methods.

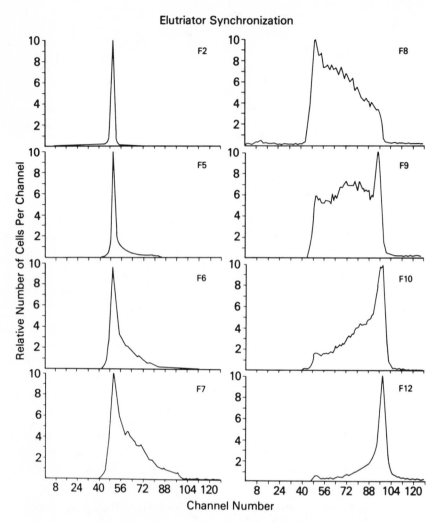

Fig. 4. A series of histograms describing the DNA content of CHO cells in the different elutriator fractions following centrifugal elutriation. The relative percents of G1-, S-, and G2M-phase cells are presented in Fig. 8. CV of G1 peaks ranged from 1.8 to 2.3%.

As shown in Fig. 5, the G1 compartment was maximally depleted of G1 cells derived by mitotic selection in 6 h. In contrast, the G1 compartment was depleted in 8 and 12 h for G1 cells derived by centrifugal elutriation (*see* Fig. 5) and automated cell sorting (*see*

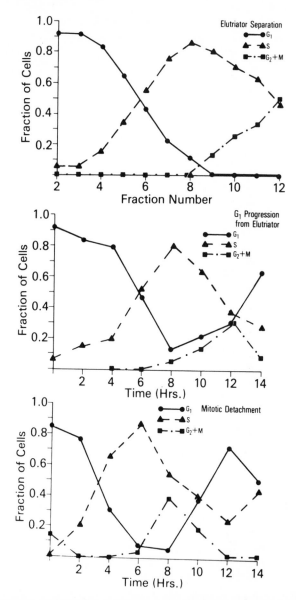

Fig. 5. The upper panel describes the relative distributions of CHO cells in G1-, S-, and G2M-phases of the cell cycle as a function of elutriator fraction number. Middle and bottom panels describe the progression of cells by FCM analysis following G1 enrichment of cells by centrifugal elutriation and mitotic detachment, respectively.

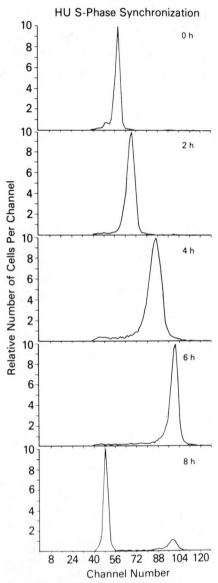

Fig. 6. A series of histograms describing the DNA content of S-phase CHO cells synchronized by mitotic selection followed by exposure to hydroxyurea (HU) as a function of time following the release of the HU block. The relative percents of G1-, S-, G2M-phase cells are presented in Fig. 5. CVs of S-phase distributions were 3, 4, 5, 2.5, 2.7, and 3.5% at the time points of 0, 2, 4, 6, 8, and 10 h, respectively.

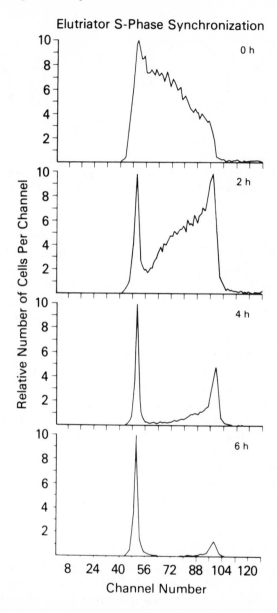

Fig. 7. A series of histograms describing the DNA content of CHO cells originally enriched in S-phase by centrifugal elutriation as a function of time following centrifugation. The relative percents of G1-, S-, G2M-phase cells are presented in Fig. 5.

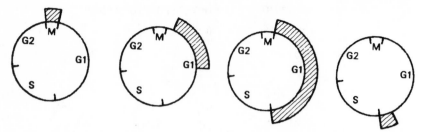

Fig. 8. Diagram describing cycle positions of cells prepared by various synchronization procedures. From left to right are: mitotic detachment, centrifugal elutriation for G1 enrichment, cell sorting, and blockade by hydroxyurea.

Fig. 10), respectively. These differences in progression through G1-phase were also reflected in the times required for progression of the G1-enriched populations through the complete cell cycle. In contrast to the 12-h progression time of mitotically selected G1-phase cells, progression of G1 cells enriched by centrifugal elutriation (fraction 3) or cell sorting through the cell cycle took at least 14 h (*see* Figs. 5 and 9).

Further studies were performed to characterize parameters that might account for the discrepancy in time required for progression through the cell cycle between G1-phase cells obtained by mitotic selection and by centrifugal elutriation. Three populations of cells— those selected by mitotic shakeoff, G1 cells selected from an asynchronous population by centrifugal elutriation, and postmitotic cells further purified by elutriation—were compared with respect to cell volume distributions and cell progression. The cells obtained by mitotic shakeoff followed by a brief incubation were slightly larger on the average and more heterogeneous in volume than the G1 cells isolated by elutriation (Fig. 11). Purification of mitotic cells by elutriation yielded a population of cells with a volume distribution identical to that of the elutriated G1 cells. Following progression of these various cell populations for up to 30 h, it was observed that the elutriated mitotic cells divided about 2 h earlier than the G1 elutriated cells, but at the same time as the unelutriated mitotic cells (Fig. 12). Thus, we conclude that elutriation alone does not induce perturbations in cell progression. The slower progression of the elutriated G1 cells must be either a property of the subpopulation of cells selected or a result of the method of harvesting cells.

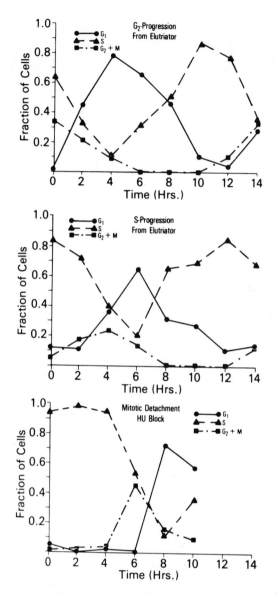

Fig. 9. The top and middle panels describe, respectively, the progression of late S- and G2M-phase and of S-phase cells, originally isolated by centrifugal elutriation, as a function of time. The bottom panel describes the progression of CHO cells following release from a HU block as a function of time.

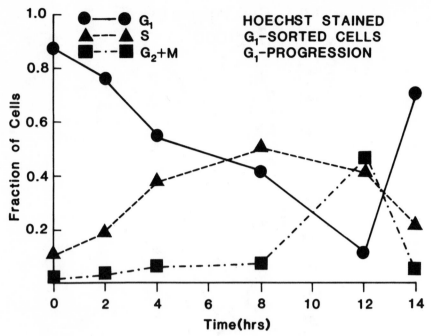

Fig. 10. Presented are data describing the progression, as determined by FCM analysis, of Hoechst-stained G1-sorted cells as a function of time.

3.2. Mathematical Analysis and Comparison of Cell Progression Parameters

As described earlier, G1-phase cells appeared to progress differently depending on the method of preparation used. Since mitotically detached cells are found at the beginning of G1 shortly after harvesting, we established this as an initial condition in our model (*see* Fig. 13). The resulting fit is presented in Fig. 14. Following sorting, however, it was assumed that the cells were initially equally distributed throughout G1 (Figs. 15 and 16). Finally from the cell progression studies, the G1 cell population isolated by centrifugal elutriation appeared to be a homogeneously sized but not a homogeneously aged group of cells. It could be assumed either that the cells are evenly spread throughout G1 or that they are composed of a homogeneous population of small cells that took a relatively long period of time to go through G1. It was this latter condition that resulted in the better fit (Figs. 17 and 18). These data are summarized for comparison in Table 1.

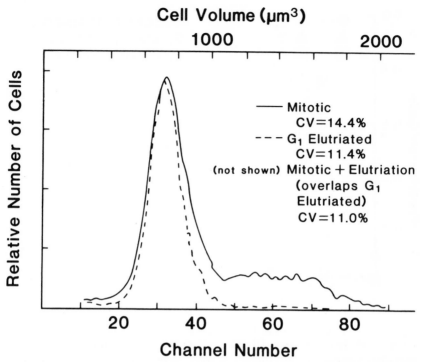

Fig. 11. Comparison of volume distributions of G1 cells separated by elutriation, cells obtained by mitotic detachment, and mitotic cells further fractionated by elutriation. Mitotic cells were incubated at 37°C for 1 h following detachment to allow for most cells to undergo cytokinesis.

In the analysis of progression data from enriched S-phase populations, we have the advantage that the entire distribution of cells in S-phase, i.e., the DNA contents of individual cells ranging from 2C to 4C, is displayed by FCM. Using S-phase cells, we can determine the magnitude of deviation of the assumed initial distribution from the actual initial distribution that is allowed before significant alterations are detected in the estimated transit times and fits to progression data.

The DNA profile describing the initial distribution of cells in fraction 8 following elutriation is presented in Fig. 19. This population was estimated to have 84% of the cells in S, with most of the cells being in early S. Five initial distributions were considered (even though some were not realistic). These were chosen to test the sensitivity of the results to using likely and unlikely assumptions. A likely condition assumed was a steadily decreasing number of cells

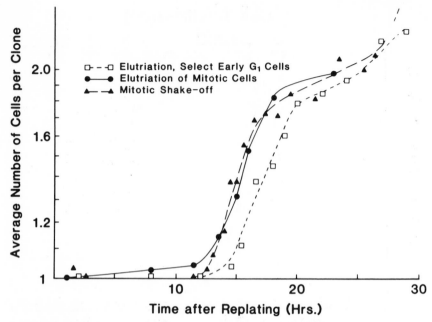

Fig. 12. Comparison of the progression of two mitotic populations and a G1 population, each obtained by mitotic selection only, mitotic selection plus elutriation, or elutriation only, respectively. Cell progression, measured as cell division, was analyzed by counting the number of cells per clone at various times after plating.

starting with a maximum number in early S and decreasing linearly to zero in late S (*see* Fig. 20). Such a distribution differs from the actual histogram by overestimating the proportion of cells in early S. Two other initial distributions of lesser plausability were tested, including one in which cell number increases from zero at the beginning of S to a maximum at the end of S. Finally two outrageous distributions were selected in which all of the cells were either at the beginning or end of S-phase.

The estimated mean transit times for the five different conditions tested are given in Table 2. The first two conditions yielded the same calculated mean transit times to within an estimated standard error. Mathematically, there is little reason to choose between the two, and the observed behavior of the cells lies between the two assumed distributions. Placing none at the start of S and increasing linearly to a maximum at the end gave a longer S-phase

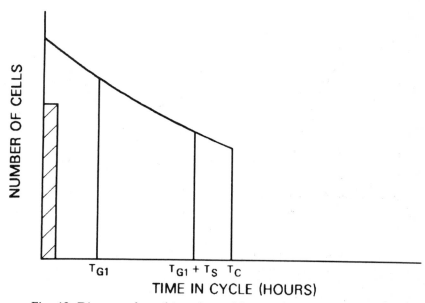

Fig. 13. Diagram describing the stable age distribution of cells (solid lines) as estimated by simulating an initial distribution of cells (hatched lines) as observed by flow cytometry following mitotic detachment. For ease of comparison to other methods of synchronization the total number of cells in the stable age distribution are normalized to unity. Numerical values for T_{G1}, T_{G1}, and T_S and T_c are found in Table 1.

than expected. The experimental and fitted data for the first condition are plotted for comparison in Fig. 21. Fitted data describing the second and third models are not presented because they appear nearly identical to those shown in Fig. 21. In contrast to the relatively good fit using either of the first three initial conditions described, the models involving all of the cells either at the beginning or at the end of S did not give rise to acceptable fits. As shown in Fig. 22, when all the cells were placed at the beginning of S, the best fit to the empirical data required that the apparent transit time through S had to be reduced from 7 to 4 h. This necessitated a reduction of the total cycle time from 12 to 7 h, a value clearly not consistent with empirical data. Similarly placing cells at the end of S led to an unacceptably long cycle time and G2M time.

Progression data for a late S plus G2M population separated by elutriation was also modeled (Fig. 23). An FCM profile of the

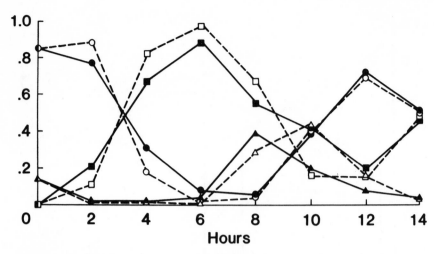

Fig. 14. Observed and fitted progression data from a population prepared by mitotic detachment. Cells were assumed to be initially at the beginning of G1. Estimated transit times are T_{G1} = 3.2 h, T_S = 5.6, and T_{G2M} = 2.4 h.

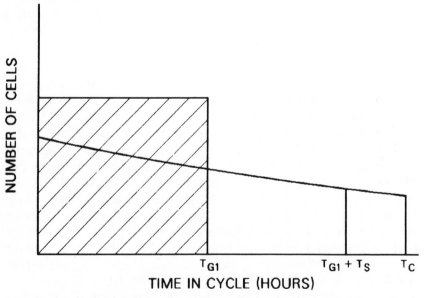

Fig. 15. Distribution of cells assumed for fitting of data following G1 phase enrichment by fluorescent activated cell sorting.

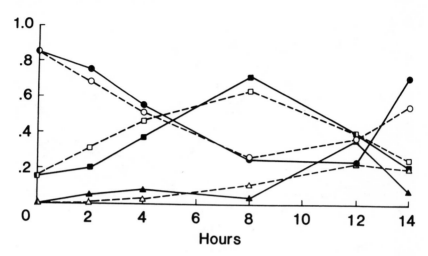

Fig. 16. Observed and fitted progression data of a G1 population prepared by cell sorting were assumed to be initially distributed uniformly over G1. Estimated transit times are $T_{G1} = 9.8$, $T_S = 8.0$, and $T_{G2M} = 3.6$ h.

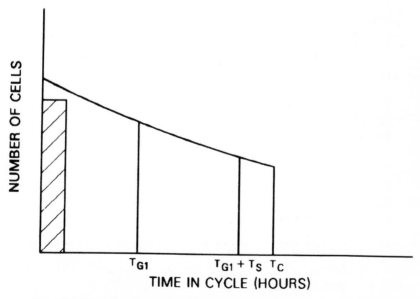

Fig. 17. Distribution of cells assumed for fitting data following G1-phase enrichment by centrifugal elutriation.

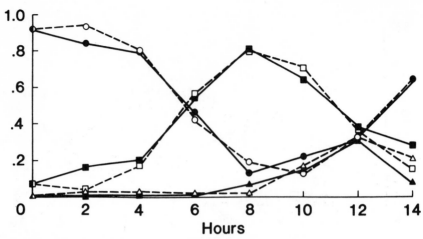

Fig. 18. Observed and fitted progression data of a G1 population prepared by centrifugal elutriation. The cells were assumed to be initially distributed at the start of G1. Estimated transit times are T_{G1} = 5.7, T_S = 6.0, and T_{G2M} = 2.3 h.

Table 1
Estimated Transit Times for a Set of Initially Pure G1 Populations

Method of preparation	Initial condition	T_{G1}, h, $\pm$ SE	T_S, h, $\pm$ SE	T_{G2M}, h, $\pm$ SE	T_c, h, $\pm$ SE
Mitotic detachment	All cells start at G1 or in mitosis	3.2 $\pm$ 0.2	5.6 $\pm$ 4.8	2.4 $\pm$ 0.7	11.2 $\pm$ 4.9
Centrifugal elutriation	Cells uniform over G1	9.8 $\pm$ 0.9	8.1 $\pm$ 0.9	2.1 $\pm$ 1.6	20. $\pm$ 2.0
Centrifugal elutriation	All cells at start of G1	5.7 $\pm$ 0.2	6.0 $\pm$ 0.4	2.3 $\pm$ 0.6	14. $\pm$ 0.8
Sorting	Cells uniform over G1	9.8 $\pm$ 1.1	8.0 $\pm$ 1.0	3.6 $\pm$ 1.6	21.4 $\pm$ 2.2

initial population is presented in Fig. 24. The initial distribution of cells in S was taken to be a linearly increasing level of S. In practice, the duration of G2M is so short that the relative distribution of cells within this compartment makes very little difference in the

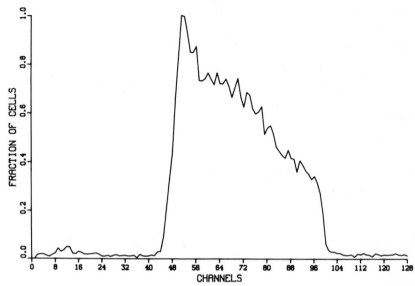

Fig. 19. The observed initial DNA histogram of S-phase population enriched by centrifugal elutriation.

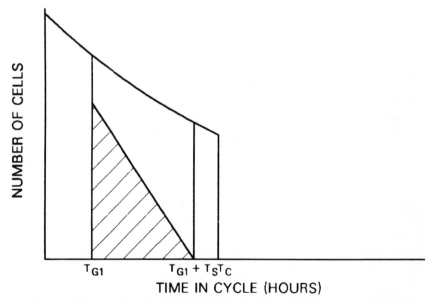

Fig. 20. Distribution of cells assumed for fitting of data following S-phase enrichment by centrifugal elutriation.

Table 2
Estimated Transit Times for Various Initial Distributions
of an S-Phase-Enriched Population

Initial distribution of cells in S	$T_{G1'}$ h, ±SE	$T_{S'}$ h, ±SE	$T_{G2M'}$ h, ±SE	$T_c = T_{G1} + T_S + T_{G2M'}$ h, ±SE
Maximum at start, decreasing to none at end	2.7 ± 0.6	5.9 ± 0.6	1.6 ± 0.3	11.2 ± 0.9
Uniform	2.9 ± 0.6	6.8 ± 0.8	2.5 ± 0.8	12.2 ± 1.2
None at start, increasing to maximum at end	3.0 ± 0.4	8.4 ± 1.0	2.9 ± 0.6	13.3 ± 1.2
All at start of S	1.6 ± 0.8	4.0 ± 0.5	1.5 ± 3.1	7.1 ± 3.2
All at end of S	2.6 ± 0.9	7.6 ± 1.4	4.7 ± 0.6	14.9 ± 1.6

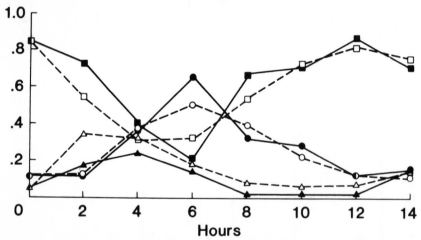

Fig. 21. Observed (solid) and fitted (dash) progression data fractions of cells of an initially enriched S-phase population. Circles represent G1 fractions, squares represent S fractions, and triangles represent G2 + M fractions. The initial conditions have cells maximally at the beginning of S, decreasing to none at the end of S. The estimated transit times are: T_{G1} = 2.7, T_S = 5.9, and T_{G2M} = 1.6 h.

model chosen. Results are presented in Fig. 25, in which a flat distribution of cells through out G2M was assumed. Transit times

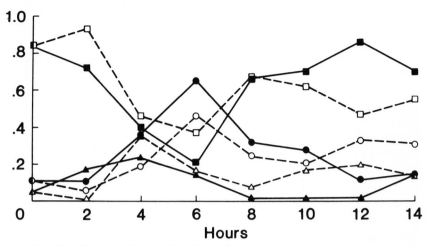

Fig. 22. Observed and fitted S phase progression data with cells assumed to be entirely at the beginning of S phase initially. The symbols are the same as those used in Fig. 16. Estimated transit times are T_{G1} = 1.6, T_S = 4.0, and T_{G2M} = 1.5 h.

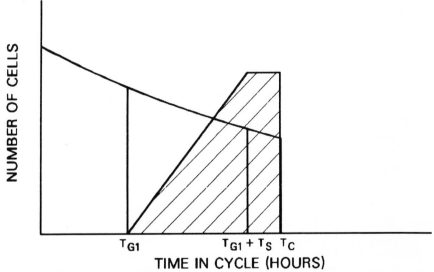

Fig. 23. Distribution of cells assumed for fitting of data following G2 + M enrichment by centrifugal elutriation.

were calculated and were in general agreement with values obtained with other populations (Table 3).

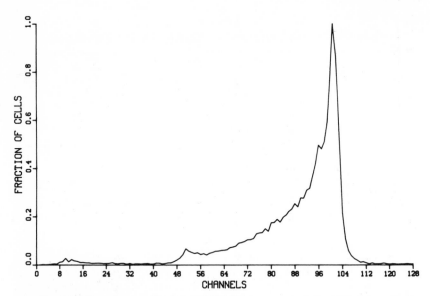

Fig. 24. The observed initial DNA histogram of the G2M population enriched by centrifugal elutriation.

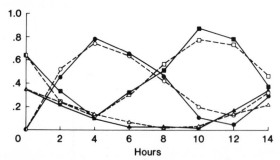

Fig. 25. Observed and fitted progression data from a population initially enriched in G2M cells. The S-phase cells were assumed to be initially distributed increasing from none at the beginning of S to a maximum at the end of S. The symbols are the same as those used in Fig. 16.

4. DISCUSSION

4.1. Comparison of Cell Synchronization Methods

The ultimate goal in the development of cell synchrony techniques is the acquisition of a large number of cells with a homogeneous age distribution capable of subsequent unperturbed progres-

Table 3
Estimated Transit Times for an Enriched G2M Population

Initial distribution	T_{G1}, h, $\pm$SE	T_S, h, $\pm$SE	T_{G2M}, h, $\pm$SE	T_c, h, $\pm$SE
Increasing from none at start of S	4.9 $\pm$ 0.5	6.7 $\pm$ 1.0	2.1 $\pm$ 0.4	13.2 $\pm$ 1.2

sion. Although mitotic selection has historically approached such a goal, it can provide cells only at the beginning of G1 and its applicability is restricted to only a few cell lines. Recently, the methods of centrifugal elutriation and automated cell sorting have been developed and applied as cell synchrony techniques. In this chapter we have evaluated and compared each of these methods to the mitotic selection technique, either alone or with HU, using the CHO cell line.

The three procedures used to obtain G1-enriched populations can be compared to the purity of the initial population obtained, as well as to the subsequent progression parameters observed. With respect to the former, little difference in the purities among the initial G1 populations was observed. If cells isolated by each of these procedures exhibited the same transit time, then after some time, T_{G1}, there would be no cells remaining in G1. In fact, 6 h after mitotic detachment, only 8% of the cells were still in G1, and 2 h later this value fell to 6%. In contrast, 8 h following elutriation and cell sorting, the numbers of cells in G1 were found to be 13 and 25%, respectively. These data suggest that G1 cells isolated by the latter two methods are either progressing more slowly than those obtained by mitotic selection or a fraction of the cells is not progressing at all. For mitotically selected G1 phase cells, cell cycle progression appeared to be complete within 12 h. G1-enriched cell populations obtained either by centrifugal elutriation or automated cell sorting required about 14 h to complete progression through the cell cycle.

The delay in progression of elutriated G1 cells may have been a result of the rigors of cell separation. An alternative explanation, however, is that elutriation selected a population of G1 cells with a volume smaller than that of the average-size mitotic cell. As a result these cells might require a longer time to traverse the cell cycle (39). Finally it is possible that the trypsinization of cells, a step required to suspend interphase cells grown in monolayer cultures for elutriation, resulted in a delay of progression. From the data presented in Figs. 10 and 11, it appears that cell size is not a major determining factor. In addition, mitotic cells appeared to progress at the same rate regardless of whether they were elutriated or not. Thus, the most likely explanation for the longer time required for division of G1 cells selected by elutriation is that a period of time is required for the resynthesis of cell surface components removed by trypsin before cells can progress in their division cycle. The delay in progression of Hoechst-stained sorted cells may have been a result of the rigors of the staining and destaining process and the passage through an Argon-ion laser beam.

Progression studies of cell populations enriched in S or G2 + M were more gratifying. The same transit time estimates were obtained for cells synchronized in G2M and S as those obtained by mitotic selection. Centrifugal elutriation as described earlier produced substantially longer G1 transit times, but not longer S and G2M times. It also appeared that the actual initial distribution of cells did not materially affect the estimated transit times. The exception to this was if "outrageous" initial distributions were chosen, such as all of the cells at the beginning or end of S phase.

The dispersion of cell synchrony as a function of time appeared to be similar for mitotically selected and sorted cells (12). In contrast to mitotically selected cells that were in early G1, sorted cells were initially spread throughout the G1 compartment. A modal G1-channel 40 ($\pm$5 channel-wide sort "window") was used (see Fig. 3) in an attempt to obtain enough enriched G1 cells for subsequent subculturing, and cell cycle traverse rate estimates. The selection of this 10-channel-wide G1 sort window represented a compromise between achieving the best possible G1 synchronized cohort (narrower window) and the practical problem of sort time required to harvest at least 2×10^6 cells/sample at 10^6 cells/mL sample concentration and a flow rate of 2–2.5×10^3 cells/s. A comparable 10-channel-wide S-window would require significantly larger sort times, but synchrony should be substantially better.

The FCM profiles in Fig. 4 represent the DNA distributions of cells that can be obtained following centrifugal elutriation. The separations described here are comparable to those previously obtained (26), except for the lower purity of the G2 cells as a result of not continuing the collection of fractions at high enough flow rates to minimize contamination of late S-phase cells. The general procedure used here for separation by elutriation represents the result of extensive attempts to optimize the elutriation conditions (*24*). We have investigated alternative procedures based on the "long collection method" (*18*), but that method did not result in improved separation of CHO cells. No other methods for improvement of separation of CHO cells by elutriation have been reported. The volume dispersion of the elutriated fractions of cells obtained here (CVs of 10% for fraction 3, 14% for fraction 8, 11% for fraction 11) represent among the lowest for any cell type reported in the literature by centrifugal elutriation (*18,24*). Regardless of the measures taken to optimize the technical aspects of the separation system, there are cellular factors that can limit the purity of cell fractions separated by elutriation. The volume dispersion of cells at a particular age in the generation cycle is sufficiently large that all cells of the same volume may not be at precisely the same position in the cell cycle (*39*). For example, CHO cells selected at mitosis have a volume dispersion with a coefficient of variation of 18% (*1*). Therefore in our experiments, elutriator-selected synchronized cells, even with a volume CV of 10%, represent cells with a range of cell ages from the previous mitosis. Although this can preclude the complete simultaneous synchronization of cells in G1, S, or G2 by centrifugal elutriation, the enrichment of cells in the various phases of the cell cycle is comparable to that obtained by mitotic selection and subsequent progression.

As described earlier, it is of considerable interest to be able to synchronize and isolate cells in other phases of the cell cycle besides G1 phase. Regardless of the method used to isolate G1 cells, cell dispersion during progression will give rise to less than optimally enriched populations of S- or G2-phase cells. Of the procedures followed here, the method of mitotic selection followed by a HU block gave rise to a more enhanced synchrony of S-phase cells than was obtainable by any other technique. However, the use of a cytotoxic/cytostatic agent such as HU may give rise to perturbed cells. It is unclear what effects on cell viability, cell kinetics, and biochemistry this agent may induce during the subsequent growth of these cells. Although centrifugal elutriation alone was not sufficient to obtain the degree of synchrony observed with HU, neither was the use of mitotic selection alone. If an agent such as HU is to be used to obtain S-phase synchrony, it is reasonable to assume it can be used as effectively with elutriated G1 populations as it is with those that are mitotically selected.

4.2. Advantages and Disadvantages of Each Synchronization Method

Each of these methodologies possess advantages and disadvantages. With respect to minimal perturbation to the cells, mitotic selection appears to be the method of choice, followed by centrifugal elutriation and then cell sorting. The major disadvantages of mitotic selection are its limited applicability to only selected cell lines growing in monolayer culture and its low yield of cells. For example, after setting up 10^8 cells, only a total of 2–8 × 10^6 mitotic cells are routinely recovered for use. This accounts for a recovery of only 2–8%. Also, a method such as mitotic selection, which relies on selection of cells in one phase of the cell cycle, is limited by the dispersion of cell synchrony as a function of time following the isolation of the target population, which is an intrinsic characteristic of each cell line. A low cell yield by sorting results from limitations on the sort rate obtainable. Since cell sorting was limited to a sort rate of 2 × 10^3 cells/s, 20 min was required to accumulate only 2 × 10^6 G1-phase cells. A sort rate of 5 × 10^3 cells/s can be achievable on the EPICS V instrument. However, these higher rates might compromise the sort fidelity, quality, and reproducibility of the sorted samples. Further instrument development and improved signal processing have recently provided sorting rates of up to 5 × 10^4/s. Hence, an order of magnitude increase in sorting rate for cells should eventually be feasible. Furthermore, the separation and synchronization of cells by automated cell sorting has to date been a costly process requiring sophisticated instrumentation and computer backup. The degree of synchrony, however, is no better than that obtained by other less costly methods. In addition, this procedure requires staining of cells with DNA-specific dyes such as Hoechst 33342, giving rise to the possibility of loss of viability or mutagenicity. The advantage of this technique, however, is its ability to sort and isolate cells with specific cellular parameters. Criteria such as cell size, nuclear/cytoplasmic ratio, metabolic, and/or enzymatic activities or specific cell surface markers can be used to sort and separate cells.

The method of centrifugal elutriation offers advantages with respect to the speed of separating cells. The method is rapid, in that large numbers of asynchronously growing cells (3 × 10^8) can be separated within 30 min into subpopulations enriched in the various phases of the cell cycle, with the exception of mitosis. The cells are separated at room temperature in culture medium and thus are not subjected to cytotoxic or cytostatic agents. Its major disadvantage, however, is the limited purity achievable in the S- and G2M-phase-enriched populations. This appears to be limited by the inherent heterogeneity of size within each phase of the cell cycle. A summary of the relative advantages and disadvantages of each of these methods for synchronizing cells is presented in Table 4.

5. CONCLUSIONS

The application of new technologies to the development of cell synchrony procedures has been described and compared to the

Table 4
Advantages and Disadvantages of Each Method

	Advantages	Disadvantages
Mitotic selection	Easy, high-purity, minimal perturbation, low cost, wide variety of cultured cell lines	Low yield (2–8% of population, 10^7 cells), rapid synchrony decay through G1
HU + mitotic selection	Easy, high S-phase purity, low cost	Metabolic perturbation, low yield
Elutriation	Speed, minimal perturbation, high viability, large cell numbers (3×10^8), high recovery (~90%), homogeneous cellular volumes, all cycle phases simultaneously, cells treated before or after separation	Cost ($20,000 elutriator equipment), need channelyzer for size analysis, recommend FCM capability, low mitotic index, maximum cells in S-phase only 80%, purity of some fractions less than achieved by other methods
Cell sorting	DNA markers more specific for cell age, capability of isolating unique subpopulations based on multiple cellular features, capability of sorting and isolating rare events	High cost (~$85,000 to 200,000[a]), large computer capability required, highly trained support staff required, slow throughput of cells (30,000–300,000 total cells/min), possible mutagenicity or loss of viability

[a]Technological advances may reduce cost to $50,000 or less by 1987.

standard mitotic selection technique using CHO cells. Each procedure offers advantages and disadvantages. However, with respect to the degree of synchrony obtainable other than for S-phase, the various techniques are comparable. Optimum methods for cell synchrony will, no doubt, be determined by cellular parameters unique to each cell system. It is most probable that improved synchrony techniques will arise through the integration of techniques such as those described here.

APPENDIX

Mathematical Methods

A wealth of mathematical models can be used to describe the movement of cells through the cycle. The present work is based on methods introduced by Hahn (*15*) and by Roti Roti and Dethlefsen (*31*). For a full description see Thames and White (*36*). Alternate methods are described in (*5*), (*12*), and (*40*), and an overview of the mathematical analysis are found in (*20*) and (*21*). Briefly, the progression of cells is simulated by a set of K difference equations of the form:

$$x_i\,(t\,+\,\Delta t)\,=\,\alpha_i x_i(t)\,+\,\beta_{i\,-\,1} x_{i\,-\,1}(t)\,+\,\gamma_{i\,-\,2} x_{i\,-\,2}(t)$$

$$\alpha_i\,+\,\beta_i\,+\,\gamma_i\,=\,1$$

for $K > 2$ and

$$x_1\,(t\,+\,\Delta t)\,=\,\alpha_1 x_1(t)\,+\,2\,\beta_K x_K(t)\,+\,2\,\gamma_{K\,-\,1} x_{K\,-\,1}(t)$$

$$x_2\,(t\,+\,\Delta t)\,=\,\alpha_2 x_2(t)\,+\,\beta_1 x_1(t)\,+\,2\,\gamma_K x_K(t)$$

where the x_i are a set of states through which the cells move, and α, β, and γ are the probability of advancing zero, one or two states in time Δt. The number of states K in the cycle and the probabilities of advancement β and γ are computed through the use of two quantities, a mean time T and variance σ^2. These quantities are very close but not identical to the mean and variance of the transit time T and σ^2 (*36*). These are related by the following:

Let
$$\rho\,=\,\frac{T^2}{\sigma^2\,+\,T\Delta t}$$

Then

$$\rho \leqslant K \leqslant 2\rho$$

$$\beta = K\Delta t(2 - K\rho^{-1})/T$$

$$\gamma = K\Delta t(K\rho^{-1} - 1)/(2T)$$

Thus far we have described the methods necessary to model the cycle without reference to phases. The extensions to separate phases is straightforward and consists of introducing T_{G1}, T_S, T_{G2M}, and their corresponding values of σ^2, from which specific values of K_{G1}, ρ_{G1}, γ_{G1}, and so on, can be computed.

The fitting of data is then done by choosing a set of initial conditions, i.e., $X_1(0)$, and simulating the movement of cells. Simulation of data showed that the model is relatively insensitive to the variance in S and G2M, so that these were set equal to 1. Fitting was done by a least squares Levenberg-Marquardt algorithm (22) using a program by Chandler (7) that allows for bounded fitting regions. An estimate of the standard error was computed using the usual inverse of the information matrix. It was through this procedure that the fact that the variance could not be accurately estimated was demonstrated.

REFERENCES

1. Anderson, E. C., Bell, G. I., Peterson, D. F., and Tobey, R. A. Cell growth and division. IV. Determination of volume growth rate and division probability. Biophys. J., 9: 246–263, 1969.
2. Aqad, S. R., Fox, M., and Winstanley, D. The use of ficoll gradient centrifugation to produce synchrous mouse lymphoma cells. Biochem. Biophys. Res. Comm., 37: 551–558, 1969.
3. Arndt-Jovin, D. J. and Jovin, T. M. Analysis and sorting of living cells according to deoxyibonucleic acid content. J. Histochem. Cytochem., 25: 585–589, 1977.
4. Barlogie, B., Spitzer, G., Hart, J. S., Johnston, D. A., Buchner, T., Schumann, J., and Drewinko, B. DNA histogram analysis of human haemopoietic cells. Blood, 48: 245–258, 1976.
5. Beck, H. P. A new analytical method for determining duration of phases, rates of DNA synthesis and degree of synchronization from flow cytometric data on synchronized cell populations. Cell Tissue Kinet., 11: 139–148, 1978.

6. Bootsma, D., Budke, L., and Vos, O. Studies on synchronous division of tissue culture cells initiated by excess thymidine. Exp. Cell Res., *33*: 301–309, 1964.

7. Chandler, J. D. Quantum chemistry program exchange. Program 307, Indiana University, Evansville, Ind.

8. Dean, P. N. Data analysis in cell kinetic research. *In*: (J. W. Gray and Z. Darzynkiewicz, eds.), Techniques in Cell Cycle Analysis, New Jersey: Humana 1986.

9. Flow Cytometry and Sorting, (M. Melamed, P. Mullaney, and M. Mendelsohn, eds.) New York: John Wiley, 1979.

10. Glick, D., von Redlich, D., Yuhos, E. T., and McEwen, C. R. Separation of mast cells by centrifugal elutriation. Exp. Cell Res., *65*: 23–26, 1979.

11. Grabske, R. J. Separating cell populations by elutriation. Fractions, *1*: 1–8, 1978.

12. Gray, J. W. Cell-cycle analysis of perturbed cell population. Computer simulation of sequential DNA distribution. Cell Tissue Kinet., *9*: 499–516, 1976.

13. Gray, J. W., Dolbeare, F., Pallavicini, M., and Vanderlaan, M. Flow cytokinetics. *In*: (J. W. Gray and Z. Darzynkiewicz, eds.) Techniques in Cell Cycle Analysis. New Jersey: Humana 1986.

14. Grdina, D. J., Meistrich, M. L., Meyn, R. E., Johnson, T. S., and White, R. A. Cell synchrony techniques. I. A comparison of methods. Cell Tissue Kinet., *17*; 223–236, 1984.

15. Hahn, G. M. State vector description of the proliferation of mammalian cells in tissue culture. I. Exponential growth. Biophys. J., *6*: 275–290, 1966.

16. Howard, A. and Pelc, S. R. Synthesis of deoxyribonucleic acid in normal and irradiated cells and its relation to chromosome breakage. Heredity (suppl.), *6*: 261–273, 1953.

17. Johnston, D. A., White, R. A., and Barlogie, B. Automatic processing and interpretation of DNA distributions: comparison of several techniques. Comp. Biomed. Res., *11*: 393–404, 1978.

18. Keng, P. C., Li, C. K. N., and Wheeler, K. T. Characterization of the separation properties of the Beckman elutriator system. Cell Biophys., *3*: 41–56, 1981.

19. MacDonald, H. R. and Miller, R. C. Synchronization of mouse L-cells by a velocity sedimentation technique. Biophys. J., *10*: 834–842, 1970.

20. Macdonald, P. D. M. Statistical inference from the fraction labelled mitoses curve. Biometrika, *57*: 489–511, 1970.

21. Macdonald, P. D. M. Age distributions in the general cell kinetic model. *In*: (A.-J. Valleron and P. D. M. Macdonald, eds.), Biomathematics and Cell Kinetics, Amsterdam: Elsevier/North-Holland Biomedical Press, 1978.

22. Marquardt, D. W. and Snee, R. D. Ridge regression in practice. Amer. Statistician, *29*: 3–30, 1975.

23. Meistrich, M. L., Meyn, R. E., and Barlogie, B. Synchronization of mouse L-P59 cells by centrifugal elutriation separation. Exp. Cell Res., *105*: 169–177, 1977.

24. Meistrich, M. L. Experimental factors involved in separation by centrifugal elutriation. *In*: (T. G. Pretlow and T. P. Pretlow, eds.), Cell Separation: Methods and Selected Applications, vol. 2, Florida: Academic, 1983.

25. Meyn, R. E., Hewitt, R. R., and Humphrey, R. M. Evaluation of S phase synchronization by analysis of DNA replication in S-bromodeoxyuridine. *In*: (D. M. Prescott, ed.), Methods in Cell Biology, vol. IX, San Francisco: Academic, 1975.

26. Meyn, R. E., Meistrich, M. L., and White, R. A. Cycle-dependent anticancer drug cytotoxicity in mammalian cells synchronized by centrifugal elutriation. J. Natl. Cancer Inst., *64*: 1215–1219, 1980.

27. Miller, R. G. and Phillips, R. A. Separation of cells by velocity sedimentation. J. Cell. Physiol., *73*: 191–202, 1967.

28. Morris, N. R., Cramer, J. W., and Reno, D. A simple method for concentration of cells in the DNA synthetic period of the mitotic cycle. Exp. Cell. Res., *48*: 216–218, 1967.

29. Rajewsky, M. F. Synchronization in vivo: Kinetics of a malignant cell system following temporary inhibition of DNA synthesis with hydroyurea. Exp. Cell Res., *60*: 269–276, 1970.

30. Romsdahl, M. M. Synchronization of human cell lines with colcemid. Exp. Cell Res., *50*: 463–467, 1968.

31. Roti Roti, J. L. and Dethlefsen, L. A. Matrix simulation of duodenal crypt cell kinetics. I. The steady state. Cell Tissue Kinet., *8*: 321–351, 1975.

32. Sinclair, R. and Bishop, A. H. L. Synchronous culture of strain-L mouse cells. Nature (Lond.), *205*: 1272–1273, 1965.

33. Sinclair, W. K. Hydroxyurea: Effects on Chinese hamster cells grown in culture. Cancer Res., *27*: 297–308, 1967.

34. Stearns, B., Losee, K. A., and Bernsterin, J. Hydroxyurea. A new type of potential antitumor agent. J. Med. Chem., *6*: 201, 1963.

35. Terasima, T. and Tolmach, L. J. Changes in X-ray sensitivity of HeLa cells during the division cycle. Nature, *190*: 1210–1211, 1961.

36. Thames, H. D. and White, R. A. State-vector models of the cell cycle. I. Parametrization and fits to labeled mitoses data. J. Theor. Biol., *67:* 733–756, 1977.

37. Tobey, T. A. and Ley, K. D. Regulation of initiation of DNA synthesis in Chinese hamster cells. I. Production of a stable, reversible G1-arrested population in suspension culture. J. Cell Biol., *46:* 151–157, 1970.

38. White, R. A., Grdina, D. J., Meistrich, M. L., Meyn, R. E., and Johnson, T. S. Comparison of cell synchrony techniques II. Cell Tissue Kinet., *17:* 237–245, 1984.

39. Yen, A., Fried, J., Kitahara, T., Strife, A., and Clarkson, B. D. The kinetic significance of cell size. I. Variation of cell cycle parameters with size measured at mitosis. Exp. Cell Res., *95:* 295–302, 1975.

40. Zeitz, D. S. FPI Analysis. I. Theoretical outline of a new method to analyze time sequences of DNA histograms. Cell Tissue Kinet., *13:* 461–471, 1980.

Index

403